Sustainable Civil Infrastructures

About this Series

Sustainable Infrastructure impacts our well-being and day-to-day lives. The infrastructures we are building today will shape our lives tomorrow. The complex and diverse nature of the impacts due to weather extremes on transportation and civil infrastructures can be seen in our roadways, bridges, and buildings. Extreme summer temperatures, droughts, flash floods, and rising numbers of freeze-thaw cycles pose challenges for civil infrastructure and can endanger public safety. We constantly hear how civil infrastructures need constant attention, preservation, and upgrading. Such improvements and developments would obviously benefit from our desired book series that provide sustainable engineering materials and designs. The economic impact is huge and much research has been conducted worldwide. The future holds many opportunities, not only for researchers in a given country, but also for the worldwide field engineers who apply and implement these technologies. We believe that no approach can succeed if it does not unite the efforts of various engineering disciplines from all over the world under one umbrella to offer a beacon of modern solutions to the global infrastructure. Experts from the various engineering disciplines around the globe will participate in this series, including: Geotechnical, Geological, Geoscience, Petroleum, Structural, Transportation, Bridge, Infrastructure, Energy, Architectural, Chemical and Materials, and other related Engineering disciplines.

More information about this series at http://www.springer.com/series/15140

Hadi Khabbaz · Heejung Youn
Mounir Bouassida
Editors

New Prospects in Geotechnical Engineering Aspects of Civil Infrastructures

Proceedings of the 5th GeoChina International Conference 2018 – Civil Infrastructures Confronting Severe Weathers and Climate Changes: From Failure to Sustainability, held on July 23 to 25, 2018 in HangZhou, China

Springer

Editors
Hadi Khabbaz
School of Civil and Environmental Engineering
University of Technology Sydney
Sydney, NSW, Australia

Mounir Bouassida
National Engineering School of Tunis
University of Tunis El Manar
Tunis, Tunisia

Heejung Youn
Hongik University
Seoul, Korea (Republic of)

ISSN 2366-3405 ISSN 2366-3413 (electronic)
Sustainable Civil Infrastructures
ISBN 978-3-319-95770-8 ISBN 978-3-319-95771-5 (eBook)
https://doi.org/10.1007/978-3-319-95771-5

Library of Congress Control Number: 2018948643

This Springer imprint is published by the registered company Springer Nature Switzerland AG
The registered company address is: Gewerbestrasse 11, 6330 Cham, Switzerland

Contents

Introduction

Earth is experiencing a global climate change, world's population is rising fast and the rate of urbanization is very high. Advancement in technologies to incorporate the impact of these parameters on civil infrastructures is vital. Technological and engineering advances have been crucial to efficient and economical civil infrastructures development. The articles presented in this volume are attempts made by the researchers and practitioners to address many geotechnical challenges, based on the state-of-the-art practices, innovative technologies, new research results and case histories in construction and design towards safer infrastructures. This volume covers a wide range of topics, which are of direct relevance to people within the broad field of geotechnical engineering, including consultants, contractors, academics, materials suppliers and the owners and operators of civil infrastructures. This volume specifically presents:

- The advancements in the field of geotechnical engineering.
- Various creative methods for improvement of weak soils.
- Recent findings on the characterization of treated soils.
- Insights on various geotechnical issues attracting everyone's attention involved in civil infrastructures.

This volume is part of the proceedings of the 5th GeoChina International Conference on Civil Infrastructures Confronting Severe Weathers and Climate Changes: From Failure to Sustainability, HangZhou, China 2018.

Numerical Simulation and Parametric Study of a Single Pile in Clay Layer to Examine the Effect of Loading on Settlements and Skin Friction Distribution

Salma Al Kodsi(✉) and Kazuhiro Oda

Geotechnical Division, Department of Civil Engineering, Graduate School of Osaka University, Osaka, Japan
salma_1987k@hotmail.com

Abstract. Skin friction is a shear stress distributes along pile's shaft as a reason of the movement between pile and adjacent soil. Applying loads on the pile head lead the pile to move downward and the adjacent soil to move upward resisting this movement, and a positive skin friction distribution will occur. On the other hand, applying a surcharge load on the surface next to the pile will cause extra settlements in the soil layers which may be larger than the pile settlements, and a negative skin friction (NSF) will be distributed along the pile's shaft. Load combination on both; pile and the adjacent soil, is a common reason of NSF phenomenon. To study the loading effect widely, a parametric study was carried out in this paper for different cases including pile and surcharge loading. This parametric study was held by a numerical simulation using elastic-plastic soil model defined by Matsui-Abe and the elastic-viscoplastic soil model defined by Sekiguchi-Ohta model to study the case of a single pile driven in a loaded soft clay layer. The model was first validated by comparing the results obtained from the primary consolidation with field test measurements. The comparison between two soil constitutive models was important to examine the effect of loading in different cases of soil behavior. Viscosity effect on NSF distribution was studied and shown in the graphs. As NSF may lead to pile's material fracture and structural failure, the aim of this paper is to examine the effect of loading on NSF, and clarify the viscosity behavior during primary consolidation stage. Viscous effect, surcharging and pile load combination have played a major rule in changing the skin friction distribution.

1 Introduction

A relative movement between pile and the soil leads to mobilize shear stresses along the interface between pile and adjacent soil. The main reason behind such movement is the application of surcharge load on the ground surface next to pile. Terzaghi and Peck (1948) assumed full mobilization of shear strength along the pile-soil interface up to the pile toe for a single pile, or along the perimeter of pile group. Therefore, the neutral point (point of zero shear stress) is assumed to be located at the bearing stratum of the

H. Khabbaz et al. (eds.), *New Prospects in Geotechnical Engineering Aspects of Civil Infrastructures*, Sustainable Civil Infrastructures,
https://doi.org/10.1007/978-3-319-95771-5_1

pile. Indraratna et al. (1992) suggested that in order to minimize NSF, piles might be driven few week, or a month after applying the surcharge load on the ground surface.

Bipul et al. (2003) presented a constitutive relationship for one-dimensional consolidation of clays. They recognized the importance of structural viscosity on clay consolidation. The viscous effect through consolidation is less at initial stage, and likely to increases gradually along with the progress of consolidation. Hanna and Sharif (2006) conducted a study on piles driven into clay and subjected to indirect loading through the surcharge applied symmetrically on the surrounding area. The study was based on a numerical model using a finite element technique and the soil was assumed to follow a linear elastic-perfectly plastic stress-strain relationship, which defined by Mohr-Coulomb failure criterion. They concluded that negative friction is developed on the pile's shaft during the undrained period and continues to exist until the consolidation of the surrounding soil be fully completed under surcharge loads. Tung-Lin (2009) studied the effect of viscosity on consolidation of poroelastic soil due to groundwater table depression. He showed that the displacement and pore water pressure of clay stratum are strongly related to the viscosity effect. The overestimation of soil displacement will occur only when the viscosity effect is being neglected. Werjuan et al. (2012) studied NSF of the super long pile caused by soil settlement under large-scale surcharge loading. They showed the impact of NSF and to what extent the soil parameters surrounding pile affect the NSF distribution. In their series of model tests on NSF on single pile and pile groups under different surcharge loading conditions, Ting et al. (2016) found that pile head settlement and NSF under side load were smaller compared with the result of the test under uniform loading, while the location of neutral plane (LNP) was lower. Arpan and Sujit (2017) developed a 3-D consolidation apparatus and performed a series of 3-D consolidation test under different surcharge pressures. They concluded that the consolidation characteristics were largely affected by the surcharge pressure. Increasing the surcharge on the surrounding soil makes it denser to reduce both the lateral movements of soil particles, and the lateral pore water pressure.

Many researchers studied the effect of surcharging on the NSF distribution, however the objective of this paper is to study the effect of the combination of surcharging and pile load on the distribution of shear stresses along the pile length besides the effect of viscosity during primary consolidation compared to elastic-plastic behavior. An elastic-viscoplastic and elastic-plastic soil models were represented herein to carry on the parametric study. A field test of a single pile was used to validate the model and conduct the numerical study.

2 Methodology

2.1 Field Loading Test

Matsui (1993) carried out a full-scale loading test of a cast-in-place RC bored pile in Osaka bay, Osaka, Japan. The soil profile consisted of both Holocene Deposits and Upper Pleistocene Deposits. The Holocene layers consisted of loose and soft silts and clays. A sandy gravel layer was adopted as a bearing layer of the tested pile. The tested pile was a cast-in-place RC pile with 1.5 m in diameter and 28.5 m in length. The vertical loading test was carried out, in accordance with the Standard of the JGS (JGS 1972). Figure 1 shows the field test soil section.

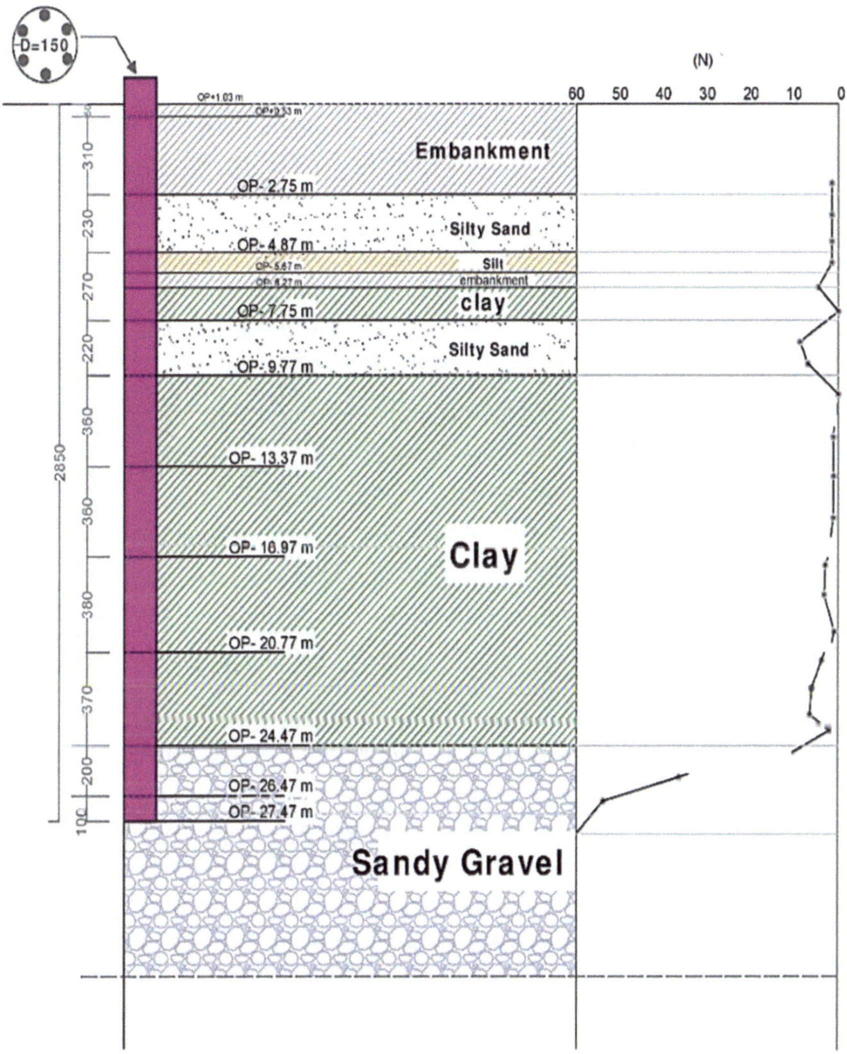

Fig. 1. Field test section

2.2 Numerical Model

2.2.1 Soil Constitutive Models

Finite element analysis was performed in order to compare results obtained from field test measurements with those from the numerical model. The model is an axisymmetric model assumes to behave as elastic-plastic material to represent the correlation between the stresses and the strains due to Matsui and Abe (1981) constitutive soil model for the main clay layer. This elastic-plastic soil model will be compared with the elastic-viscoplastic soil constitutive model represented by Sekiguchi and Ohta (1977). The different between these two soil models is the time dependency. Although elastic-plastic soil model is a time-dependent stress strain model but the viscous effect has been noticed during primary consolidation which affected the soil settlements. Elastic-plastic model is represented by Matsui-Abe in Eqs. (1) and (2). The Sekiguchi-Ohta model is a Cam-Clay type effective stress model for the behavior of clay-type soils and the yield surface (f) is given by Eq. (3). The bearing layer behavior was represented by Yasufuku model in Eq. (4).

$$e_0 - e = \lambda \ln\left(\frac{p'}{p'_0}\right) + (\lambda - \kappa)\left(\frac{\alpha_a}{\alpha_a - 1}\right) \ln\left\{\frac{M_a + (\alpha_a - 1)\eta}{M_a + (\alpha_p - 1)\eta_{k0}}\right\} \quad (1)$$

For the active state

$$e_0 - e = \lambda \ln\left(\frac{p'}{p'_0}\right) + (\lambda - \kappa)\left(\frac{\alpha_p}{1 - \alpha_p}\right) \ln\left\{\frac{M_p + (1 - \alpha_p)\eta_{k0}}{M_p + (1 - \alpha_p)\eta}\right\} \quad (2)$$

For the passive state

where e is the void ratio, e_0 is the initial value of e, λ is the compression index, κ is the swelling index and α_a, α_p are the strain increment ratio.

$$f = \mu \ln\left|\frac{1}{\delta}\left[\left\{1 - \exp\left(-\frac{\delta \dot{v}_r^v}{\mu} t\right)\right\} \exp\left(\frac{v^p}{\mu}\right) + \delta \exp\left(-\frac{\delta \dot{v}_r^v}{\mu} t\right)\right]\right| - v^{vp} = 0 \quad (3)$$

where μ is the coefficient of secondary consolidation, $\dot{v}_r^v$ is the reference viscous volume strain rate, δ is a material constant and v^p is the plastic volumetric strain.

$$f = \ln\frac{p}{p_0} + \frac{C}{2(C-1)} \ln\left[\frac{(1-C)(2\alpha - \eta)\eta + \{N - (2-C)\alpha\}N}{(1-C)\alpha^2 + \{N - (2-C)\alpha\}N}\right] = 0 \quad (4)$$

model in Eq. (4)

Where N and C are experimental parameters, α is an internal parameter to reflect the influence of the proportional loading path on the yield surface and $\eta = \frac{q}{p'}$ is the stress ratio.

2.2.2 Model Boundary, Mesh and Initial Conditions

The surface of the top layer is assumed to be permeable while the low and side boundaries of the model are supposed to be impermeable. The level of the ground water is at 0.80 m down from the surface. The low model boundary is being fixed, while the boundaries at the axis of symmetry and sides are free to move vertically.

The mesh was divided into 13 blocks. The first block -the pile- is divided into 360 elements behave as elastic material. The main clay layer includes block (2) and block (3) which both are divided into 234 elements. The abruption between pile shaft and

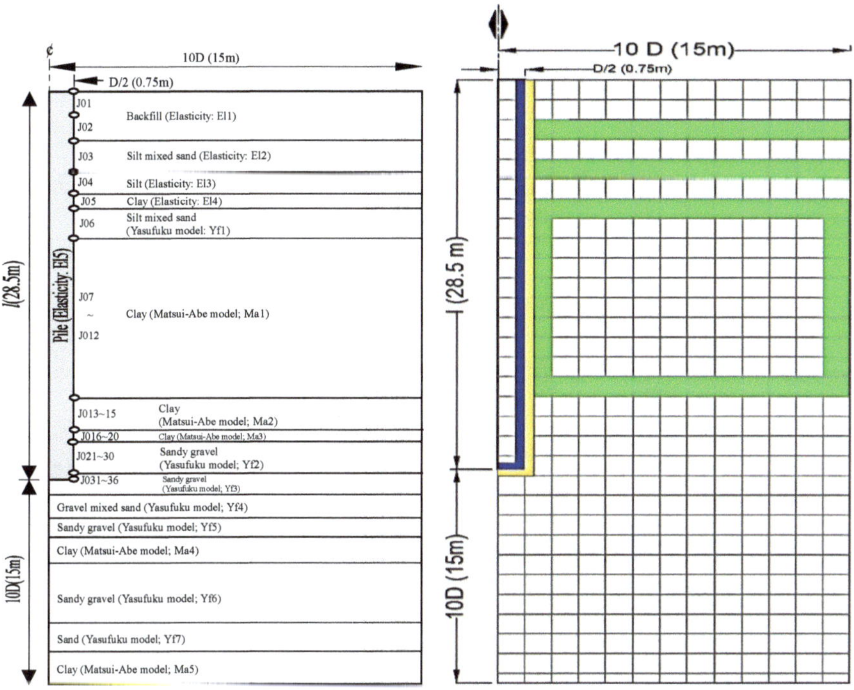

Fig. 2. Numerical model and the mesh

surrounding soil is represented by 36 joint elements (Goodman et al. 1968) in which the main clay layer includes 6 joint elements. Figure 2 shows the analytical study model and mesh boundary.

2.3 Model Validating

Numerical modelling using the finite elements technique was used herein to validate the used model. Table 1 shows the analytical elastic parameters for the pile material. Table 2 describes the basic mechanical constants for clay in Elastic-plastic behavior by Matsui-Abe. The main parameters for Sekiguchi-Ohta model are mentioned in Table 3 Yasufuku basic parameters for the bearing soil layer are shown in Table 4. Finally, the parameters of joint elements for the main clay layer are shown in Table 5. All parameters were determined through soil investigations including SPT and CPT (after Oda 2012).

A comparison between the measured axial forces and predicted values by the present numerical model is shown in Fig. 3. It can be noticed from the comparison results that the numerical model is valid to carry on the parametric study.

Table 1. Analytical parameters of elasticity

No.	E (kN/m²)	υ
E11	8.40E+04	0.47
E12	2.70E+04	0.47
E13	3.00E+04	0.35
E14	1.50E+04	0.30
E15	2.20E+07	0.22

Table 2. Analytical parameters of Matsui-Abe model

No.	λ	κ	M
Ma1	0.2480	0.0124	1.41
Ma2	0.4950	0.0248	1.26
Ma3	0.4000	0.0248	1.26
Ma4	0.5800	0.0243	1.26
Ma5	0.4480	0.0224	1.32

Note λ is the slope of normally consolidation line, κ is the slope of elastic swelling line and M is the frictional constant. Poisson ratio (υ) = 0.333

Table 3. Analytical parameters of Sekiguchi-Ohta model

No.	λ	κ	M
Se1	0.1024	0.01240	1.47
Se2	0.2475	0.01240	1.41
Se3	0.2475	0.01240	1.41
Se4	0.4950	0.02480	1.41
Se5	0.5800	0.02430	1.26

Note $\mu = 0.002, \delta = 1.0 \times 10^{-10}$ *and* $v_0 = 1.0 \times 10^{-10} (1/\text{min})$

Table 4. Analytical parameters of Yasufuku model

No.	N	M
Yf1	1.90E−02	0.75
Yf2	2.82E−03	0.80
Yf3	2.82E−03	0.75
Yf4	2.82E−03	0.75
Yf5	3.27E−03	0.80
Yf6	2.82E−03	0.8
Yf7	3.27E−03	0.75

Note α, Cg and Cf are constants. α = 0.4, Cg = 2.0, Cf = 1.5

Table 5. Analytical parameters of joint elements

No.	Kn (kN/m^3)	Ks (kN/m^3)	C0 (kN/m^2)
J07	9.8E+09	9.8E+09	5.64
J08	9.8E+09	9.8E+09	17.29
J09	9.8E+09	9.8E+09	15.90
J10	9.8E+09	9.8E+09	11.52
J11	9.8E+09	9.8E+09	15.01
J12	9.8E+09	9.8E+09	10.97

Note Kn, Ks are the slope of elastic line in compression state and shear deformation state, respectively, C0 is cohesion factor between pile shaft and surrounding soil. θ is the internal angle of friction = 0

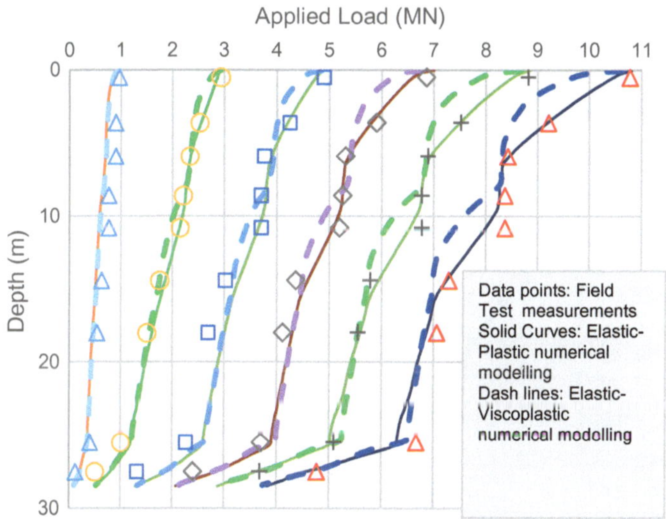

Fig. 3. Pile axial load distribution

3 Parametric Study

The aim of this parametric study is to examine different types of loading, to observe the behavior of both soil and pile, and to study the loading cases effect on settlements and skin friction distribution. The study is based on two types of loading;

① S: First, applying the pile load and then applying the surcharge load on the ground surface next to pile head (i.e. no surcharging on pile).

② P + S: First, applying the pile load and then applying the surcharge load on both ground surface and pile head (i.e. all model surface is under surcharging).

Table 6. Loading cases for the parametric study

Case	PL[a]	SL[b]	S	P + S	Case	PL[a]	SL[b]	S	P + S
1	100	2.0	✓		5	100	8.0	✓	
2	100	2.0		✓	6	500	2.0		✓
3	100	4.0		✓	7	1100	2.0		✓
4	100	6.0	✓						

[a]Pile load (tf), [b]Surcharge load (tf/m^2)

In the light of these two types of loading, Table 6. was found to clarify the parametric study of loading cases.

The following figures present the effect of loading and viscosity on the settlements, axial force and skin friction distribution. Figure 4 shows the pile axial load distribution for elastic-viscoplastic model. When applying a small value of surcharging, larger distribution was examined along pile's shaft in regarding to comparing viscosity with elastic-plastic. In the contrary, in the light of large surcharging, the difference in the distribution in two models did not occurred. A comparison of the case of small and large surcharging in two soil models is shown in Fig. 5.

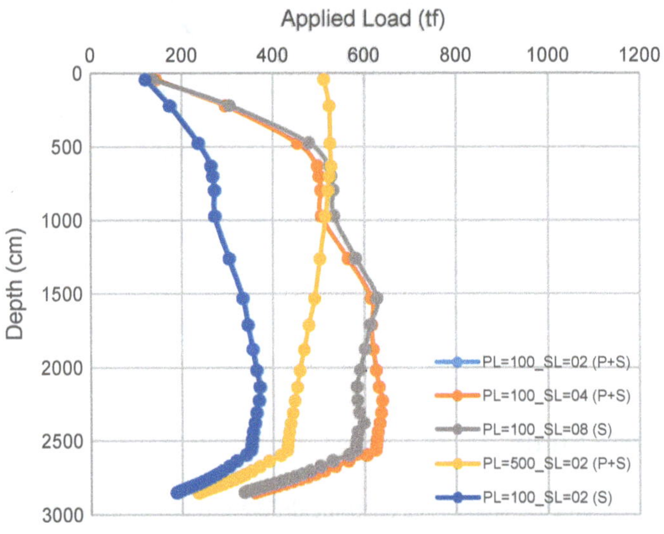

Fig. 4. Pile axial load distribution/elastic-viscoplastic model

The ground settlements with time for the two soil models and different cases of loading are shown in Figs. 6 and 7. Viscous effect had a major influence on soil settlements, where this influence played a rule in case of applying a small surcharging while bigger surcharge load leaded to a small different in the settlements values between elastic-plastic and elastic-viscoplastic soil models and that is clarified in

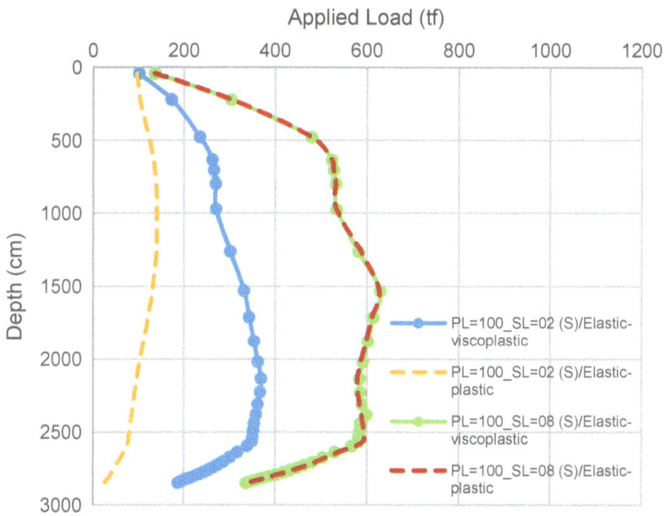

Fig. 5. Pile axial load distribution/comparison between elastic-plastic and elastic-viscoplastic model

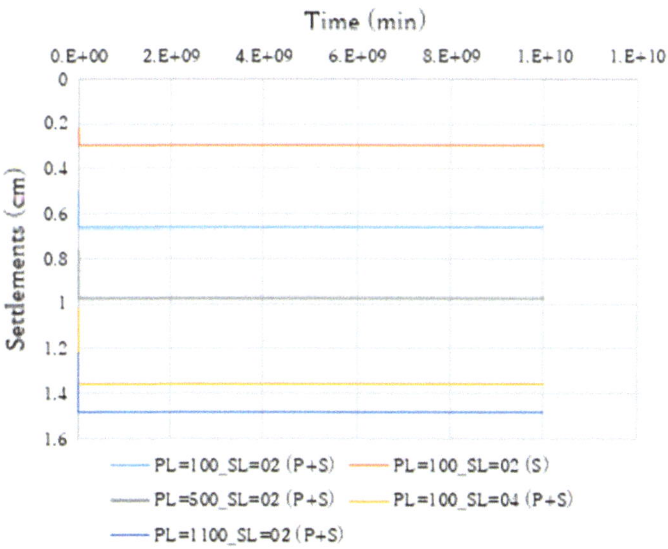

Fig. 6. Ground settlements with time/elastic-plastic model

Fig. 8. Changing the applied pile load to a specific value occurred the pile to settle more than the ground, whereas this conclusion can be shown in Figs. 9 and 10 for two cases of pile loading.

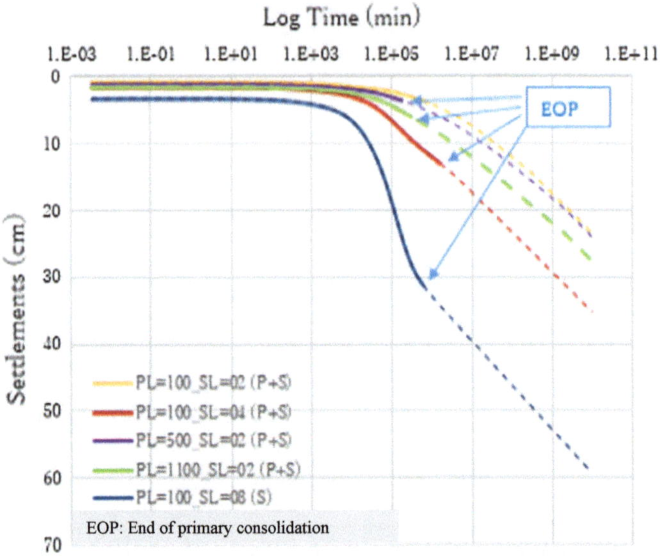

Fig. 7. Ground settlements with time/elastic-viscoplastic model

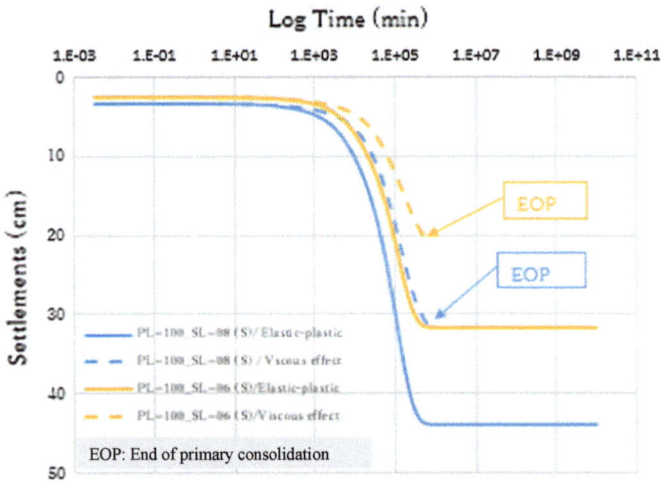

Fig. 8. Ground settlements with time/comparison between elastic-plastic and elastic-viscoplastic model

As the distribution of skin friction is related to pile and soil settlements, same conclusion was observed in Figs. 11 and 12, whereas applying a small surcharge load leaded to a bigger skin friction distribution in case of viscosity. Figure 13 shows the case of applying a larger pile load which changed the distribution of NSF into a positive skin friction.

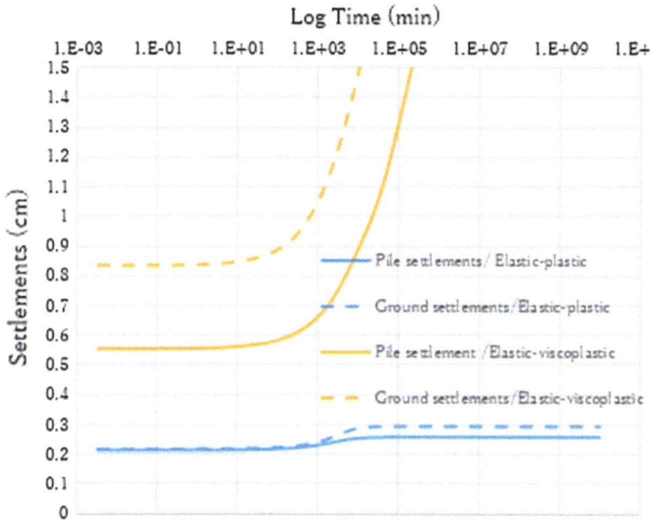

Fig. 9. Ground and pile settlements with time/PL = 100_SL = 02 (S)

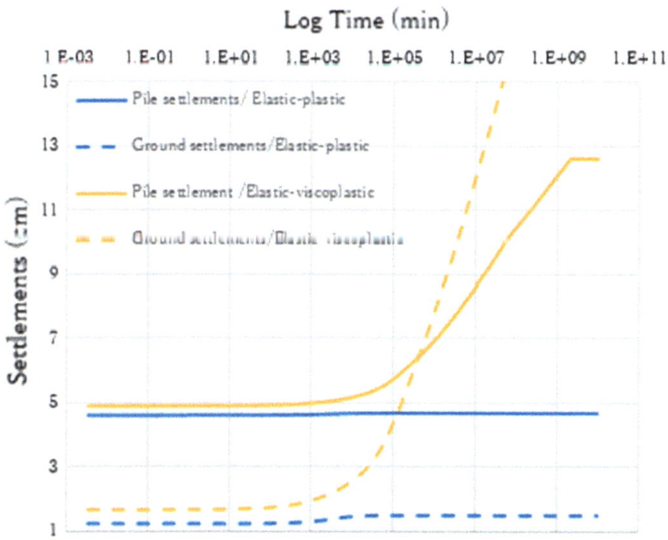

Fig. 10. Ground and pile settlements with time/PL = 1100_SL = 02 (P + S)

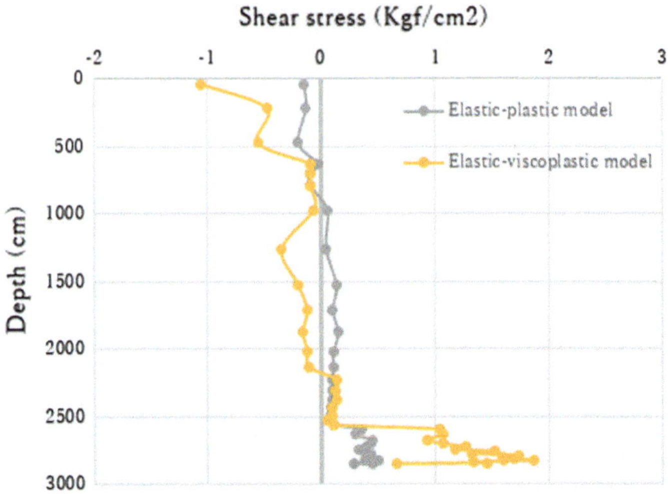

Fig. 11. Shear stress distribution/comparison between elastic-plastic and elastic-viscoplastic model (PL = 100_SL = 02(S))

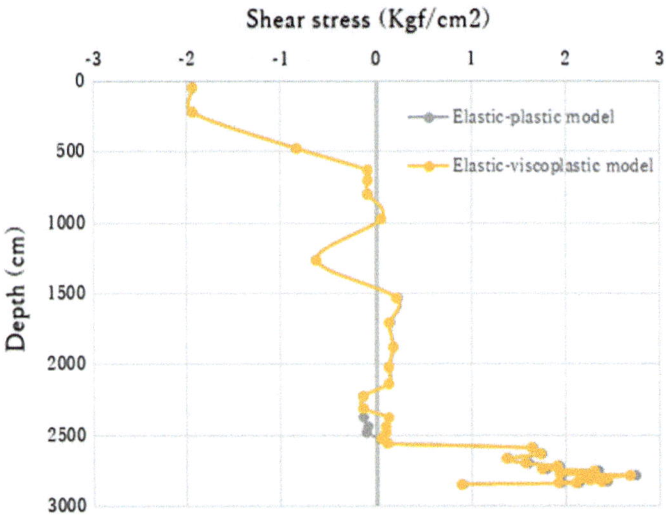

Fig. 12. Shear stress distribution/comparison between elastic-plastic and elastic-viscoplastic model (PL = 100_SL = 08 (S))

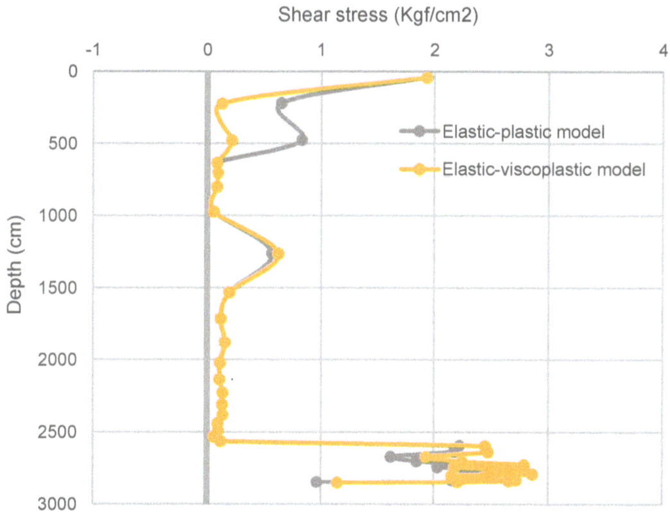

Fig. 13. Shear stress distribution/comparison between elastic-plastic and elastic-viscoplastic model (PL = 1100_SL = 02(P + S))

4 Conclusions

Surcharging and pile load play a vital role in mobilizing shear stresses along the pile's shaft (negative and positive friction) regardless the used soil model in analysis. In this paper two different soil models were used and a parametric study for different loading cases was mentioned. It can be concluded that:

- Surcharging is the main reason of soil settlements and NSF mobilizing and its effect will be more significant when it's applied only on the ground surface next to the pile.
- When the whole model surface is under surcharging and the pile head is loaded, the NSF will be smaller comparing with the case of surcharging is applied on the ground surface only. Adding an extra load to the pile, leads the pile at specific point to settle equal or more than the adjacent soil which reduces NSF.
- Viscous effect has a major role in increasing the soil settlements and skin friction distribution even though the study has been done during the primary consolidation stage. This effect will be noticed clearly in case of small values of surcharging but when the surcharge load exceeded a specific amount, the viscous influence was approximately similar to elastic-plastic.
- It is important to take the viscous effect into consideration when designing pile foundations. Carrying on this study to examine the soil behavior and skin friction distribution during secondary consolidation due to structural viscosity is also important and must be covered through future researches.

References

Arpan, L., Sujit, K.P.: The effect of different surcharge pressure on 3-D consolidation of soil. Int. J. Appl. Eng. Res. **12**(8), 1610–1615 (2017). ISSN 0973-4562

Awwad, T., Kodsi, S.A.: A comparison of numerical simulation models to determine the location of neutral plane. In: Proceedings of the 19th International Conference on Soil Mechanics and Geotechnical Engineering, pp. 1947–1950 (2017)

Awwad, T., Donia, M., Awwad, L.: Effect of a stiff thin Foundation soil layer's depth on dynamic response of an embankment dam. Proc. Eng. **189**, 525–532 (2017)

Bipul, C.H., Balasingam, M., Goro, I.: Viscosity effects on one-dimensional consolidation of clay. Int. J. Geomech. **3**(1) (2003)

Hanna, A., Sharif, A.: Negative skin friction on single piles in clay subjected to direct and indirect loading. Int. J. Geomech. (2006)

Indraratna, B., Balasubramaniam, A. S., Phamvan, P., Wong, Y.K.: Development of negative skin friction on driven piles in soft Bangkok clay. Can. Geotech. J. **29**, 393–404 (1992)

Matsui, T.: Case studies on cast-in-place bored piles and some considerations for design. Deep Foundations on Bored and Auger Piles. Ghent, pp. 77–101 (1993)

Matsui, T., Abe, N.: Multi-dimensional elasto-plastic consolidation analysis by finite element method. Soils Found. **21**(1), 79–95 (1981)

Oda, K.: Numerical simulations of field loading tests of cast-in-place bored piles with large diameter. In: International Conference(Proceedings) on Testing and Design Methods for Deep Foundations, pp. 859–866 (2012)

Poorooshasb, H.B., Alamgir, M., Miura, N.: Negative skin friction on rigid and deformable piles. Comput. Geotech. **18**(2), 109–126 (1996)

Sekiguchi, H., Ohta, H.: Induced anisotropy and time dependency in clays. In: 9th ICSMFE, Tokyo, Constitutive Equations of Soils, pp. 229–238 (1977)

Terzaghi, K., Peck, R.B.: Mechanics in Engineering Practice. Wiley, New York (1948)

Ting, H., Liangde, H., Lijun, H.: The influence of surcharge load on the negative skin friction on pile groups. In: Proceedings of Twenty-Sixth (2016) International Ocean and Polar Engineering Conference, Rhodes, Greece, 26 June–1 July 2016

Tung-Lin, T.: Viscosity effect on consolidation of poroelastic soil due to groundwater table depression. Environ. Geol. **57**, 1055–1064 (2009)

Werjuan, Y., Yimin, L., Jun, C.: Characteristics of negative skin friction for super long piles under surcharge loading. Int. J. Geomech. **12**(2) (2012)

Study on Low-Strength Biocemented Sands Using a Temperature-Controlled MICP (Microbially Induced Calcite Precipitation) Method

Yang Wang, Hanlong Liu, Zhichao Zhang, Peng Xiao, Xiang He, and Yang Xiao(✉)

School of Civil Engineering, Chongqing University, Chongqing, China
trey@cqu.edu.com, {hliuhhu,zhangzhichaopt,xpcqdx2012, medihe,hhuxyanson}@163.com

Abstract. MICP (Microbially Induced Calcite Precipitation) technique is recently known as a new research area in geotechnical engineering. This technique provides a more environmentally way to enhancing the soil strength and stiffness by the MICP process in the soil pores. However, the spatial uniformity of MICP in the treated sands, which determines the effectiveness of MICP technique, remains a challenging issue even in the laboratory tests, especially for low-strength biocemented sands. Noting that the MICP process could be greatly inhibited under low temperatures before the homogeneous conditions of MICP reactions is achieved in sands, a temperature-controlled MICP method is proposed in this paper to improve the MICP uniformity in low-strength biocemented sands. A series of temperature-controlled MICP tests are made and the results are compared with the MICP tests under a constant temperature.

Keywords: MICP technique · Uniformity · Temperature · ISO sands

1 Introduction

MICP (Microbially Induced Calcite Precipitation) technique (DeJong et al. 2006; Mitchell and Santamarina 2005; van Paassen et al. 2010; Xiao et al. 2017), is recently known as a new research area in geotechnical engineering. In the technique, a ureolytic bacteria (Al Qabany and Soga 2013; Chu et al., 2013; DeJong et al., 2006; Montoya et al., 2013; Mortensen and DeJong 2011; van Paassen et al., 2010; Whiffin et al. 2007), is used to induce the precipitation of calcite among the pores in soils or sands to improve the soil mechanical properties. To enable the MICP process in soils, the bacterial fluid and the mixed solution of urea and calcium chloride are first grouted into the soil pores. The urea is then hydrolyzed into ammonium and carbonate ions, and the calcite precipitation is finally induced in the solute environment with carbonate ions. Thus, the main chemical reactions involved are as follows:

H. Khabbaz et al. (eds.), *New Prospects in Geotechnical Engineering Aspects of Civil Infrastructures*, Sustainable Civil Infrastructures,
https://doi.org/10.1007/978-3-319-95771-5_2

$$CO(NH_2)_2 + H_2O \xrightarrow{ureolytic\ bacteria} 2NH_4^+ + CO_3^{2-} \tag{1}$$

$$CO_3^{2-} + Ca^{2+} \rightarrow CaCO_3(s) \downarrow \tag{2}$$

Many laboratory tests had been made by different researchers to study the MICP process in sands. Experimental results (Cheng et al. 2013) indicated that larger agglomerated crystal clusters of calcite could more effectively fill the gaps between the sand grains, leading a higher compressive strength improved by the MICP process. Meanwhile, the clusters, acting as calcite bridges, could increase the thermal conductivity of soil, especially for soils with low saturation (Venuleo et al. 2016). Centrifuge models test (Montoya et al. 2013) also illustrated the greatly improved dynamic behavior of MICP-treated soils, which showed lower pore pressures and shaking-induced settlements.

The uniformity of MICP-treated sand specimen was studied by Feng and Montoya (2016), Jiang et al. (2016), Lin et al. (2016) and Martinez et al. (2013). The results showed that more $CaCO_3$ content was found at the bottom, especially in the heavily treated specimen. In a 240 L soil microcosm, the tip resistance at a depth of 40–50 mm, measured by cone-penetration tests, was several times higher than that at other depth ranges (Burbank et al., 2013), as shown in Fig. 1. These test results indicated a significant heterogeneities of both the MICP precipitation (see Fig. 1a) and the improved soil strength, depending on many factors such as the grouting methods as well as the particle content, the gap ratio and the particle size distribution (Jiang et al. 2016). In general, the calcium carbonate precipitation content and its spatial uniformity, which determine the effectiveness of MICP technique, still need to be improved even under laboratory conditions. Especially, due to the convection and diffusion process in the grouting, the spatial differences in the concentrations of the bacteria solution and urea-$CaCl_2$ mixed solution may be one of the most important resources of the heterogeneous calcium carbonate precipitation in sands. To improve the uniformity of MICP, a temperature-controlled MICP method is proposed in this paper. A series of temperature-controlled MICP tests are made and the results are discussed.

2 Methods and Materials

As discussed in the above section, the convection and diffusion of the bacteria solution and the urea-$CaCl_2$ mixed solution in the treated soils may be the major factor leading to the bad uniformity of MICP. For example, for the simple one-dimensional MICP process in which the bacteria and reagent solutions are grouted from the top to the bottom of a sand column, the MICP reactions at the top always start earlier than that at the bottom, which means more calcium carbonate precipitations at the positions close to grouting inlet. Moreover, in such a heterogeneous MICP reaction case, the reduction in the permeability of the sand close to the grouting inlet furtherly hinder the transport of bacteria and reagent solutions to the sand pores far away from the grouting inlet.

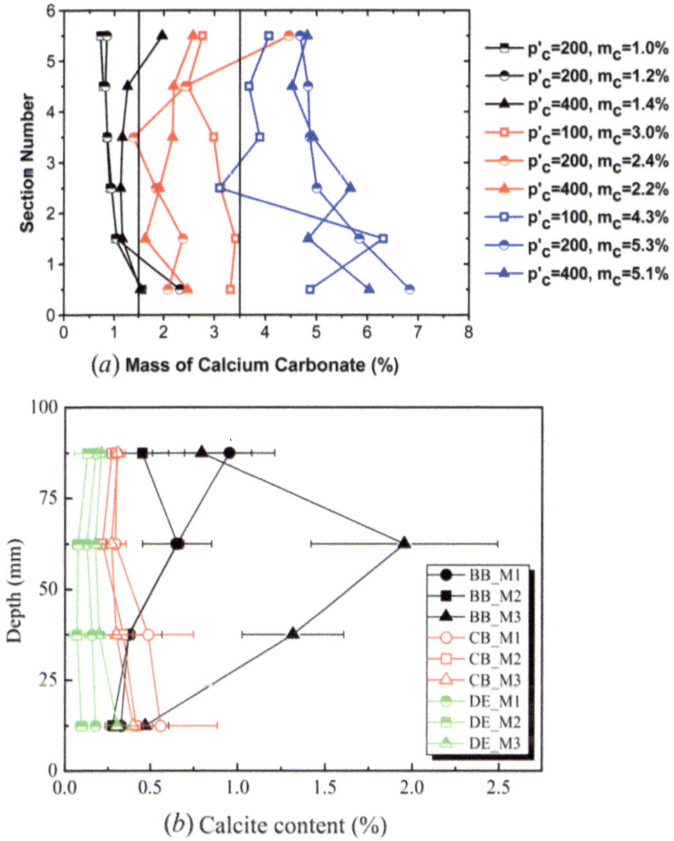

Fig. 1 **a** Cementation variation along the height of a treated sand specimen [p'_c, in kPa, is the effective confining pressure and m_c is the mass content of calcite cementation (Feng and Montoya 2016)]; **b** profiles of carbonate precipitation along the depth of the MICP treated soils (Jiang et al. 2016)

Correspondingly, remarkable nonhomogeneous MICP and improved strength distributions can be observed, significantly resulting in the heterogeneity of the treated specimen.

One of the possible approaches improving the MICP uniformity is to ensure the homogeneous initial conditions of the MCIP reactions in the completely sand sample, including the uniform initial concentrations of bacteria and reagent solutions. To this end, the MICP reaction must be inhibited before such homogeneous initial conditions are reached in the convection and diffusion process of the grouted solutions. It is known that the microbially induced urea hydrolysis rate can be greatly decreased at low temperatures, due to the low urease activity. Therefore, the MICP uniformity can be expected to be improved by a temperature-controlled MICP method. Based on such a concept, we designed an improved grouting apparatus as shown in Fig. 2.

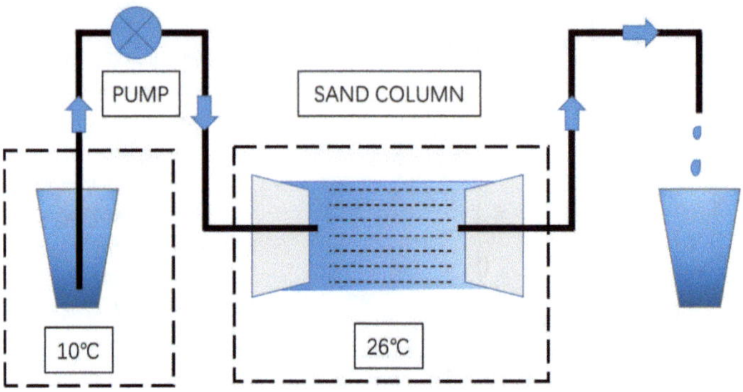

Fig. 2. Test design of the temperature-controlled MICP method

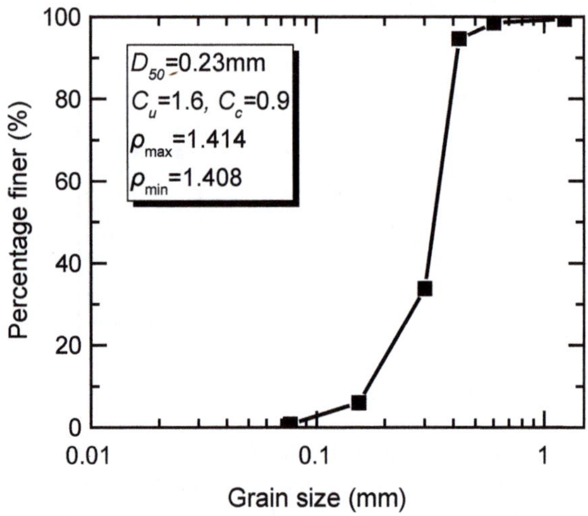

Fig. 3. Grain size distribution curve & properties of Ottawa 50-70 sand

Two steps are designed in the temperature-controlled MICP tests: (1) the grouting step at low temperature and (2) the reaction step at high temperature. In the step (1), the reagent and bacteria solutions, mixed at 10 °C almost without any MICP reactions, are first grouted into the sample by a pump until the steady seepage is reached, so that the relatively uniform concentration distributions of the reagent and bacteria solutions can be ensured and thus the MICP reactions develops simultaneously within the treated sample. Then, in the step (2), the temperature in the sand column is allowed to be increased to 26 °C to activate the MICP reactions. The temperature-controlled grouting procedure described above can be repeated for several times in order to get different contents of calcite precipitated. This temperature-controlled MICP method can be useful especially for the ground improvement in regions where the environmental temperature varies remarkably within one day. After the treatments, the improved triaxial shear strength and the spatial distribution of MICP are measured.

Sporosarcina pasteurii (DSM 33; ATCC 11859), stored at 4 °C, is also used in this study (Al Qabany and Soga 2013; Chu et al. 2013; DeJong et al. 2006; Montoya et al. 2013; Mortensen and DeJong 2011; van Paassen et al. 2010; Whiffin et al. 2007). The reagent solution is composed of $CaCl_2$ and $CO(NH_2)_2$ with a fixed initial concentration. Ottawa 50-70 sand is used and packed into a PVC tube, forming a sand column with a height of 80 mm and a diameter of 39 mm. The detailed information of the main materials used can be found in Table 1.

Table 1. Materials used in the temperature-controlled MICP tests

Materials	Notes
Bacteria: *Sporosarcina pasteurii*	OD_{600} = 1.628–1.821
Reagent: $CaCl_2$, $CO(NH_2)_2$	Concentration: 0.25 mol/L
Silica sand: Ottawa 50-70 sand	Details were shown as Fig. 3

3 Experimental Test Program

In order to determine the temperature at which the urease activity is inhibited adequately, the relationship between the urease activity, which is measured by the electrical conductivity of the bacteria solution used, and the temperature should be analyzed. In this section, the urease activity is therefore presented firstly, and then the detailed test procedures are introduced.

3.1 Urease Activity Test

Keeping all other factors unchanged, the urease activity tests are implemented at different temperatures.

Table 2 shows the parameters in the urease activity tests. The tests includes the following three steps:

(1) Preparing 6 beakers filled with water at different temperatures: 10, 15, 20, 25, 30 and 40 °C;
(2) Putting five tubes with 9 ml reagent into each beaker and letting stand for 3 min so that the reagent reaches the same temperature as the water filled in the beaker.
(3) Pouring 1 ml bacteria into every tube and then measuring the urease activity by the conductivity method (Whiffin et al. 2007).

Table 2. The parameters in urease activity tests

Materials	Notes
Bacteria: *Sporosarcina pasteurii*	OD_{600} = 1.801, 1 ml per tube
Reagent: $CO(NH_2)_2$	1.1 mol/L 9 ml per tube
Temperature	10, 15, 20, 25, 35, 40 °C
Temperature compensation coefficient	0.02

The measured urease activities at different temperatures are shown in Fig. 4. Following with the increase in temperature, the urease activity is gradually increased to a peak value at about 33 °C, and a stable stage, in which the urease activity is approximately independent on the temperature, can be found within the temperature range 15–24 °C. Obviously, the urease activity is almost completely inhibited at the temperature below 10 °C. Therefore, 10 °C can be an appropriate temperature for the reagent in the grouting step.

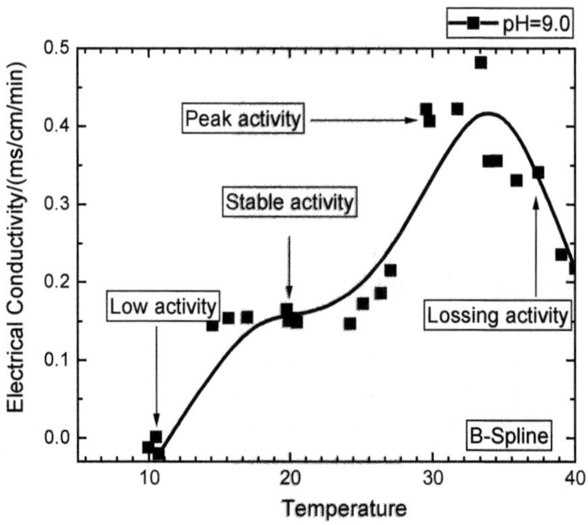

Fig. 4. The relationship curve of EC and temperature

3.2 MICP Test in Sand Column

Firstly, a standard method for sand reconstitution is necessary. Dry pluviation method was used here according to the studies in Raghunandan et al. (2013). The sand is pouring into the tube at a mass-flow rate of 8.5 g/s from a funnel with an opening size of 9.4 mm, and the drop height is 300 mm. The reconstituted sand columns with void ratios e = 0.680–0.694 and relative densities D_r = 50–55% are finally prepared. Each sand sample, reconstituted from Ottawa 50-70, weight about 135 g.

The volume proportion of mixture grouted into the samples is shown in Eq. (3). In this paper, the low-strength biocemented sands, in which the MICP uniformity is more difficult to be achieved, are expected. To this end, two total volumes of grouted mixture, 250 and 500 ml, are chosen in the tests. In the tests with 500 ml total mixture volume, two batches of mixture, 250 ml per batch, are grouted successively. The comparative tests on untreated sand samples are also implemented. Detail information of the test scheme can be found in Table 3. Figure 5 shows the photograph of the MICP test apparatus and samples.

$$CO(NH_2)_2{:}CaCl_2{:}Bacteria = 5{:}5{:}1 \qquad (3)$$

Table 3. Scheme of the temperature-controlled MICP tests

No.	Condition	Mixture/ml	Notes
1	10 °C	250	(a) Sand: 135 g
		500	(b) Sand column: $D \approx 39$ mm, $H \approx 80$ mm
2	Untreated sand		(c) Bacteria: OD_{600} = 1.628–1.821
			(d) Reagent ($CO(NH_2)_2$, $CaCl_2$): 0.25 ml/L
			(e) Mixture: $CO(NH_2)_2$:$CaCl_2$:Bacteria = 5:5:1

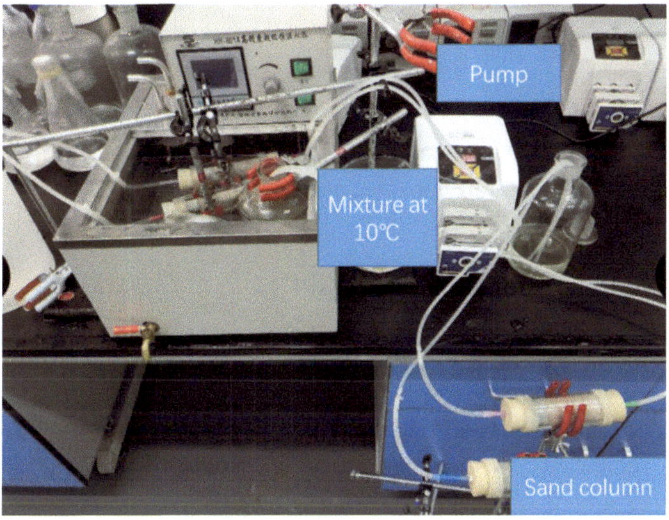

Fig. 5. Photograph of the MICP test apparatus and samples

3.3 Evaluation of Treatment Results

To examine the mechanical properties of the treated sand samples, consolidated drained triaxial tests were carried out. Considering the low-strength of the biocemented sand samples, the effective cell pressure was less than 10 kN/m^2 during the saturation step. The finally pore pressure coefficient, B, was up to 0.96.

The calcite precipitation distribution was measured along the height of the samples to estimate the uniformity of MICP (Fig. 6). In this study, the carbonate content was measured by the dissolving method (Jiang et al. 2016; Lin et al. 2016; Whiffin et al. 2007). Samples from different positions of each sand column were mixed with hydrochloric acid, and the content was calculated by Eq. (4)

$$CaCO_3 \text{ content } (\%) = \frac{m_0 - m_1}{m_0} \times 100\% \tag{4}$$

where m_0 is the total original dry weight of samples; m_1 is the total dry weight of samples after dissolving and washing.

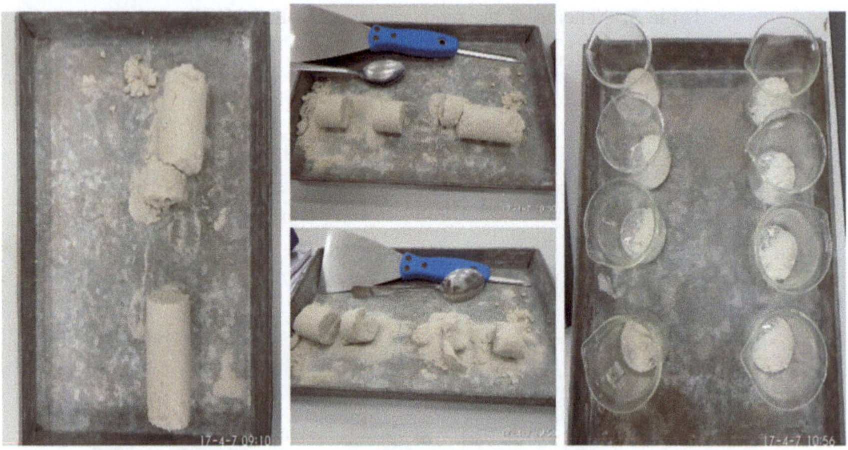

Fig. 6. Photographs of the treated sand columns and the sectional samples used for the carbonate content measurement by dissolving method

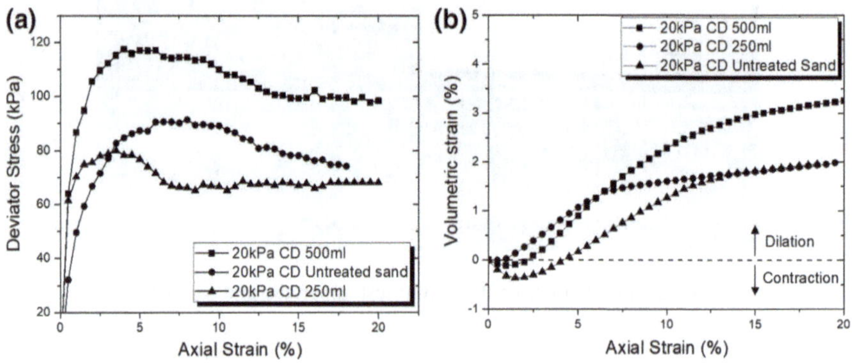

Fig. 7. Consolidated drained triaxial test results: **a** stress-strain curves; **b** volumetric-strain curves

4 Results and Discussion

4.1 Onsolidated Drained Triaxial Tests

The stress-strain and volumetric-strain relation curves of treated and untreated sand columns are presented in Fig. 7. More details about the mechanical properties of the biocemented sand samples was showed in Table 4.

Comparing with the responses of untreated sand sample, the results indicate the significant increase in the stiffness of both the sand samples treated by 250 and 500 ml grouting mixtures. However, the effects of MICP on the peak and residual strengths seem to be nonlinearly sensitive to the precipitated calcite contents for low-strength biocemented sands. For the 500 ml-treated sand sample, both the two strengths are

Table 4. Parameters of biocemented sand

Parameters	Untreated sand	250 ml	500 m
Peak Strength/kPa, q_u	91.339	80.016	117.415
Residual Strength/kPa, q_s	70.350	67.703	99.632
Initial Elastic Modulus/MPa, E_0	2.337	3.121	3.1901
Max Dilation angle, φ'_{max} (°)	12.4	16.2	17.8

largely enhanced with more volume dilation. On contrary, for the 250 ml-treated sand sample, the peak strength decreases comparing with the untreated case, while the residual strength is almost unchanged by the treatment. Such a conclusion is supported by the repeated tests in our study, but had not been seldom reported in the existing literatures. It may be attributed to the breakage of weak MICP-bonding among particle contacts in extremely low-strength biocemented sands. In such a weakly biocemented case, the vulnerable MICP-bonding increases the initial stiffness of the sand, while are liable to be destroyed under a small shear stress, resulting in a decreased peak strength. It may be then inferred that the peak strength of MICP-treated sands is controlled by the MICP-induced contact bonding.

4.2 Calcite Precipitation Distribution

Using the dissolving method, the carbonate content distributions along the height of all the treated samples are reported in Fig. 8. Overall, compared with the results shown in Fig. 1, the temperature-controlled MICP method provides an improvement of MICP uniformity in sands. The MICP of 250 ml-treated and 500 ml-treated samples are to some extent distributed uniformly within the samples, with two small slopes of the linear regression lines (k(500 ml) = −0.0206 and k(250 ml) = −0.00694). The correlation coefficients of the regression lines, R^2, are at a low level less than 0.4, indicating a relatively low dispersion of the calcite precipitation in the samples. Moreover, the MICP uniformity seems to be fairly well within a large height range from 20 mm to the end of sand columns.

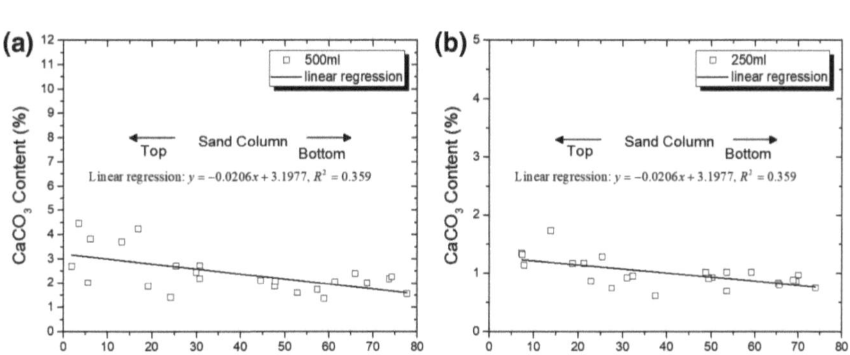

Fig. 8. Calcite precipitation distribution **a** mixture volume: 500 ml; **b** mixture volume: 250 ml

It should be noted that the uniformity here is discussed in a statistical sense and can be studied by analyzing the MICP distribution rules from the statistical perspective. Figure 9 shows that the distribution of $CaCO_3$ content in biocemented sands well meets the Gamma distribution. Calculated using the t-test method, the two fitting probabilities are 75.3% (500 ml) and 85% (250 ml), respectably. The mean square error (MSE) of the Gamma distribution can be used to analyze the uniformity of MICP. More detailed discussions on this issue will be made in our future experimental studies.

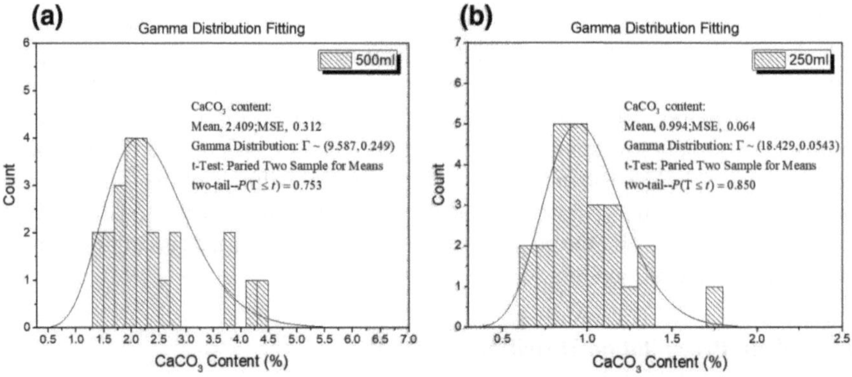

Fig. 9. Histograms of $CaCO_3$ content and Gamma distribution fitting (**a** mixture volume: 500 ml; **b** mixture volume: 250 ml) (MSE represents the mean square error and the two values in brackets are the two parameters of Gamma distribution)

5 Conclusion

A temperature-controlled MICP method is proposed in this study to improve the uniformity of calcite precipitation in MICP-treated sands, which is still a challenging issue of the MICP technology even under laboratory conditions. The urease activity tests show that the urease activity is almost completely inhibited at the temperature below 10 °C. Based on this temperature dependent property of urease activity, studies in this paper show that a better MICP uniformity can be achieved by the microbial grouting at a low temperature, which ensures the homogeneous distribution of bacteria-reagent mixture solution in sand columns before the MICP reactions are activated at a higher temperature. Therefore, the temperature-controlled MICP method includes the grouting step at low temperature and the reaction step at high temperature.

The mechanical properties and MICP distributions of MICP-treated Ottawa sand columns prepared under temperature-controlled conditions are then studied in this study. The results indicate that the effects of MICP on the peak and residual strengths seem to be nonlinearly sensitive to the precipitated calcite contents for low-strength biocemented sands, dependent on the contact bonding induced by MICP. The distribution of $CaCO_3$ content in biocemented sands well meets the Gamma distribution.

Moreover, the measured $CaCO_3$ content distributions along the height of sand columns show that the temperature-controlled MICP method provides an improvement of MICP uniformity in sands.

References

Al Qabany, A., Soga, K.: Effect of chemical treatment used in MICP on engineering properties of cemented soils. Géotechnique **63**(4), 331–339 (2013). https://doi.org/10.1680/geot.SIP13.P.022

Burbank, M., Weaver, T., Lewis, R., Williams, T., Williams, B., Crawford, R.: Geotechnical tests of sands following bioinduced calcite precipitation catalyzed by indigenous bacteria. J. Geotech. Geoenviron. Eng. **139**(6), 928–936 (2013). https://doi.org/10.1061/(ASCE)GT.1943-5606.0000781

Cheng, L., Cord-Ruwisch, R., Shahin, M.A.: Cementation of sand soil by microbially induced calcite precipitation at various degrees of saturation. Can. Geotech. J. **50**(1), 81–90 (2013). https://doi.org/10.1139/cgj-2012-0023

Chu, J., Ivanov, V., Stabnikov, V., Li, B.: Microbial method for construction of an aquaculture pond in sand. Géotechnique **63**(10), 871–875 (2013). https://doi.org/10.1680/geot.SIP13.P.007

DeJong, J.T., Fritzges, M.B., Nüsslein, K.: Microbially induced cementation to control sand response to undrained shear. J. Geotech. Geoenviron. Eng. **132**(11), 1381–1392 (2006). https://doi.org/11(1381)

Feng, K., Montoya, B.M.: Influence of confinement and cementation level on the behavior of microbial-induced calcite precipitated sands under monotonic drained loading. J. Geotech. Geoenviron. Eng. **142**(1), 9 (2016). https://doi.org/10.1061/(ASCE)GT.1943-5606.0001379

Jiang, N., Kuo, M., Soga, K.: Microbially induced carbonate precipitation for seepage-induced internal erosion control in sand-clay mixtures. J. Geotech. Geoenviron. Eng. (2016). https://doi.org/10.1061/(ASCE)GT.1943-5606.0001559

Lin, H., Suleiman, M.T., Brown, D.G., Kavazanjian, E.: Mechanical behavior of sands treated by microbially induced carbonate precipitation. J. Geotech. Geoenviron. Eng. **142**(2), 4015066 (2016). https://doi.org/10.1061/(ASCE)GT.1943-5606.0001383

Mitchell, J.K., Santamarina, J.C.: Biological considerations in geotechnical engineering. J. Geotech. Geoenviron. Eng. **131**(10), 1222–1233 (2005). https://doi.org/10(1222)

Montoya, B.M., DeJong, J.T., Boulanger, R.W.: Dynamic response of liquefiable sand improved by microbial-induced calcite precipitation. Géotechnique **63**(4), 302–312 (2013). https://doi.org/10.1680/geot.SIP13.P.019

Mortensen, B.M., DeJong, J.T.: Strength and stiffness of MICP treated sand subjected to various stress paths. Geo-Frontiers 4012–4020 (2011). https://doi.org/10.1061/41165(397)410

Raghunandan, M., Juneja, A., Hsiung, B.: Preparation of reconstituted sand samples in the laboratory. Int. J. Geotech. Eng. **6**(1), 125–131 (2013). https://doi.org/10.3328/ijge.2012.06.01.125-131

van Paassen, L.A., Ghose, R., van der Linden, T.J.M., van der Star, W.R.L., van Loosdrecht, M.C.M.: Quantifying biomediated ground improvement by ureolysis: large-scale biogrout experiment. J. Geotech. Geoenviron. Eng. **136**(12), 1721–1728 (2010). https://doi.org/10.1061/(ASCE)GT.1943-5606.0000382

Venuleo, S., Laloui, L., Terzis, D., Hueckel, T., Hassan, M.: Microbially induced calcite precipitation effect on soil thermal conductivity. Geotech. Lett. **6**(1), 39–44 (2016). https://doi.org/10.1680/jgele.15.00125

Whiffin, V.S., van Paassen, L.A., Harkes, M.P.: Microbial carbonate precipitation as a soil improvement technique. Geomicrobiol. J. **24**(5), 417–423 (2007). https://doi.org/10.1080/01490450701436505

Xiao, Y., He, X., Liu, H.: New lightweight geomaterials: biocemented sand mixed with expanded polystyrene beads. Sci. China Tech. Sci. **60**(7), 1118–1120 (2017)

Curing Behaviour of Lightly Solidified Clays Monitored with Bender Element and Electrical Conductivity Measurements

Chee-Ming Chan(✉) and Mohammed Mansoor Mofreh Gubran

Faculty of Engineering Technology, Universiti Tun Hussein Onn Malaysia, Johor, Malaysia
chan@uthm.edu.my

Abstract. Soft, weak soils are usually subjected to pre-treatment prior to application in load-bearing to avoid long term stability problems. A similar approach can be adopted for reclaimed land backfilled with dredged marine sediments recovered from the surrounding waters. Pre-treatment like solidification is a time-dependent process, where the improved stiffness over curing or rest period determines how soon the reclaimed land can be utilized. In the present study, a sample of dredged marine sediment was lightly solidified with cement and examined in a large oedometer incorporated with bender element (BE) and electrical conductivity (EC) measuring devices. Comparisons were made with an artificial clay, kaolin. Simultaneous acquisition of BE and EC data enabled real-time monitoring of the improved compressibility of the treated sediment during the 7-day curing period, providing insights to the cementation mechanisms taking place. The measurements revealed the physical and chemical transformation underwent by the treated soil with time, though the results are found to be very much dependent on the soil type. Reported in changes of moisture content (MC), shear wave velocity (Vs) by the BE tests and EC measurements throughout the 7-day curing time lapse, reduction in MC was in general accompanied by a rise in Vs and initial rise followed by decline of EC. The recorded patterns of the parameters with time was mainly attributed to the different soil properties, mineralogy and pore water chemistry inherent in the soil samples examined.

1 Introduction

Marine dredging is a process of excavating sediments from the bottom of water bodies by mechanical or hydraulic machineries, with the common purposes of creating waterways for the navigation of marine vessels, coastal development and removal of contaminated sediments. For instance, sediments clearance for rivers, lakes and shorelines are necessary to maintain the water level within the desired level. According to Bosch (2009), many harbours, cities and past civilisations like Memphis in Egypt, Effuses and Millets in Greece declined and died. In addition, Aiguesm-Mortes in France lost their prominent role due to siltation. Several hundred million cubic meters of sediment were dredged in the U.S from waterways, ports and harbours each year to maintain and enlarge the nation's navigation system for commercial, national defence,

H. Khabbaz et al. (eds.), *New Prospects in Geotechnical Engineering Aspects of Civil Infrastructures*, Sustainable Civil Infrastructures,
https://doi.org/10.1007/978-3-319-95771-5_3

and recreational purposes (U.S Environmental Protection Agency 2004). In addition, according to EUDA (2006), the volume of the dredged sediments in Europe was estimated to be 200 million cubic meters. Zeng and Hong (2016) reported more than 100 million cubic meters of fine-grained clay were dredged yearly in China. On a smaller scale, in year 2012 over 3.5 m^3 were generated from the rehabilitation of Lumut waters of Perak, Malaysia, to maintain and enlarge the navigation channel for commercial, fishing and national fishing purposes (Chan et al. 2013).

2 disposal methods are available for the displaced sediments: open water dumping and inland disposal. However, there are harmful environmental consequences associated with each method. The former operation generates not only a considerable amount of sediments with poor geotechnical properties, but also various contaminants that may cause pollution to the environment (Bray 2013). For instance, coral reefs, flora-fauna and benthic organism could be affected directly by burial or indirectly by smothering (Katsiaras et al. 2015). The sediment physical alteration may also induce distribution patterns of contaminated sediments due to the presence of heavy metals i.e. Cu, Pb, Zn, Mn, and Ni (Homayoun et al. 2015). Moreover the suspended sediments has the potential to harm the fish in short term, causing fish mortality and deposited egg and larvae of eggs (Kjelland et al. 2015). On the other hand, constructing a disposal facility inland often raises the concern of contamination leaching to the groundwater.

Instead of dumping, a more sustainable approach would be to reuse these materials for beneficial purposes in civil engineering construction, e.g. land reclamation and beach restoration works. However, these fine-grained dredged marine soils are of high moisture content, and the presence of a considerable amount of organic matters make direct use impossible due to the high compressibility and low shear strength. Therefore, appropriate treatment must be adopted to reduce DMS compressibility. Solidification technique could be adopted to reduce the compressibility. Some common solidifying agents include cement and lime as adopted by Dong et al. (2012) and industrial by-products such as coal ash and steel slag experimented by Wang et al. (2016).

The present study incorporated bender element (BE) and electrical conductivity (EC) measuring devices in an oedometer to provide real-time monitoring of the time-dependent improvement in compressibility. The complementing measurements enabled the simultaneous mechanisms of solidification and consolidation to be captured for further analysis.

2 Instrumented Oedometer: BE and EC Measurements

The instrumented oedometer cell is a modified oedometer designed as a fixed ring system (Fig. 1). The standard oedometer cells are most commonly used for conducting the 1-D consolidation test on undisturbed saturated soils. For instance, a standard ring of 75 mm diameter and 20 mm high is described in the BS1377-5 (1990). Nonetheless, modified oedometer cells have been fabricated in some past studies as summarised in Table 1.

The oedometer in the present study was designed to provide enough space for installation of the BE and EC probes. In addition, the slurry nature of the soil sample requires a special cell with reasonable travel distance of the shear waves between the

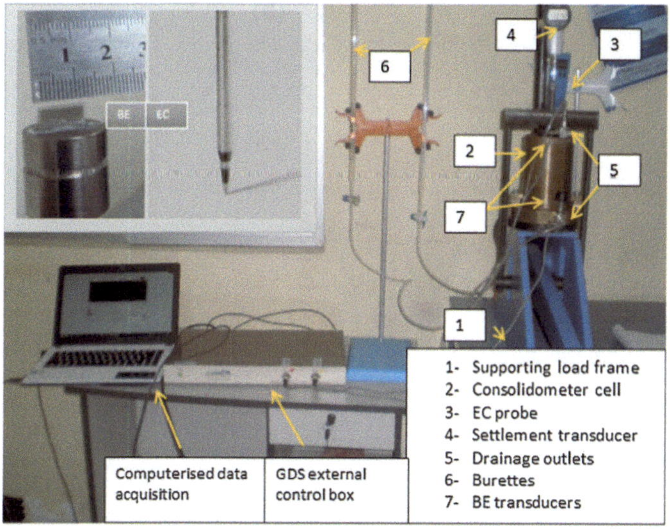

Fig. 1. Test setup for the instrumented oedometer

Table 1. Modified oedometer cell from previous studies

Height (mm)	Diameter (mm)	H:D	Measured parameters	Reference
80	100.6	1:1.26	Compressibility, electromagnetic (EM) properties and BE	Fam and Santamarina (1996)
90	100	1:1.11	Compressibility and BE	Chan (2006)
60	130	1:2.17	Compressibility, EC and BE	Comina et al. (2008)
101	114	1:1.13	Compressibility and BE	Kang et al. (2014)
63	100	1:1.59	Compressibility and BE	Lee and Santamarina (2005) and (Choo et al. (2016)

BE, as well as to be watertight to keep the sample drained via specified routes only. BS1377-5 (1990) states that "the inside diameter shall not be less than 50 mm and not greater than 105 mm; the height of the ring shall be not less than 18 mm and not more than 0.4 times the internal diameter, or H:D ratio of 1:3.75". In addition, ASTM D2435-04 requires cell diameter of not less than 50 mm and height not less than 12.5 mm, or H:D ratio of 1:2.5. Details of the oedometer are presented in Fig. 2 and described in Table 2.

For the BE measurements, a sinusoidal wave form was selected due to its simplicity where the received peaks are easier to be identified compared to other wave forms. The amplitude was limited in GDS system between 5 and 14 V. For this study, the amplitude was fixed at 10 volts. The selection of the wave periods, t (msec) were based on the loading stage. The frequency is calculated as inversion of the input single wave

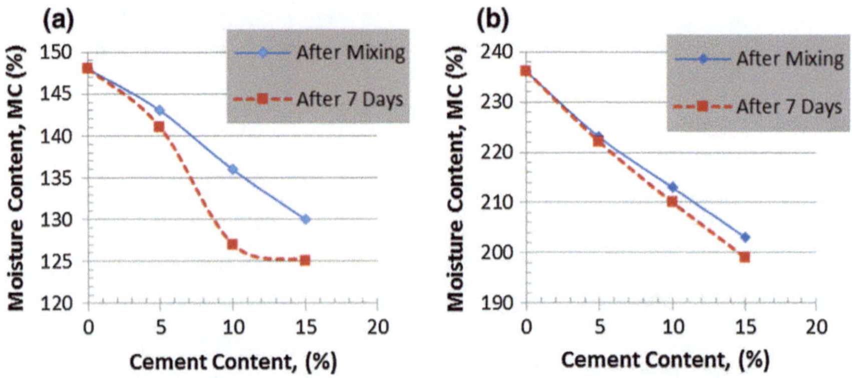

Fig. 2. Moisture content reduction after mixing and curing against cement content; **a** kaolin **b** DMS

Table 2. Main components of the instrumented oedometer

Components	Descriptions	Functions
Mould 	• Dimensions: H × D × W 146 × 100 × 4 (mm) • Weight = 1868.9 (g)	• Contain soil specimen
Top cap 	• Dimensions: H × D 36 × 99 mm • Weight = 2400.3 g • Accommodate three holes for BE, EC and drainage path	• Contain a central seating to support the applied vertical stress • BE S-wave source transducer • EC probe inserted 50 mm inside the specimen • Top drainage path
Base platen 	• Base inserted in mould 10 mm • Weight = 4719.6 g • Accommodate two holes for BE and drainage path	• Support the mould and the cap • BE S-wave receiver transducer • Bottom drainage path

period (T), i.e. f = 1/T. Frequencies lower than 1 kHz was used for the smaller stresses where the stiffness of the sample is low and higher frequencies (up to 100 kHz) were used for higher stresses.

Note that there were some inevitable limitations to the EC measuring probe though. The natural DMS was diluted with distilled water to reduce the salt concentration to obtain lower EC. However, in order to achieve EC of 9 mS/cm DMS, MC should be 600% which is approximately 5 liquid limits of the tested soil. In this study, all specimens were mixed at 2LL. Based on that, the current probe was applicable to measure the EC for kaolin specimens only which had EC below 7 mS/cm.

3 Soil Samples

Refined Malaysian kaolin grade was used in this study and its physical properties and chemical composition are shown in Table 3. Kaolin (K) was selected as a control soil to be compared with the DMS at the same cement dosage. The dredged marine sediment (DMS) was collected from a dredged site and was classified as a high plasticity clay associated with high moisture content and high amount of fine grain particles in addition to the dissolved salt in water. Physical properties of DMS can be found in Table 3. In addition, XRF analysis was conducted to identify the concentration of the DMS minerals at the soil's original state. The average concentration level of the minerals are determined: silica (SiO_2) is the dominant mineral 41.63% followed by aluminum oxide (Al_2O_3) 14.63% and sodium oxide (Na_2O) 8.07%, while the other minerals are found to be of less significant levels.

Table 3. Physical properties of DMS and kaolin

Physical properties	DMS (%)	Kaolin (%)
Natural moisture content	149	–
Liquid limit	118	74
Plastic limit	57	36
Plastic index	61	–
Particle density	2.46	2.60
LOI	9.97	10.5–13.0

4 Preparation of Test Samples

Both soil samples were pre-mixed at twice the respective liquid limit prior to addition of ordinary Portland cement as solidification agent. The cement dosages were predetermined as 0, 5, 10 and 15% per dry mass of the soil, where the test samples were termed simply as 5 K and 5DMS for kaolin and DMS respectively with the prefix denoting the cement dosage. Mixed in a conventional kitchen mixer, the soil, cement and water were thoroughly blended to from a uniform mixture prior to being placed in the oedometer. Once deposited in the cell, the sample was left to cure for 7 days, where simultaneous BE and EC measurements were carried out.

5 Results and Discussions

Moisture Content (MC). The addition of hydraulic binders to clay slurry soil normally reduces the MC due to the chemical reactions (i.e. hydration products) between clay-water-cement mixtures. From Fig. 2 it is obvious that MC reduced linearly after mixing. Interestingly, the reduction in MC for kaolin and DMS showed a typical value for each soil. In other words, when the cement content increased from 5 to 10%, MC in kaolin reduced by 6.5%, and when the cement content increased from 10 to 15% the same percentage was obtained. Similarly, the MC of DMS reduced by 10% for each 5% cement dosage increment. Thus, the reduction, immediately after mixing, in DMS was higher by 3.5% compared to kaolin.

After 7 days of curing, MC of the DMS specimens showed no major reduction. However, the difference became slightly bigger as the cement content increased, as can be observed in Fig. 2b. This indicates curing induced no major reaction, hence no significant cementation in DMS compared to kaolin. On the other hand, the reduction in MC for 7-day curing of the kaolin specimens was higher. Apparently 5% cement addition resulted in no significant loss of moisture after a week. However, it is noted that with $\geq 10\%$ cement addition kaolin underwent more significant MC loss than DMS, as depicted by deviation of the plots between 'after mixing' and 'after 7 days'. This indicates that 10% is the minimum required to produce meaningful MC reduction in both cases, which affects the hydration products and hence producing effective cementation.

Similar observations have been reported by Chew et al. (2004), where much reduction in MC after 7 days compared to after mixing was reported. It was further explained that the dry cement caused the immediate reduction whereas the major reduction took place within 7 days curing as a result of the absorbed water that was transferred to hydrated CSH and CASH due to hydration and pozzolanic reactions. Nonetheless the reduction in MC after mixing seemed to be higher in this study for all tested soils. A possible explanation could be mineralogical effects, such as the total surface area which might contribute to immediate consumption of the water when mixed with soil during mixing.

Shear Wave Velocity (Vs). Figure 3 depicts the cement dosage effects on V_s during the curing time for kaolin specimens. Generally, Vs increased with the prolonged curing time. During the first 12 h, Vs increased sharply up to 50 m/s almost at the same rate and magnitude for all the kaolin specimens regardless of the cement content. This suggests that the early stages of hydration do not contribute much to the formation of distinguishable rigid and cemented network yet (Fam and Santamarina 1996). This is obvious when 15 K showed significant and continuous Vs increment over time, while the Vs of 5 and 10 K continued increasing with the same magnitude up to 24 h. After that there was no significant increment for 5 K. Vs of 10 K continued increasing but with lower magnitude compared to the 15 K specimen. Vs for 5 K remained almost constant after the first day of curing. This shows that the cement dosage was not sufficient to make further enhancement in terms of small strain stiffness beyond that time. On the other hand, specimens 10 and 15 K cement exhibited further increment which seemed to continue increasing beyond the specified curing time.

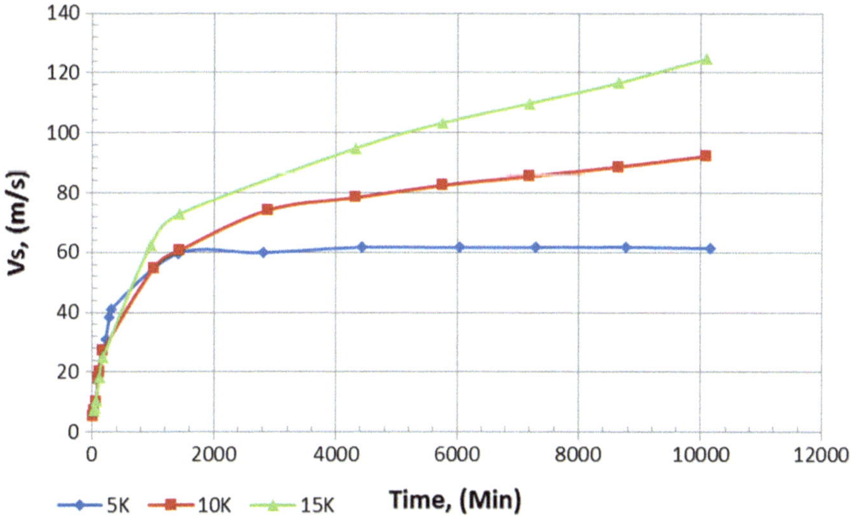

Fig. 3. Relationship between Vs and elapsed curing time for solidified Kaolin specimens

Seemingly, a distinguishable improvement in the small strain stiffness based the cement content was delayed until certain time. According to Gartner et al. (1985), upon adding water to cement the solution becomes saturated with calcium hydroxide and the calcium ions were released in the solution. After 12 h, a drop in calcium content is attributed to the setting of cement. Therefore, a substantial amount of calcium and water are consumed to form CSH and $Ca(OH)_2$ (Prusinski and Bhattacharja 1999). This indicates that the mixture was still in the cation exchange phase causing shrinkage in the water layers to formulate the flocculation and agglomoration which creates yet unstabilised clay particle's formation. After that the hydration and pozzolanic reactions products (CSH and CAH) started creating the bonding among the particles which contributed to the rigidity and stiffening the mixture as shown in Fig. 2.3c, d. The maximum Vs obtained for 5 K, 10 K and 15 K at the end of the curing period (7 days) were 61.5, 97.6, 130.5 m/s respectively.

The relationship between Vs and curing time for solidified DMS specimens are shown in Fig. 4. Generally, Vs increased as time and cement content increased. Unlike kaolin, Vs for DMS increased slightly at almost the same rate and magnitude for the first 2 h. After that, 10 and 15 K showed sharp increment during the second and third day of curing and moderately for the remaining time. The 5 K specimen showed continuous increase but with insignificant increment (5–9.5 m/s only) over 7 days of curing.

Although it can be said that the 5% cement almost doubled Vs comparing with the control specimen but it is far smaller than Vs for the kaolin or even the 10DMS and 15DMS. It is apparent that the cement effects on Vs for kaolin specimens was more significant than DMS. This indicates that the stiffness gain in DMS was less compared to kaolin. According to Sasanian and Newson (2014), the mineralogical differences, clay contents, liquid limits and moisture chemistry parameters govern the behaviour of

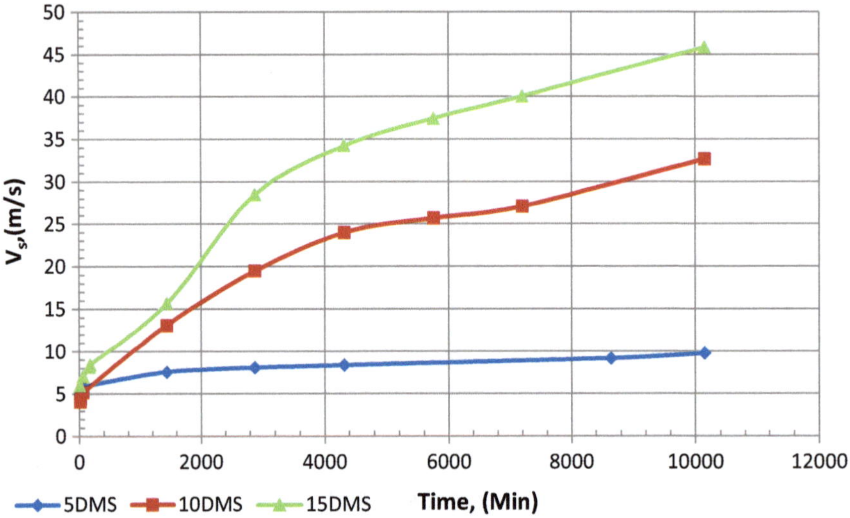

Fig. 4. Relationship between Vs and elapsed curing time for solidified DMS specimens

cementation behaviours of solidified soil. The maximum Vs obtained on the seventh day of curing for DMS were 9.5, 32.7 and 45.9 m/s for 5DMS, 10DMS and 15 DMS respectively.

Although the first reading of the Vs for kaolin and DMS prior to the cement addition was almost identical at 4 m/s, generally cement addition to the soil increased Vs significantly. It is noted that Vs was higher in kaolin compared to DMS. Moreover, the initial major increment for kaolin was sharp within shorter time. This is identical with Vs measurement during the curing period for concrete conducted by Zhu and Bate (2014). However, the increment in DMS was gradual and required a longer time. It can be concluded that Vs increased as the cement content and curing time increased, a relationship which could be used to assess the solidification process in terms of small strain shear stiffness.

It was also observed that Vs at the end of the specified curing time increased with further curing time, which is in parallel with the perception that stiffness of soil improves with curing time. The relationship between the final Vs and the cement content for kaolin and DMS specimens at the end of the specified curing time is shown in Fig. 5a. As anticipated, the gain in Vs is linearly proportionate to the cement content, although both kaolin and DMS showed good relationship with cement content, but it seemed more noticeable in kaolin. Moreover, it was found that the magnitude of Vs was higher in kaolin compared to the DMS specimens. The hydration and pozzolanic reactions apparently transferred the absorbed water into cementation products, i.e. CSH and CASH, which was not ejected again by reheating (Chew et al. 2004). Thus, these products form bonding among the soil particles which contributed to the stiffness of the

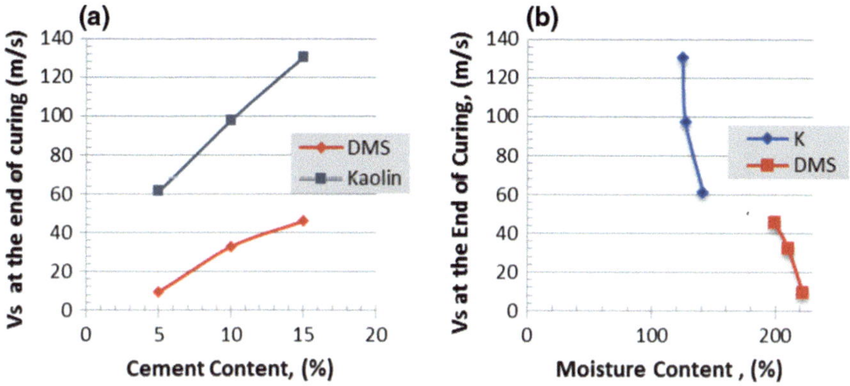

Fig. 5. Effects of **a** cement content and **b** moisture content on the final Vs at the end of curing time

solidified soils and hence increased Vs. Likewise, the final Vs at the end of the specified curing time was plotted against the MC Fig. 5b. It is obvious that Vs increased when the MC was reduced which in turn was consumed to form the cementation products.

Electrical Conductivity (EC). As mentioned earlier, EC readings were not possible in DMS due to limitation of the measuring range of the EC probe. Figure 5 shows the behaviour of the electrical conductivity (EC) of kaolin specimens during the curing time. Generally, EC values increased as the time and cement content increased. EC increased sharply during the first 20 h for all specimens. The 5 K specimen reached the maximum EC of 5.97 mS/cm on the third day of curing while 10 and 15 K reached their maximum EC of 6.73 mS/cm on the first day curing. After that, EC started decreasing continuously for all specimens. While decreasing, note that the plots for all specimens shown in Fig. 5 merged on the third day of curing onward.

Similar trend was observed on the Vs where the EC values have the same trend for all specimens at this time. It can be noted that the EC for all samples peaked at the 17th hour, after which EC started to decline. The increment of EC is due to the active hydration process as mentioned previously. The increase of the EC at the first day can be attributed to the presence of the free ions in the solution due to the cation exchange. For instance, within 12 min, upon adding water to cement, the solution became saturated with calcium hydroxide and the calcium ions (Ca^{2+}) were released in the solution. After that the chemical reactions in terms of emitting new ions slowed down and the bound water around the particles was reduced by the hydration process which is attributed to the setting of the cement where Ca^{2+} was consumed to produce CSH and CASH products (Prusinski and Bhattacharja 1999). Therefore, EC started to decrease. In other techniques i.e. the dielectric permittivity measurements in microwave frequency range can provide a good indication of free water volume fraction, while EC can indicate the availability and mobility of ions in soil (Lee et al. 2008) (Fig. 6).

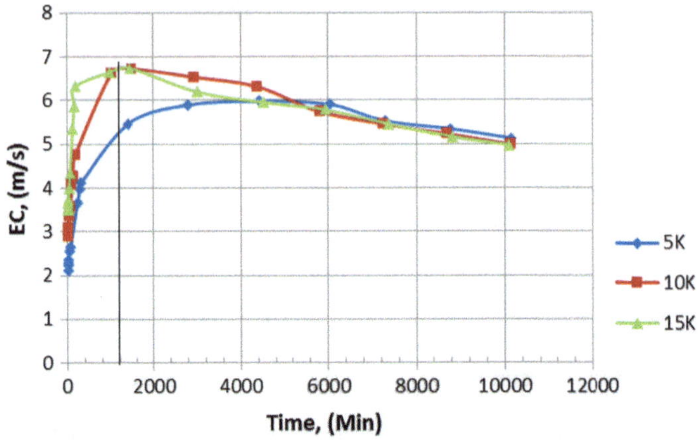

Fig. 6. Curing effects on EC for solidified kaolin specimens

6 Conclusions

The instrumented oedometer described in the present paper was used to capture changes in Vs and EC over the 7-day curing time of lightly solidified kaolin and DMS. Overall the same amounts of cement added to the soils, i.e. 5, 10 and 15%, were found to be more effective in kaolin than DMS, primarily due to the differences in the inherent properties, mineralogy as well as pore water chemistry. Nonetheless a general pattern of change in the parameters monitored was observed: reduction of moisture with time due to hydration and subsequent pozzolanic reactions between soil-cement took place in conjunction with increasing Vs (hence stiffness) and EC, though the latter showed gradual decline after 24 h. This suggests stiffness improvement attributed to solidification, a result of chemical changes and formation of cementation bonds in the soil-cement mixture, albeit the one week lapse of time.

Acknowledgements. Technical assistance by the laboratory personnel at RECESS and Geotechnical Lab of UTHM are duly acknowledged. Funds for the second author's doctoral work was provided by GIPS (ORICC), UTHM. The registration fee is owed to IGSP (U676) by ORICC, UTHM.

References

ASTM International: ASTM D2435-04 Standard Test Methods for One-Dimensional Consolidation Properties of Soils Using Incremental Loading. http://www.astm.org/Standards/D2435.htm

Bosch, R.V.D. Dredging Technology. Hydraulic Structures and Water Data Acquisition Systems, vol. II (2009)

Bray, R.N.: Environmental Aspects of Dredging. CRS Press, Taylor & Francis Group, London (2013)

British Standards Institution: BS1377-5 methods of tests for soils for civil engineering purposes: compressibility, permeability and durability tests (1990)

Chan, C.M., Pun, K.H., Ahmad, F.: A fundamental parametric study on the solidification of malaysian dredged marine soils. World Appl. Sci. J. **24**(6), 784–793 (2013)

Chew, S.H., Kamruzzaman, A.H.M., Lee, F.H.: Physicochemical and engineering behavior of cement treated clays. J. Geotech. Geoenviron. Eng. **130**(7), 696–706 (2004). Available at: http://ascelibrary.org.libproxy1.nus.edu.sg/doi/abs/10.1061/(ASCE)1090-0241(2004)130:7(696)

Dong, X.W., et al. Solidification/stabilization of dredged marine sediments for road construction. Environ. Technol. 37–41. (November 2012)

European Dredging Association (EUDA): Dredging Creates Space: Annual Report (2006)

Fam, M.A., Santamarina, J.: Study of clay-cement slurries with mechanical and electromagnetic waves. J. Geotech. Geoenviron. Eng. **122**(May), 365–373 (1996)

Gartner, E., Tang, F., Weiss, S.: Saturation factors for calcium hydroxide and calcium sulfates in fresh Portland cement pastes. J. Am. Ceram. Soc. **68**(12), 667–673 (1985). Available at: http://onlinelibrary.wiley.com/doi/10.1111/j.1151-2916.1985.tb10122.x/abstract

Homayoun, F., et al.: Significance of environmental dredging on metal mobility from contaminated sediments in the Oskarshamn Harbor, Sweden. Chemosphere **119**(January), 445–451 (2015)

Katsiaras, N., et al.: Impacts of dredged-material disposal on the coastal soft-bottom macrofauna, Saronikos Gulf, Greece. Sci. Total Environ. **508**, 320–330 (2015)

Kjelland, M.E., et al.: A review of the potential effects of suspended sediment on fishes: potential dredging-related physiological, behavioral and transgenerational implications. Environ. Syst. Decis. **35**(3), 334–350 (2015)

Lee, J.Y., Santamarina, J.C., Ruppel, C.: Mechanical and electromagnetic properties of northern Gulf of Mexico sediments with and without THF hydrates. Mar. Pet. Geol. J. **25**, 884–895 (2008)

Prusinski, J., Bhattacharja, S.: Effectiveness of Portland cement and lime in stabilizing clay soils. In: Transportation Research Record, 1652, pp. 215–227 (1999). Available at: http://trb.metapress.com/index/D106M01163UG21V5.pdf

Sasanian, S., Newson, T.A.: Basic parameters governing the behaviour of cement treated clays. Soils Found. **54**(2), pp. 209–224 (2014). Available at: http://dx.doi.org/10.1016/j.sandf.2014.02.011

U.S Environmental Protection Agency: Report of the dredging status in United States (2004)

Wang, D., Abriak, N.E., Zentar, R.: Dredged marine sediments used as novel supply of filling materials for road construction. Mar. Georesour. Geotechnol. **618**(June), 1–9 (2016). Available at: https://www.tandfonline.com/doi/full/10.1080/1064119X.2016.1198945

Zeng, L., Hong, Z.-C.: Time-dependent compression behaviour of dredged clays at high water contents in China. Appl. Clay Sci. **123**, 320–328 (2016). Available at: http://linkinghub.elsevier.com/retrieve/pii/S0169131716300400

Zhu, J., Bate, B.: Using shear wave velocity to monitor the curing process of self-consolidating concrete by bender element. NUTC R339, Project #00042503, p. 34 (2014)

Effect of Grain Content on the Sandstone Properties Using Biconcave Bond Model of DEM

Chia-Chi Chiu[1], Shi-Feng Chen[1], and Meng-Chia Weng[2(✉)]

[1] Department of Civil Engineering, National Taiwan University, Taipei, Taiwan
[2] Department of Civil Engineering, National Chiao Tung University, Hinchu, Taiwan
mcweng@nctu.edu.tw

Abstract. This study adopted the biconcave bond model of DEM (Discrete element method) to investigate the mechanical behavior of sandstone. The biconcave bond model is capable of simulating the cementation between non-contact grains of sandstone with high porosity. It also links the micro parameters and macro material properties. The biconcave bond model incorporated with grain-matrix structure was used to investigate the influence of grain and matrix on sandstone behavior. The simulation shows that the stress-strain curves fit experiment result well. In addition, the sensitivity analysis of GAR (Grain area ratio) and porosity shows that the simulation of UCS and Young's modulus agree well with the empirical equation.

1 Introduction

Due to the complicated constitution of rock, understanding how its intrinsic microscopic properties, including texture and cementation, affect its macroscopic mechanical properties, such as strength and deformation, is essential in rock mechanics research. Numerous investigations have revealed that the microscopic characteristics that influence the uniaxial compressive strength of rock include mineral composition (Bell 1978; Bell and Lindsay 1999; Gunsallus and Kulhawy 1984; Smart et al. 1982), cementation (Clough 1981), grain size (Haztor and Palchik 1997, 1998; Singh, 1988), packing, and porosity (Bell and Culshaw 1998; Dobereiner and Freitas 1986; Dyke and Dobereiner 1991; Ersoy and Waller 1995; Jeng et al. 2004; Palchik 1999; Sabatakakis et al. 2008; Tiryaki 2008; Weng et al. 2005). Jeng et al. (2004) and Weng et al. (2005) performed mechanical experiments and petrographic analyses on 13 different sandstone layers obtained from the Western Foothill Range in Taiwan. They found that the strength of weak sandstone is negatively correlated to grain content, which means the strength of rock is mainly provided by matrix rather than grain. Figure 1 shows two microscopic images of sandstones, MS sandstone and TK sandstone, in Taiwan. MS sandstone exhibits high grain area ratio (GAR = 67.5%) and the matrix content (particle size less than 0.076 mm) is quite low (Fig. 1a). TK sandstone has a rather small grain ratio (GAR = 28.2%), which indicates that more than a half of the area is filled with matrix and porosity and most of mineral grains do not contact each other (Fig. 1b). The

H. Khabbaz et al. (eds.), *New Prospects in Geotechnical Engineering Aspects of Civil Infrastructures*, Sustainable Civil Infrastructures,
https://doi.org/10.1007/978-3-319-95771-5_4

uniaxial compressive strength of MS sandstone and TK sandstone are 37.13 and 69.0 MPa, respectively. The sandstone with higher grain area ratio (on the other hand, lower matrix content) results in lower strength. This special characteristic can be founded in the Tertiary sandstones in western Taiwan, which are not strong enough to be classified as hard rock due to relatively short rock forming period. These weak rock caused lots of engineering problem in Taiwan, such as squeezing, shear dilation, creeping and weakened by water (Weng and Li 2012).

(a) MS sandstone (*GAR*=67.5%, *n*=14.1%) (b) TK sandstone (*GAR*=28.2%, *n*=12.8%)

Fig. 1. Textural composition of sandstones with different grain content

Based on Jeng et al. (2004) and Weng et al. (2005), they adopted multivariable regressive analysis to investigate the relationship between macroscopic mechanical behavior and microscopic parameters (i.e. petrographic parameters). As a result, the empirical relation of the uniaxial compressive strength (UCS) can be expressed in terms of porosity n and GAR, as Eq. 1 and Fig. 2 shown.

$$UCS = (133.7 \cdot \mathrm{e}^{-0.107n})(3.2 - 0.026\,GAR) \tag{1}$$

Note that the porosity n typically ranges between 5 and 25%, and GAR ranges between 20 and 75% for the studied sandstone. The results show that lower porosity n and less GAR result in stronger sandstone. Similar tendency is also found in Young's modulus (E). The relationship between Young's modulus (E), porosity n, and GAR can be expressed as:

$$E = 36.3\mathrm{e}^{-0.106n}(0.354 + 0.017\,GAR) \tag{2}$$

Though it has been shown that the microscopic factors do affect the mechanical behavior of sandstone, the mechanism of relative mechanical properties between grain and matrix is difficult to be studied either by laboratory experiments or by petrographic analysis. Given this difficulty, a numerical model, provided that it is capable of accounting for the discrete packing nature of grains and the bonding strength, can serve as convenient means to study the influence of microscopic properties on the macroscopic behaviour of rock. Among the existing numerical model, discrete element

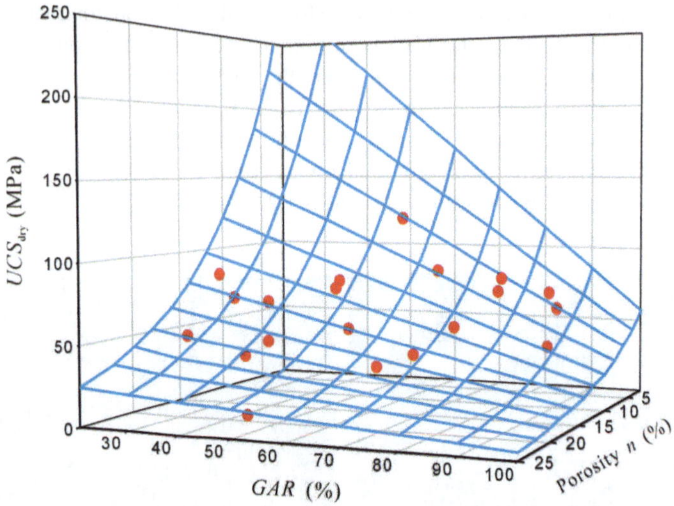

Fig. 2. Empirical relationship between uniaxial compressive strength (UCS) with GAR and porosity n. This figure plots the actual data of USC's in red solid points

method (DEM) possesses the above-mentioned requirements, and this model is adopted in this research. By using DEM, the distribution of the grain and the pore can be systematically varied and the corresponding variation of macroscopic mechanical behaviour can be observed. As a result, clarification on the effects of the porosity on strength and deformation behaviours of sandstones is accordingly obtained. The purpose of this study is to adopt a new developed numerical model of DEM to investigate the mechanical behavior of rocks with different grain content. To check its validity, the simulation is compared with the experimental results.

2 Methodology

Hsieh et al. (2008) simulated the mechanical behavior of rock with different grain contents through a modified bonded-particle model (BPM) of DEM. However, one of the major difficulties in simulating nature sandstone is to model sandstone with high porosity. Using BPM to model sandstone with high porosity may result in "floating particles", i.e. particles not supported due to the limited number of particles that can be used. Therefore, previous work introduced an additional type of particle representing the "porous matrix" to model high-porosity sandstones without leading to unsupported, floating particles. As a result, their model contains three types of particles representing grain (GP), matrix (MP) and porous matrix (PP). The consideration provides an enlightening approach that links the relation of numerical model to petrographic parameters. However, the porous particle is virtual in real environment, it is a simulation skill that remains the stability of model under high porosity, so its parameters cannot refer to any physical properties. Since the simulation technique significantly progresses in recent year, the use of porous particle is unnecessary. In this study, we adopt an advanced bond model, biconcave bond model (Chiu et al. 2015, 2016), to simulate the cementation between non-contact grains. The adopted model can simulate sandstone with high porosity.

2.1 Biconcave Bond Model

The concept of biconcave bond model is originated from the observation of rock thin section, which shows grains are bonded with biconcave shape cementation (Fig. 3). The basic theory of biconcave bond model is the Dvorkin theory (Dvorkin et al. 1991), which describe the elastic reaction of bond layer under external displacement, to calculate forces of bond under given particle displacement. Additionally, the biconcave bond model applies modifications such as motion decomposition, stress adjustment and improvement of algorithms to make model precisely and efficiently. The biconcave bond model have the above characteristics:

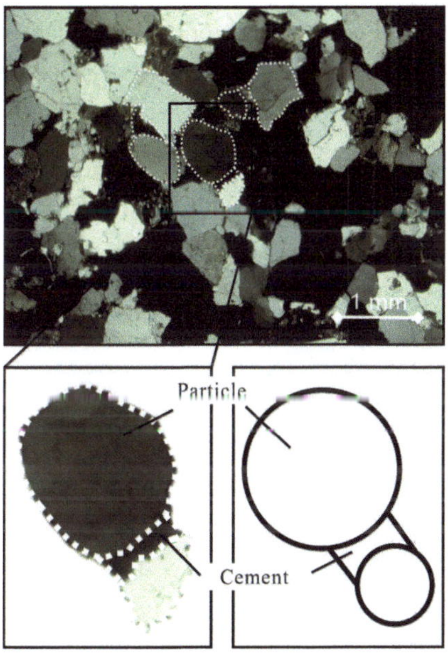

(a) Schematic particle and cement

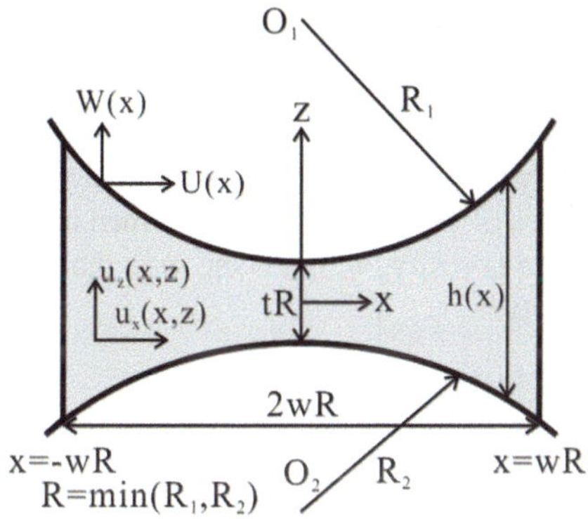

(b) Geometry and coordinates of a bond layer

Fig. 3. Illustration of biconcave bond model. The particles are treated as grains and biconcave shape cementation bond grains together

(1) It can evaluate the mechanical behavior of biconcave shape bond.
(2) Bonds can be built with a gap if distance between center of particle less than $D/\sqrt{2}$, which D is the diameter of minor particle.
(3) The parameters of biconcave bond model are material properties such as Young's modulus and Poisson's ratio rather than stiffness and stiffness ratio.

The reason that this study choose biconcave bond model as the bonding model due to its meaningful required parameters and discrete bonding feature. Since the major micro-parameters used in this study are refers to empirical equations which acquired by laboratory tests, and reproduce correct proportion of grain, matrix and pore is necessary, the biconcave bond model is the quite suitable. In summary, the biconcave bond model has the following features. Since the proposed model considers the effects of bond shape and thickness, it performs better than existing models. Especially, the proposed model can consider the bond behavior between two non-contact particles, which is common in clastic rock.

2.2 Particle Types

Compared with the particle types of Hsieh et al. (2008), this study neglects the use of porous matrix particle (PP) and only selects the two major types of particle, grain (GP) and matrix (MP), denoted as GM structure. Three types of bond, G–G, M–G and M–M, exist between the particles (Fig. 4). In this study, we combine biconcave bond model with grain-matrix structure to construct a new framework for simulate the mechanical behavior of sandstone. Since biconcave bond model can be assigned even if particles are separated, the model can be built with corresponding grain and matrix particles without using porous particle. Thus, porous particle is neglected in this study, the area of grain particle to specimen area in numerical model equals to specified GAR; and the porosity, which is the ratio that empty area to specimen area, equals to specified porosity.

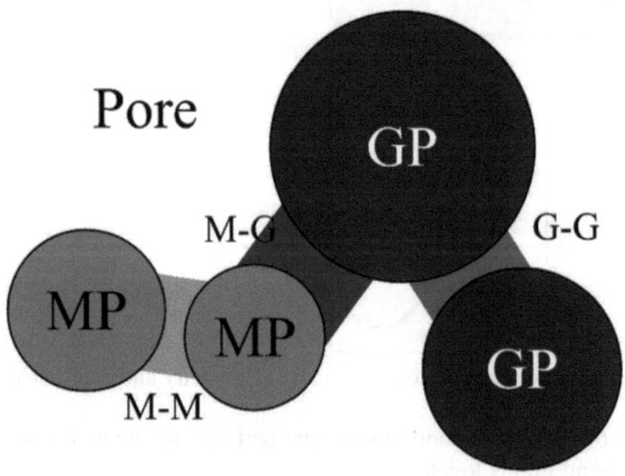

Fig. 4. Illustration of the particles and the corresponding three bonding types (G–G, M–M, M–G) adopted in this research. Symbols: GP = grain particle, MP = matrix particle. The particles have no necessary to contact each other

2.3 Required Parameters

The required parameters of simulated specimen can be divided into three parts, linear contact model, biconcave bond model, and GM structure. The parameters of linear contact model is required for handling contact reaction after bond breaks; The biconcave bond model reflects the reaction of banded particle; and the GM structure represents the distribution of grain and matrix to improve the precise of simulation.

The linear contact model requires three parameters: (1) normal stiffness; (2) shear stiffness or stiffness ratio; (3) friction coefficient. The normal and shear stiffness can be derived by Young's modulus by Potyondy and Cundall's suggestion (Potyondy and Cundall 2004). The biconcave bond model requires six parameters: (1) Young's modulus; (2) Poisson's ratio; (3) bond thickness; (4) bond width; (5) bond normal strength; and (6) bond shear strength. In addition, The GM structure, in this study, divided particle as two types (GP, MP) and bond as three types (G–G, G–M, M–M).

3 Verification of Model

To validate the application of BCB-GM framework (Biconcave bond model with Grain and Matrix framework), we prepared a verification by performing a uniaxial compression test for matrix-support rock. The uniaxial compression strength, Young's modulus, stress-strain curve and failure pattern are captured for compare to laboratory experiment results.

3.1 Model Settings

This study uses the experiment data of MS2 uniaxial compression test as the validation of BCB-GM framework. The GAR of MS2 is about 70%, the matrix is 17%, and the porosity is 13%. In PFC2D numerical model, particles are generate randomly, the gradation of specimen follows natural specimen with up-scaling. In simulations, the particles size less than 1 mm are set as matrix particle, and larger than 1 mm are defined as grain particle, thus, the ratio between GAR, matrix and porosity are sets to 70, 17 and 13% (Fig. 4).

To assign the parameters in BCB-GM framework, Eqs. (1) and (2) are necessary to obtain the properties of grain, matrix and bonding. The deformability of grain and matrix particles, which can be acquired by Eq. (1) through corresponding GAR and n value, for grain is GAR = 100% with n = 0%; for matrix is GAR = 0% with n = 0%. However, the Eqs. (1) and (2) has its applicability, thus this study assume that the Young's modulus of grain can be derived by setting GAR = 75% with n = 10%, and the parameters of Young's modulus of matrix is GAR = 20% with n = 10%. In the BCB-GM framework, the area of bond do not be counted into the area of matrix and is virtual in model, however, it is necessary for the operation of model, thus, we reduce the Young's modulus of bond for half to reduce its influence to the deformability, and its GAR and n are selected based on the settings of model which in this case is GAR = 70% with n = 13% (Fig. 5).

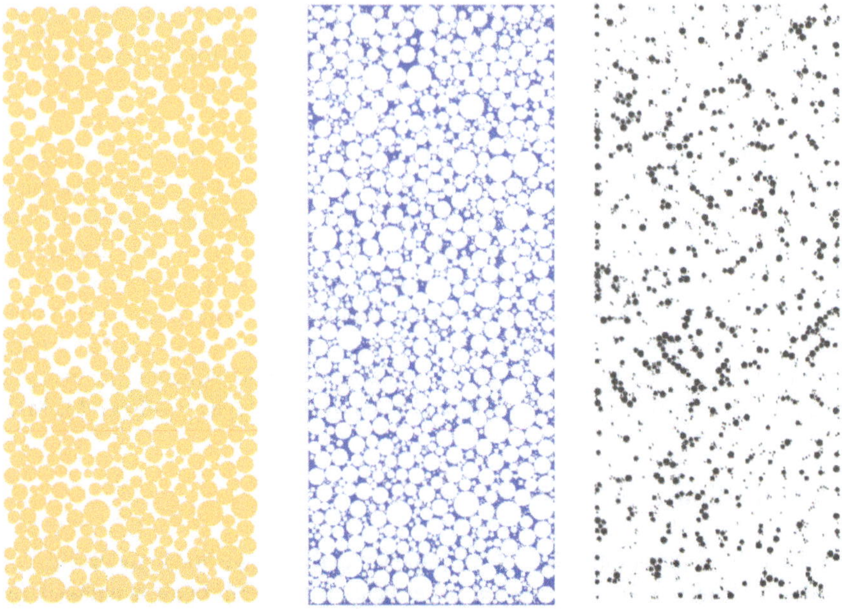

(a) Grain (70%) (b) Matrix (17%) (c) Pore (13%)

Fig. 5. Distribution of the grain, matrix, and pore of the proposed model for sandstone

The strength of bond can be assigned by Eq. (2), which is similar to the settings of Young's modulus of particle, There are three types of bond: G–G, M–G and M–M, in G–G bonding type, the parameters are set to GAR = 75% with n = 10%; in M–M bonding type, the parameters are GAR = 25% with n = 10%; at last, the parameters of M–G bonding type are decided by the settings of model.

The other parameters, such as unit weight and friction coefficient, are obtained by experimental data or back analysis and are list in Table 1. The high value of friction coefficient is for controlling post-peak behavior, since biconcave bond model is adopted, the friction coefficient is unimportant during pre-peak stage, thus this parameter do not correlate with experimental data.

3.2 Simulation Results

The stress-strain curves of simulation and experimental results are shown in Fig. 6. The uniaxial compression strength and Young's modulus of experiment data are 39.6 MPa and 7.9 GPa, respectively. The uniaxial compression strength and Young's modulus of simulation are 40.6 MPa and 8.17 GPa, respectively. The simulated compression strength and Young's modulus are capable of describing the behavior of MS sandstone. In addition, the development of cracks, including tension crack and shear crack, are also shown in Fig. 6. The tensile and shear cracks develop with a similar trend.

Table 1. Parameters of sandstone under uniaxial compression test

Parameter	Value	
Particle: linear contact model		
Types	GP (Radius > 1 mm)	MP (Radius < 1 mm)
Unit weight (kg/kgm^3)	2320	
Friction coefficient	2	
Particle Young's modulus (GPa)	$G_1(n = 10)\ G_2(GAR = 75)$	$G_1(n = 10)\ G_2(GAR = 20)$
Poisson's ratio (v)	0.174	
Stiffness ratio (k_n/k_s)	$2(1 + v)$	
Particle normal stiffness	$2tE_{GP}$, $t = 1$	$2tE_{MP}$, $t = 1$
Particle shear stiffness	$2tE_{GP}/(k_n/k_s)$, $t = 1$	$2tE_{MP}/(k_n/k_s)$, $t = 1$
Bonding: biconcave bond model		
Bond Young's modulus (GPa)	$G_1(n)G_2(GAR)/2$	
Bond poisson's ratio	0.3	
Bond thickness	0.6	
Bond width	0.8	
Bond normal strength and shear strength (MPa)	G–G	$F_1(n = 10)\ F_2(GAR = 75)$
	M–G	$F_1(n)\ F_2(GAR)$
	M–M	$F_1(n = 10)\ F_2(GAR = 25)$

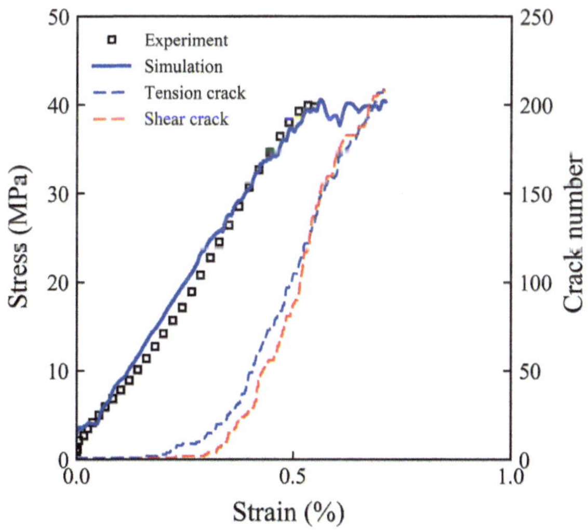

Fig. 6. Stress-strain curves of simulation (blue solid line) and experiment result (square) under uniaxial compressive loading

4 Influence of Gar and Porosity

After validating the proposed model, this study further investigates the influence of GAR and porosity on the mechanical behavior of sandstone. The GAR ranges from 35 to 75, and the porosity ranges from 10 to 20% (Fig. 7). The required parameter are same as those of MS sandstone, listed in Table 1.

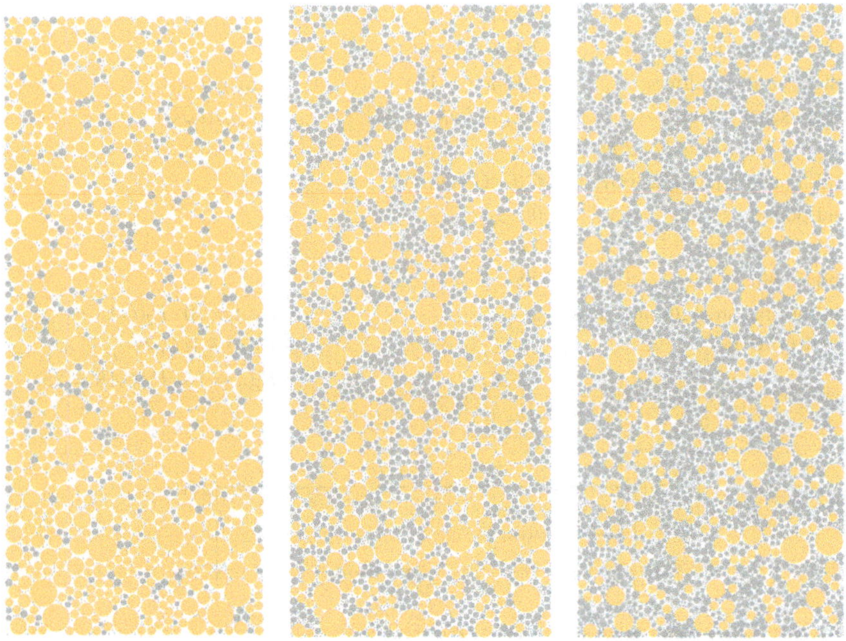

(a) GAR=75 (b) GAR=55 (c) GAR=35

Fig. 7. The simulated sandstones with different GAR and the corresponding failure patterns under uniaxial compression test (n = 15%)

Figures 8 show the stress-strain curves of sandstone with different GAR and porosity. To highlight the influence of porosity, simulations of different porosity with the same GAR are plotted in one figure. The simulation shows porosity is negatively correlated to the strength and Young's modulus of specimen, and the post-peak curve will be ductile as porosity increases. The GAR is also negatively correlated to uniaxial compression strength, however, is positively correlated to Young's modulus in general.

Figure 9 shows the results of simulation compared to theoretical values of Eq. 1. The simulation of uniaxial compression strength precisely located in the classification based on Eq. 1. The proposed model is able to reflect the influence of GAR and

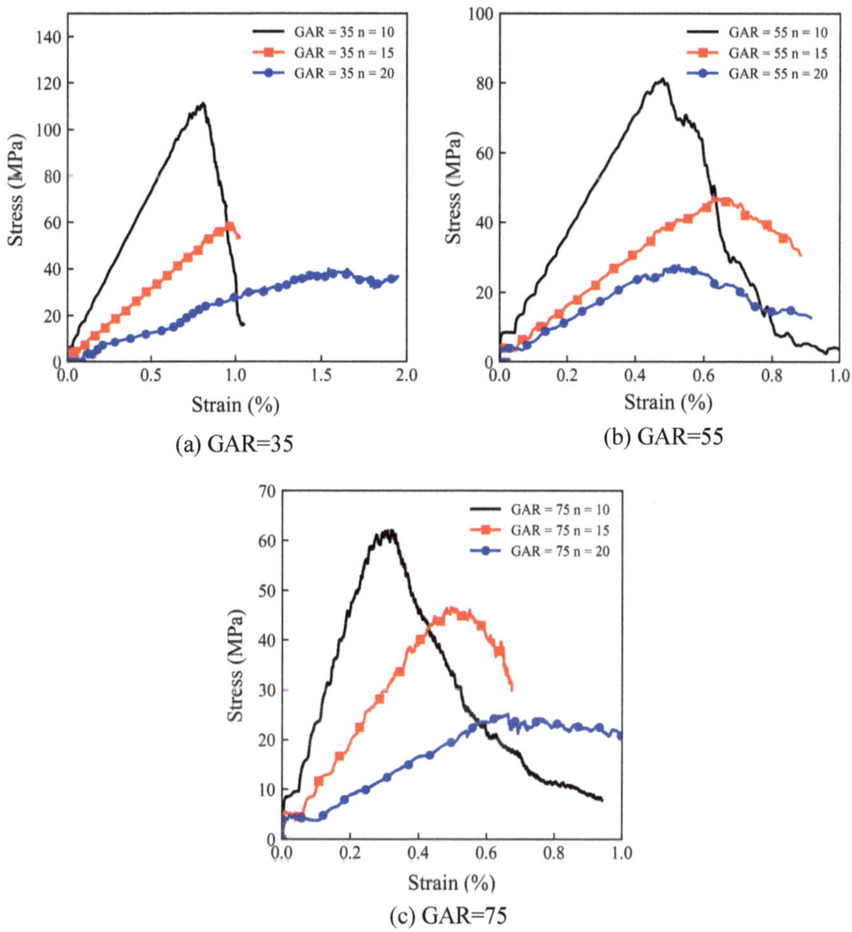

Fig. 8. Simulated stress-strain curves with different of porosity n and GAR

porosity to the uniaxial compression strength of sandstone. Furthermore, Fig. 10 shows the simulation results compared to theoretical values of Eq. 2. The Young's modulus with different GAR and porosity are plotted on specified locations. Dotted lines are theoretical GAR and porosity combination with different Young's modulus that calculated by Eq. 2. It shows the simulated Young's modulus agree with the prediction of Eq. 2 in low porosity situation, however, it is slightly underestimated in high porosity condition.

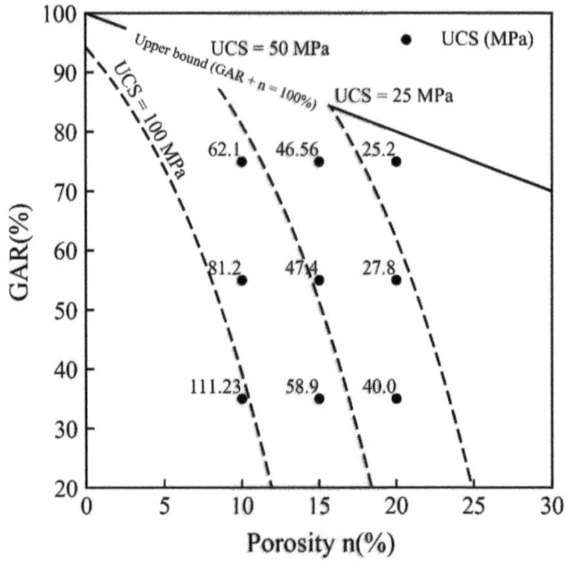

Fig. 9. Simulated relationship between uniaxial compressive strength (UCS) with different GAR and porosity n

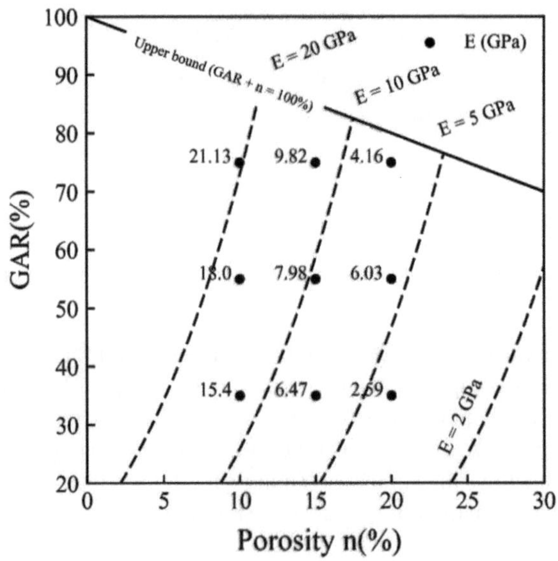

Fig. 10. Simulated relationship between Young's Modulus with different GAR and porosity n

5 Conclusions

This study adopted the biconcave bond model of DEM to investigate the mechanical behavior of sandstone. The biconcave bond model incorporated with GM structure allows grains be bonded without contact, and also links micro parameters with macro material properties. It allows the researchers to gain insight into the influence of grain and matrix on the mechanical behavior of sandstone. The simulation shows that the stress-strain curves fit experiment result. In addition, the sensitivity analysis of GAR and porosity shows that the simulation of UCS is precisely located in the classification zone based on the empirical equation. For the deformation, most simulation also agree with the theoretical values of empirical equation. The proposed model provides a useful tool to investigate the effect of grain and matrix on sandstone.

Acknowledgements. The authors would like to thank the Ministry of Science and Technology, Taiwan, for financially supporting this research under Contract No. MOST 104-2116-M-002-024.

References

Bell, F.G.: The physical and mechanical properties of the fell sandstones, Northumberland, England. Eng. Geol. (1978). https://doi.org/10.1016/0013-7952(78)90002-9

Bell, F.G., Culshaw, M.G.: Petrographic and engineering properties of sandstones from the Sneinton Formation, Nottinghamshire, England. Q. J. Eng. Geol.Hydrogeol. (1998). https://doi.org/10.1144/gsl.qjeg.1998.031.p1.02

Bell, F.G., Lindsay, P.: The petrographic and geomechanical properties of some sandstones from the Newspaper Member of the Natal Group near Durban, South Africa. Eng. Geol. (1999). https://doi.org/10.1016/S0013-7952(98)00081-7

Clough, G.W.: Cemented sands under static loading. J. Geotech. Eng. Division **107**(6), 799–817 (1981)

Chiu, C.C., et al.: A biconcave-shaped bonding model for DEM. In: The Twenty-Sixth KKHTCNN Symposium on Civil Engineering. University Town, Singapore, G-1-4 (2013)

Chiu, C.C., et al.: Biconcave bond model for cemented granular material. J. Geoeng. (2015). https://doi.org/10.6310/jog.2015.10(3).3

Chiu, C.C., et al.: Characterization of clastic rock using a biconcave bond model of DEM. Int. J. Numer. Anal. Meth. Geomech. (2016). https://doi.org/10.1002/nag.2568

Dobereiner, L., Freitas, M.H.: Geotechnical properties of weak sandstones. Géotechnique (1986). https://doi.org/10.1680/geot.1986.36.1.79

Dyke, C.G., Dobereiner, L.: Evaluating the strength and deformability of sandstones. Q. J. Eng. Geol. Hydrogeol. (1991). https://doi.org/10.1144/GSL.QJEG.1991.024.01.13

Dvorkin, J., et al.: The effect of cementation on the elastic properties of granular material. Mech. Mater. (1991). https://doi.org/10.1016/0167-6636(91)90018-U

Ersoy, A., Waller, M.D.: Textural characterisation of rocks. Eng. Geol. (1995). https://doi.org/10.1016/0013-7952(95)00005-Z

Gunsallus, K.L., Kulhawy, F.H.: A comparative evaluation of rock strength measures. Int. J. Rock Mech. Min. Sci. (1984). https://doi.org/10.1016/0148-9062(84)92680-9

Hatzor, Y.H., Palchik, V.: A microstructure-based failure criterion for Aminadav dolomites. Int. J. Rock Mech. Min. Sci. (1998a). https://doi.org/10.1016/S0148-9062(98)00004-7

Hatzor, Y.H., Palchik, V.: The influence of grain size and porosity on crack initiation stress and critical flaw length in dolomites. Int. J. Rock Mech. Min. Sci. (1997). https://doi.org/10.1016/S1365-1609(96)00066-6

Hatzor, Y.H., Palchik, V.: A microstructure-based failure criterion for Aminadav dolomites. Int. J. Rock Mech. Min. Sci. (1998b). https://doi.org/10.1016/S0148-9062(98)00004-7

Howarth, D.F., Rowlands, J.C.: Quantitative assessment of rock texture and correlation with drillability and strength properties. Rock Mech. Rock Eng. (1987). https://doi.org/10.1007/BF01019511

Hsieh, Y.M., et al.: Interpretations on how the macroscopic mechanical behavior of sandstone affected by microscopic properties-Revealed by bonded-particle model. Eng. Geol. (2008). https://doi.org/10.1016/j.enggeo.2008.01.017

Jeng, F.S., et al.: Deformational characteristics of weak sandstone and impact to tunnel deformation. Tunn. Undergr. Space Technol. (2002). https://doi.org/10.1016/S0886-7798(02)00011-1

Jeng, F.S., et al.: Influence of petrographic parameters on geotechnical properties of tertiary sandstones from Taiwan. Eng. Geol. (2004). https://doi.org/10.1016/j.enggeo.2003.12.001

Palchik, V.: Influence of porosity and elastic modulus on uniaxial compressive strength in soft brittle porous sandstones. Rock Mech. Rock Eng. (1999). https://doi.org/10.1007/s006030050050

Potyondy, D.O., Cundall, P.A.: A bonded-particle model for rock. Int. J. Rock Mech. Min. Sci. (2004). https://doi.org/10.1016/j.ijrmms.2004.09.011

Smart, B.G.D., et al.: Progress towards establishing relationships between the mineralogy and physical properties of coal measures rocks. Int. J. Rock Mech. Min. Sci. (1982). https://doi.org/10.1016/0148-9062(82)91633-3

Singh, S.K.: Relationship among fatigue strength, mean grain size and compressive strength of a rock. Rock Mech. Rock Eng. (1988). https://doi.org/10.1007/BF01020280

Sabatakakis, N., et al.: Index properties and strength variation controlled by microstructure for sedimentary rocks. Eng. Geol. (2008). https://doi.org/10.1016/j.enggeo.2007.12.004

Tiryaki, B.: Predicting intact rock strength for mechanical excavation using multivariate statistics, artificial neural networks, and regression trees. Eng. Geol. (2008). https://doi.org/10.1016/j.enggeo.2008.02.003

Weng, M.C., et al.: Characterizing the deformation behavior of tertiary sandstones. Int. J. Rock Mech. Min. Sci. (2005). https://doi.org/10.1016/j.ijrmms.2004.12.004

Weng, M.C., Li, H.H.: Relationship between the deformation characteristics and microscopic properties of sandstone explored by the bonded-particle model. Int. J. Rock Mech. Min. Sci. (2012). https://doi.org/10.1016/j.ijrmms.2012.07.003

Numerical Modeling of Embankment Settlement Over Soft Soils

Liangcai He(✉)

Wood, 121 Innovation Dr #200, Irvine, CA 92617, USA
liangcai.he@woodplc.com

Abstract. This paper presents a numerical modeling of embankment settlement over soft soils. The embankment is in the Gulf Coast of the United States. It was originally constructed over soft soils in the 1960s and has been raised and reshaped numerous times. As the construction history and corresponding loading history of the embankment were not clear, vibrating wire piezometers were installed in the underlying soft soils after the most recent re-construction to better quantify excess pore water pressures and future consolidation settlement. Based on measured pore water pressures, a large strain numerical study was performed to estimate future consolidation settlement of the embankment. A two-dimensional uncoupled analysis was performed on a 2-dimensional (2D) cross section of the embankment. A fully coupled fast flow fluid-mechanical interaction analysis and an uncoupled fluid-mechanical interaction analysis were performed on a soil column model to verify that the uncoupled model produces comparable results to the fully coupled model. The modeling yielded results consistent with the Terzaghi's one-dimensional (1D) consolidation theory, and offered insights into the deformation pattern of the embankment. The 2D cross section model may be used in subsequent soil-structure interaction analyses.

1 Introduction

The embankment modeled in this study is located in St. Bernard Parish adjacent to the Mississippi River Gulf Outlet (MRGO) between Bayou Bienvenue and Bayou Dupre. Historically, flood protection at this location has been provided by a levee. Originally constructed in the 1960s from materials excavated from the MRGO construction, the levee was raised and reshaped numerous times over the last 45 years. Based on information from the US Army Corps of Engineers (USACE 2010, 2012a, b), following Hurricane Katrina, the levee was rebuilt with additional fill placed; and an approximate 3.5-m-high T-wall was constructed on top of the levee to provide post-Katrina flood protection. The T-wall was supported on battered piles.

During design of the T-wall, it was recognized that future consolidation settlement would generate bending moments in the battered piles. Following construction, an effort was made to better estimate future settlement in selected areas by measuring pore water pressures in the foundation soils through the use of vibrating wire piezometers. Based on measured pore water pressures in the vicinity of the T-wall and site-specific field and laboratory testing data, it was estimated that future consolidation settlement in

H. Khabbaz et al. (eds.), *New Prospects in Geotechnical Engineering Aspects of Civil Infrastructures*, Sustainable Civil Infrastructures,
https://doi.org/10.1007/978-3-319-95771-5_5

one design reach would be as large as 0.8 m (m) or 31.5 in., significantly greater than estimated during design.

To bring insights into the deformation pattern of the embankment and perform soil-structure interaction analyses for the T-wall, a 2D soil model was developed for a full cross section of the embankment in this study. The model was analyzed and calibrated using uncoupled settlement analysis. Calibration was based on field pore pressure measurements and 1D consolidation hand calculation performed by USACE. A fully coupled fast flow fluid-mechanical interaction analysis and an uncoupled fluid-mechanical interaction analysis were also performed in this study on a soil column model to verify that the uncoupled model produces comparable results to the fully coupled model.

2 Site Conditions

A cross section of the levee and detailed layers of the soil profile are shown in Fig. 1. The soil layers of most interest in this study are the compressible soft clayey fill between elevations +0.91 and −3.81 m (+3 and −12.5 ft) and soft native clay layers between elevations −3.81 and −16.76 m (−12.5 and −55 ft). Also note that native layers between elevations −16.76 and −21.34 m (−55 and −70 ft) consisted of coarse-grained soils underlain by fine-grained soils between elevations −21.34 and −27.43 m (−70 and −90 ft), which in turn were underlain by coarse-grained soils to the bottom of the cross section at elevation −36.58 m (−120 ft).

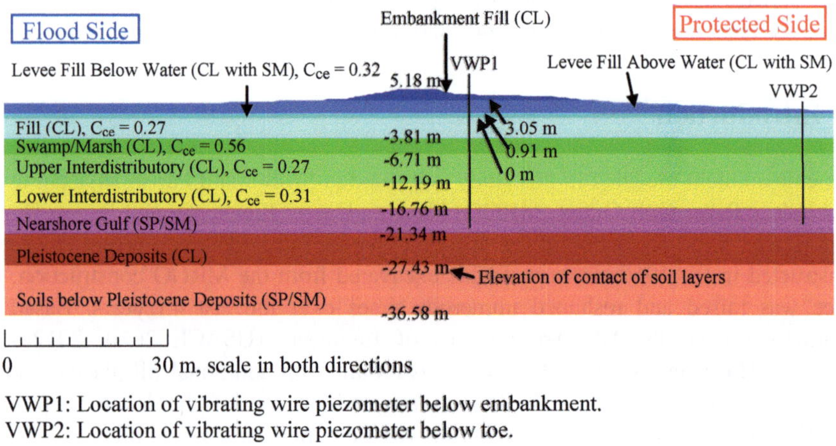

Fig. 1. Cross section and soil profile. VWP1: Location of vibrating wire piezometer below embankment. VWP2: Location of vibrating wire piezometer below toe

Groundwater at the site is at approximately elevation +0.91 m (+3 ft). Vibrating wire piezometers were installed near the centerline (VWP1) and the protected-side toe (VWP2) of the embankment (Fig. 1). Pore pressures were measured between May and

July, 2012, approximately two years after construction of the T-wall. Pore pressures with depth from VWP1 and VWP2 are shown in Fig. 2.

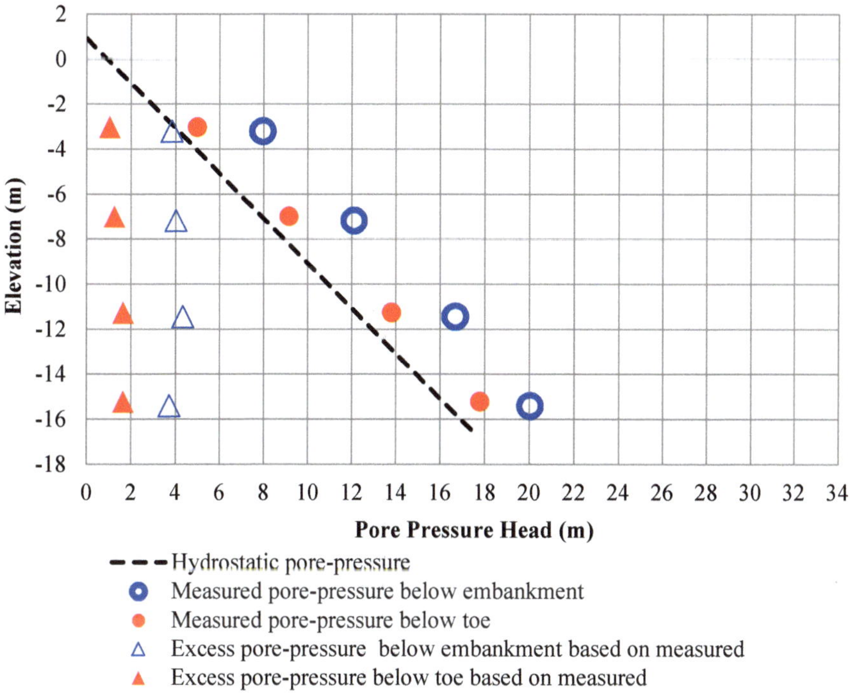

Fig. 2. Pore-pressures below embankment and toe (VWP1 and VWP2), based on measurements between May and July, 2012, approximately two years after construction of the T-wall

Material properties of the various layers required in the numerical modeling are shown in Table 1.

3 Flac Model Grid and Boundary Conditions

The analyses were performed using the 2D finite difference program FLAC (Itasca 2011). FLAC can carry out uncoupled or fully coupled fluid-mechanical analyses to capture the effects of fluid/solid interaction. The formulation of fully coupled fluid-deformation mechanisms in FLAC is based on the Biot theory of consolidation. In a fully coupled analysis, changes in pore pressure cause changes in effective stress, which affect the response of the solids (for example, changes in effective stress can produce shear and volumetric distortions, including plastic yield), and the fluid in a zone reacts to mechanical volume changes by a change in pore pressure. A fully coupled quasi-static fluid-mechanical analysis with FLAC is often time-consuming. If there is a variation of permeability across the FLAC grid, the time step of a fully

Table 1. Model material properties

Soil layer (USCS classification)	Elevation (m)	Dry density ρ_d (kg/m^3)	Porosity n	Compression ratio Cce	Poisson's ratio ν	Constrained modulus D, (kPa)	Bulk modulus K (kPa)	Shear modulus G (kPa)	Drained friction angle ϕ'(°)	Cohesion c (kPa)	Fluid mobility Coefficient k_f (m^2/Pa-sec)
Embankment (CL)	5.18–3.05	1448	0.474	–	0.35	11,712	8,112	2,702	28	4.8	0.00
Levee fill above water (CLwith SM)	3.05–0.91	1287	0.474	–	0.35	11,808–1,248	8,112–864	2,702	28	4.8	0.00
Levee fill below water (CL with SM)	0.91–0	1287	0.474	0.32	0.38	Eq. (1)	Eq. (2)	Eq. (3)	23	0	Eq. (10)
Fill (CL)	0 to −3.81	1216	0.545	0.27	0.38	Eq. (1)	Eq. (2)	Eq. (3)	23	0	Eq. (10)
Swamp/marsh (CL)	−3.81 to −6.71	584	0.778	0.56	0.35	Eq. (1)	Eq. (2)	Eq. (3)	27	0	Eq. (10)
Upper interdistributory (CL)	−6 71 to −12.19	1034	0.615	0.27	0.38	Eq. (1)	Eq. (2)	Eq. (3)	23	0	Eq. (10)
Lower Interdistributory (CL)	−12.19 to −16.76	1034	0.615	0.31	0.38	Eq. (1)	Eq. (2)	Eq. (3)	23	0	Eq. (10)
Nearshore Gulf, Pleistocene deposits, and soils below (SP/CL/SM)	−16.76 to −36.58	1546	0.375	–	0.33	30,480	20,304	7,632	30	0	0.00

coupled analysis will be dominated by the largest permeability, and a large variation may result in prohibitively long computer run time. An alternative is to perform an uncoupled fluid-mechanical interaction analysis (Itasca 2011). In cases where the model response is driven by a change in pore pressures, it is likely that the fluid flow process can be uncoupled from the mechanical process (Itasca 2011). This is the case in this study where long-term consolidation settlement is a result of dissipation of excess pore pressure. Therefore, uncoupled fluid-mechanical analyses were performed on the 2D cross section to simplify and speed up the calculation.

The FLAC model grid is shown in Fig. 3. Its top is at elevation +5.18 m. Its bottom extends to elevation −36.58 m (−120 ft). The model's left and right boundaries extend 76.2 m (250 ft) away from the center of the embankment. The model consists of 4875 grid zones. The size of the grid zones varies from 0.61 × 0.3 to 0.91 × 0.3 m^2 (2 × 1 to 3 × 1 ft^2) with aspect ratios of 3:1 or less in the main model area of interest, which is within 22.86 m (75 ft) of the center of the embankment and above elevation −27.43 m (−90 ft). Away from the main area, the size of the grid zones and the aspect ratio gradually increase. Maximum size of the grid zones is about 3.05 × 3.05 m^2 (10 × 10 ft^2) at the bottom corners. Maximum aspect ratio of the model grid is about 10:1 at the ground surface near the side boundaries.

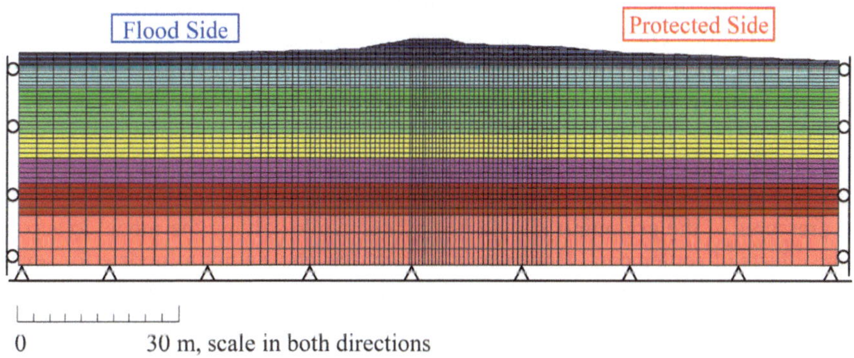

Fig. 3. FLAC model grid. *Notes* 1. See Fig. 1 for soil profile information 2. Mechanical boundary conditions: bottom fixed in both the horizontal and vertical directions; left and right boundaries free to move in the vertical direction and fixed in the horizontal direction. 3. Fluid boundary conditions: "no flow" along the left and right boundaries, and free drainage at elevations +0.91 and −16.76 m

The mechanical boundary conditions include fixed bottom in both the horizontal and vertical directions. The left and right boundaries are free to move in the vertical direction and fixed in the horizontal direction.

The fluid boundary conditions include a "no flow" boundary along the left and right sides, and free drainage boundaries at elevations +0.91 and −16.76 m (+3 and −55 ft). The free drainage boundaries are achieved by fixing pore pressure and saturation at and above elevation +0.91 m (+3 ft, the elevation of the groundwater table), and at and below elevation −16.76 m (−55 ft, the bottom of the soft clay layers).

The soils between elevations −16.76 and −21.34 m (−55 and −70 ft) were coarse-grained material, underlain by fine-grained soils between elevations −21.34 and −27.43 m (−70 and −90), which in turn were underlain by coarse-grained soils to −36.58 m. As excess pore pressure was within the layers above elevation −16.76 m (−55 ft), for simplicity, the soils between elevation −16.76 and −36.58 m (−55 and −120 ft), are modeled as coarse-grained material with constant hydrostatic pore pressure and without water flow in this study. This is considered appropriate because these materials are expected to have minimal effect on analysis results.

In the model, groundwater is assumed at elevation +0.91 m (+3 ft), and excess pore pressures are in the layers between elevation +0.91 and −16.76 m (+3 and −55 ft) only. At and below elevation −16.76 m (−55 ft), hydrostatic pore pressure is assumed, considering that between elevations −16.76 and −21.34 m (−55 and −70 ft), the layer consisted of sandy soils with relative high permeability and without excess pore pressure. It should be noted that the actual drainage boundary conditions at elevation +0.91 m (+3 ft) are difficult to simulate numerically. A free drainage boundary at elevation +0.91 m (+3 ft) is considered a reasonable assumption for two reasons. First, field investigations indicate that the clay levee embankment consists of clay but with thin layers of sandy silt, silty sand, and sand between elevation 0 and +3.05 m (0 and +10 ft). The actual drainage conditions may also include drainage in the horizontal direction. Second, it is not expected that the drainage boundary conditions will affect the final total settlement of the ground, and the consolidation time history is less critical for calculating the maximum response of structures in the subsequent soil-structure interaction analysis model.

It should also be noted that during consolidation, the gridpoints at elevation +0.91 m (+3 ft) may settle below the groundwater table. As a result, the corresponding hydrostatic pore pressure for these gridpoints will increase with consolidation. Ideally, pore pressures of the gridpoints should be updated while solving the model to account for the effects of settlements. However, this study has chosen not to update for several reasons. First, the modeling is to compare to the settlement from USACE hand calculation (USACE 2013), which was based on the Terzaghi's 1D consolidation theory. The hand calculation used small strain analysis and did not account for submergence effects. Second, a factor α is used to calibrate soil moduli in the numerical model such that total settlement at the embankment centerline in the model can match that from the hand calculation. The factor α would capture the various differences (e.g., soil stresses and moduli) between the model and the hand calculation in an average sense. Third, there are uncertainties in pore pressure measurement, unit weight of salt water, soil profiles, and material properties. These uncertainties likely have greater effects on analysis results than the gridpoint update does. Therefore, for practical purposes, the refinement in adjusting the pore pressure distribution during solving the model is not considered warranted.

The nonlinear soil stress-strain behavior is simulated using the Mohr-Coulomb constitutive model. This constitutive model is a simple elastic-yielding (bilinear) model. The elastic behavior of the constitutive model is characterized by a Young's modulus and a Poisson's ratio, or equivalent shear and bulk moduli.

FLAC model parameters include dry density ρ_d, porosity n, compression ratio C_{ce}, Poisson's ratio ν, bulk modulus K, shear modulus G, and drained shear strength for the soils, and fluid mobility coefficients k_f for water. Values of these parameters are presented in Table 1. Note that K and G for soft soil layers between elevations +0.91 and −16.76 m with excess pore pressure are derived from Poisson's ratio ν and constrained modulus D. In this study, D is computed based on true strain ε, also known as logarithmic strain, using compression ratio C_{ce}, initial vertical effective stress σ'_{v0}, and final vertical effective stress σ'_{vf}, after dissipation of excess pore pressure. Associated equations for calculations of the moduli are presented below.

$$D = \frac{\Delta\sigma'_v}{\Delta\varepsilon} = \frac{\sigma'_{vf} - \sigma'_{v0}}{-\ln\left(1 - C_{ce}\cdot\log\left(\frac{\sigma'_{vf}}{\sigma'_{v0}}\right)\right)} = \frac{\mathrm{dp}}{-\ln\left(1 - C_{ce}\cdot\log\left(\frac{\sigma'_{vf}}{\sigma'_{vf}-\mathrm{dp}}\right)\right)} \quad (1)$$

$$\mathrm{K} = D\frac{1+v}{3(1-v)} \quad (2)$$

$$\mathrm{G} = D\frac{1-2v}{2(1-v)} \quad (3)$$

$$v = \frac{\mathrm{k_o}}{1+\mathrm{k_o}} \quad (4)$$

$$\mathrm{K_o} = 1 - \sin\phi' \quad (5)$$

$$C_c = \frac{-\Delta e}{\log\left(\frac{\sigma'_{vf}}{\sigma'_{v0}}\right)} \quad (6)$$

$$C_{ce} = \frac{C_c}{1+\mathrm{e}} \quad (7)$$

In the above equations:

Ln: Denotes the natural logarithm
$\Delta\sigma'_v$: Vertical effective stress increase
Δε: Strain increment due to vertical effective stress increase
K_o: Horizontal earth pressure coefficient at rest
ϕ′: Soil friction angle
C_c: Compression index
e: Void ratio
Δe: Change in void ratio
k_H: Hydraulic conductivity
γ_w: Unit weight of water.

Based on Terzaghi's 1D consolidation theory, the hydraulic conductivity k_H or coefficient of permeability can be calculated from the coefficient of consolidation C_v, which is estimated from laboratory consolidation test data and the average vertical effective stress in the middle of each layer before and after dissipation of excess pore water pressure.

$$k_H = \gamma_w \frac{C_v}{D} \tag{8}$$

where γ_w is unit weight of water, and D is constrained modulus of the soil. In FLAC, the required "permeability" is the fluid mobility coefficient k_f defined as:

$$k_f = \frac{k_H}{\gamma_w} = \frac{C_v}{D} \tag{9}$$

Assuming isotropic permeability, the mobility coefficients (FLAC's permeability tensor) for each zone become:

$$k_{11} = k_{22} = k_f = \frac{k_H}{\gamma_w} = \frac{C_V}{D} \tag{10}$$

$$k_{12} = k_{21} = 0 \tag{11}$$

A "FISH" function is programmed to calculate D, K, and G. The calculation of k_{11} and k_{22} is implemented as part of the FISH function. Note that a factor α is used to modify D calculated from Eq. (1) to calibrate the model to match the 0.8 m (31.5 in.) of total consolidation settlement from USACE hand calculation. A value of 0.915 was obtained for α through trial and error.

4 Flac Analysis Procedure

The FLAC analyses are performed in several steps, including the build-up of the FLAC grid for the soil layers and the application of excess pore pressure. Detailed steps are described below.

1. Generate a FLAC grid for the cross section and initialize vertical and horizontal total stresses for hydrostatic conditions with the groundwater table at elevation +0.91 m (i.e., the groundwater conditions after dissipation of excess pore pressure). The vertical total stresses are initialized to the total overburden pressure under hydrostatic conditions. The horizontal total stresses are initialized to the sum of the hydrostatic pore pressure and the horizontal effective stress. The horizontal effective stress is calculated as K_o times the effective overburden pressure, where K_o is horizontal earth pressure coefficient at rest. For soils below elevation +3.05 m, $K_o = 1 - \sin(\phi')$ is assumed. Soils above elevation +3.05 m are compacted fill, and $K_o = 1$ is assumed, which is believed to have insignificant effects on analysis results.
2. Issue a "solve elastic" FLAC command to perform a gravity turn-on analysis in small strain mode; compute σ'_{vf}, the equilibrium vertical effective stresses after dissipation of excess pore pressure; recalculate the horizontal effective stresses

equal to K_o times σ'_{vf}, and re-initialize the horizontal total stresses equal to the sum of the recalculated horizontal effective stresses and hydrostatic pore pressure. The soft clay layers are initially modeled with a modulus typical for clay, which is then updated based on calculations results as outlined in step 5 below.

3. Issue a "solve" FLAC command to perform a second gravity turn-on analysis. After this gravity turn-on analysis, the model is in equilibrium representing the ground without structure elements under hydrostatic conditions (i.e., the conditions after dissipation of excess pore pressure). The major effective principal stress p' (under hydrostatic conditions) is then calculated for all clay grid zones with initial excess pore water pressure.
4. Calculate and specify the excess pore water pressure (PWP) distribution in the model based on measured pore pressures at measuring time shown in Fig. 2 (at the centerline and the toe of the embankment) and the interpolation scheme described below. Compute the vertical effective stress σ'_{v0} before dissipation of excess pore pressure by subtracting the specified PWP from the vertical total stress solved in step 3. Compute horizontal effective stresses by multiplying σ'_{v0} by K_o. Total horizontal stresses are then re-initialized to the sum of the computed horizontal effective stresses and specified PWP. With the specified PWP and new initial horizontal stresses, the model is re-balanced using small strain analysis.
   The excess pore pressure PWP distribution in the clay layers is based on the excess pore pressures, Δu_{cl} and Δu_{toe}, calculated from measured pore pressures near the centerline and the toe of the embankment, respectively. The distribution of pore pressures with depth along the centerline and toe are estimated using linear interpolation between pore pressure measurements. At other locations, the excess PWPs, Δu, are interpolated using values of Δu_{cl} (at the centerline) and Δu_{toe} (at the toe) according to the following equation:

$$\Delta u = \Delta u_{toe} + \frac{\Delta u_{cl} - \Delta u_{toe}}{\left(\sigma'_{vf}\right)_{cl} - \left(\sigma'_{vf}\right)_{toe}} \left(\sigma'_{vf} - \left(\sigma'_{vf}\right)_{toe}\right) \tag{12}$$

   where σ'_{vf} is final vertical effective stress (at the specified location) obtained from step 3.
   As discussed earlier, groundwater is assumed at elevation +0.91 m, and excess pore pressures are estimated in the layers between elevations +0.91 and −16.76 m only. At and below elevation −16.76 m, hydrostatic pore pressure is assumed due to the presence of coarse-grained soils.
5. Calculate and update soft clay moduli (D, K, and G) based on σ'_{v0}, σ'_{vf}, and compression ratio (C_{ce}), and re-balance using small strain analysis. This step first calculates the constrained modulus D of the soft clay based on Eq. (1). Bulk modulus (K) and shear modulus (G) are then calculated using Eqs. (2) and (3), respectively. Note that Eq. (1) is based on true strain, also known as logarithmic strain. Also, note that no changes in stresses or strains are calculated from the "re-balance" as the model is already in equilibrium in the previous step and the only changes in this step are modulus values.

6. Specify double drainage conditions at the groundwater table (elevation +0.91 m), and at the top of the sand layer (elevation −16.76 m), and perform a large strain uncoupled, mechanical and water flow analysis to dissipate excess pore water pressure, and record settlement and other parameters of interest. The uncoupled approach is a simplification of the coupled approach. In the uncoupled analysis, the numerical model is configured with the "config gwflow" option in FLAC. The flow calculation is performed independently (uncoupled from the mechanical analysis using "SET flow on" and "SET mech off") for a specified time step. Pore pressure changes calculated from the uncoupled flow analysis cause effective stress changes in the model which in turn induce strain or settlement if followed by a mechanical analysis. To evaluate settlement, the uncoupled flow analysis is followed by an independent mechanical analysis (uncoupled from the flow analysis using "SET flow off" and "SET mech on"). The uncoupled mechanical analysis cycles the model to equilibrium in mechanical mode. This process is repeated for a number of time steps until the dissipation of excess pore pressure is complete. It is important to note that to preserve the time history of the system, the fluid bulk modulus must be adjusted according to the following equation during flow calculation, and must be set to zero during mechanical calculation (Itasca 2011).

$$K_w^a = \frac{n}{\frac{n}{K_w} + \frac{1}{K + 4G/3}} \tag{13}$$

where K_w^a is adjusted fluid bulk modulus, K_w is fluid bulk modulus (i.e., water bulk modulus of about 2.208×10^6 kilopascals or kPa), n is porosity of the soil, K is bulk modulus of the soil, and G is shear modulus of the soil.

Steps 5 and 6 are repeated several times using trial and error to calibrate the factor α such that total settlement below the center of the embankment, after dissipation of excess pore water pressure, is approximately 0.8 m (31.5 in.) by USACE hand calculation (USACE 2013). The value of α was finally estimated to be 0.915.

5 Results of Analyses and Discussions

Key results of the modeling are presented in Figs. 4, 5, 6, 7, 8 and 9. Figures 4 and 5 show the model horizontal and vertical displacement contours, respectively, after dissipation of excess pore pressure. Settlements of the model are in response to dissipation of the excess pore pressures, and are greater near the embankment centerline. Horizontal displacements indicate that the embankment moves 2.5 to 5 cm toward the flood side after dissipation of excess pore pressure. The observed horizontal movements are consistent with what is typically seen in slope stability studies, considering that the embankment slope on the flood side is higher and steeper than on the protected side. At depth or further away from the embankment, the ground on both sides of the embankment moves toward the embankment centerline. This is due to a larger ground volume reduction as a result of dissipation of larger excess pore pressures near the centerline. Figure 4 further indicates that the protected side moves more toward the centerline compared to the flood side. This is due to the asymmetrical ground surface profile, and the larger initial excess pore pressures on the protected side near the embankment.

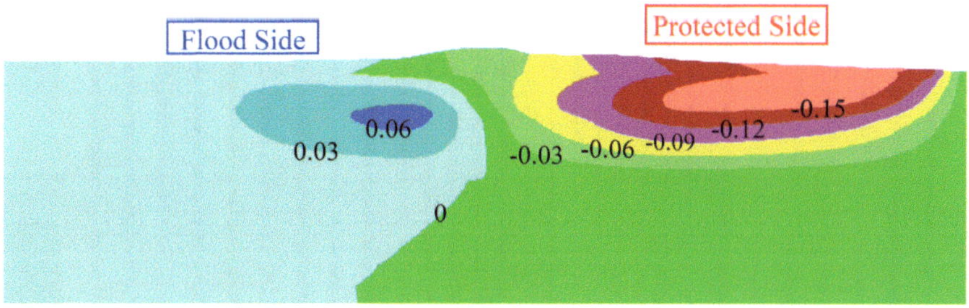

Fig. 4. Horizontal Displacement Contours after Dissipation of Excess Pore Pressure. *Notes* 1. Positive displacement is to the right. 2. Contour interval = 0.03 m

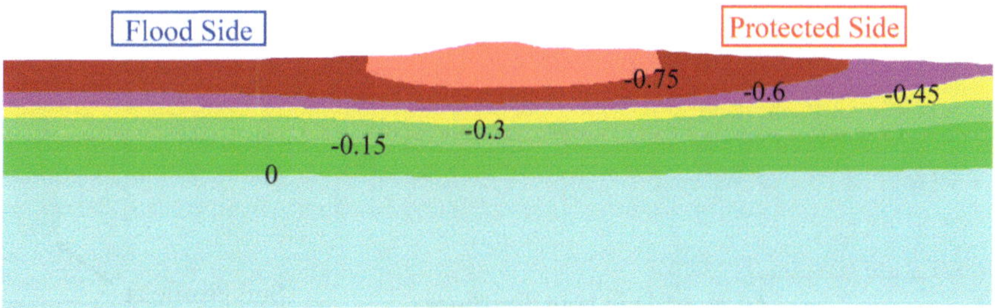

Fig. 5. Vertical displacement contours after dissipation of excess pore pressure. *Notes*: 1. Negative displacement indicates settlement. 2. Contour interval = 0.15 m

Figure 6 shows displacement vectors near the embankment. It shows displacement patterns similar to those seen in Figs. 4 and 5. Figure 6 indicates that deformation of the ground is mostly vertical settlement with slight lateral displacement toward the embankment centerline.

Figure 7 shows settlement time histories at three locations in the model. It can be seen that the centerline has the largest settlement of about 0.8 m (31.5 in.) after dissipation of excess pore pressure. It takes about 100 years for the dissipation and total settlement to complete. USACE (2013) indicated in emails that the time rate compares favorably to the results of its 1D average time rate calculation.

Figure 8 shows the settlement profile of the soft clay layers between elevations +0.91 and −16.76 m (+3 and −55 ft) beneath the embankment centerline compared to the USACE hand calculation. It indicates that FLAC model with a coefficient α of 0.915 for soil moduli reasonably simulates the 1D settlement computations.

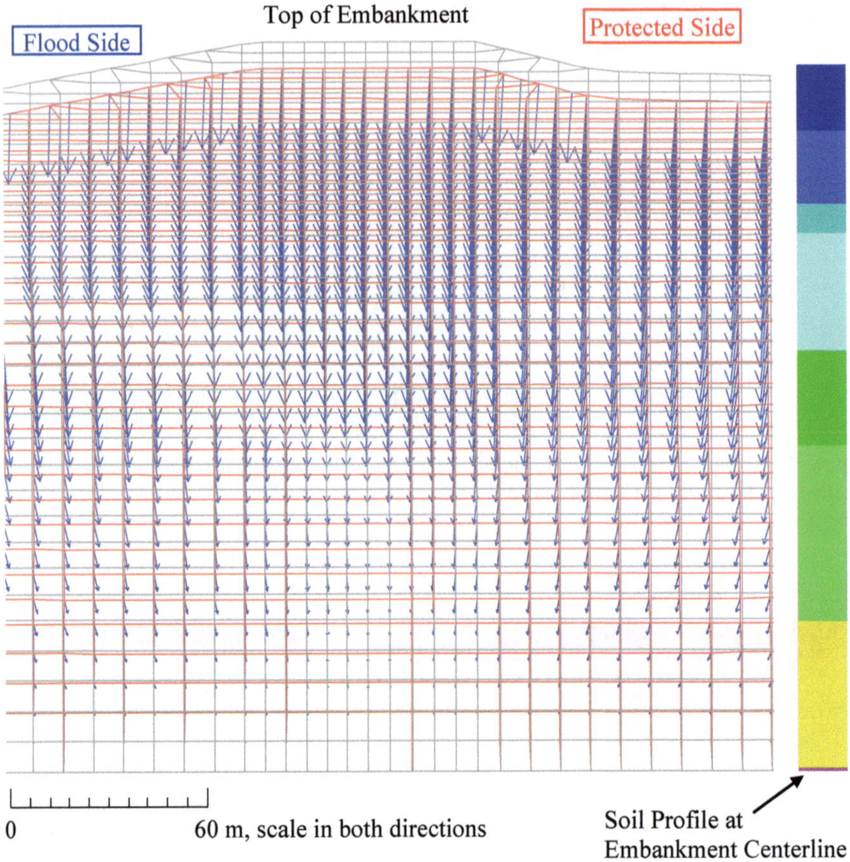

Fig. 6. Deformed and un-deformed grids and displacement vectors near the embankment after dissipation of excess pore pressure. *Notes* 1. See Fig. 1 for soil profile information. 2. Gray grid is undeformed grid, and red grid is deformed grid. 3. Blue arrows are displacement vectors, and the maximum vector is about 0.81 m

In performing the analyses, a soil column model was also developed representing a column of the ground below the center of the embankment between elevations +5.18 and −16.76 m (+17 and −55 ft). A fully coupled fast flow analysis and an uncoupled analysis were performed on the soil column model to verify that the uncoupled approach produces comparable results to the fully coupled model. The soil properties, mechanic and flow boundary conditions, and value of α (i.e., 0.915) in the coupled and uncoupled analyses for the soil column were the same as those in the 2D model. Figure 9 compares the settlement time histories of the soil column (for the fully coupled and uncoupled analyses) with that of the uncoupled 2D model. It shows that total

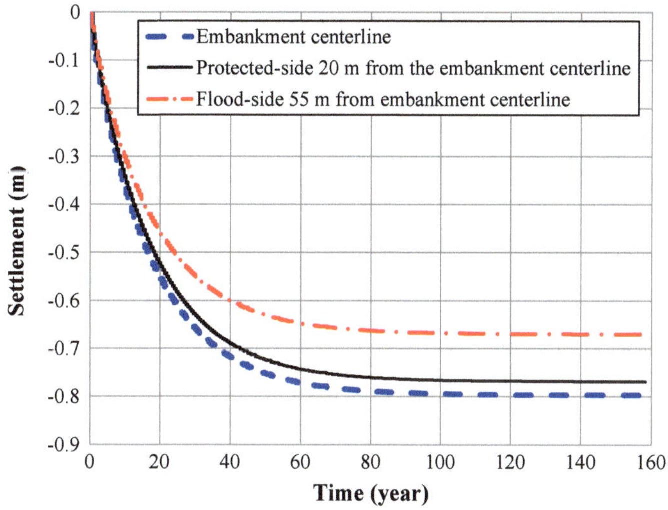

Fig. 7. 2D model settlements due to dissipation of excess pore pressure

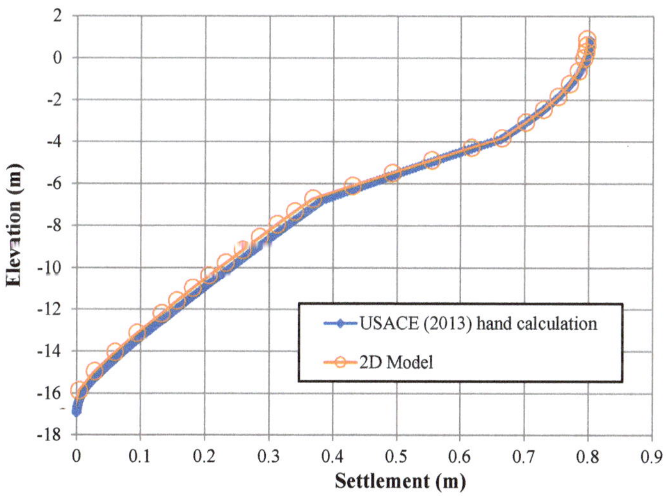

Fig. 8. Comparison of settlements due to dissipation of excess pore pressure

settlements in the soil column models are up to 2.5% larger than that of the 2D model. This is probably because the stresses at the embankment centerline in the 2D model are slightly smaller than those in the soil column models (which represent a half space). Nevertheless, Fig. 9 indicates that the uncoupled approach used for this study is adequate for practical purposes.

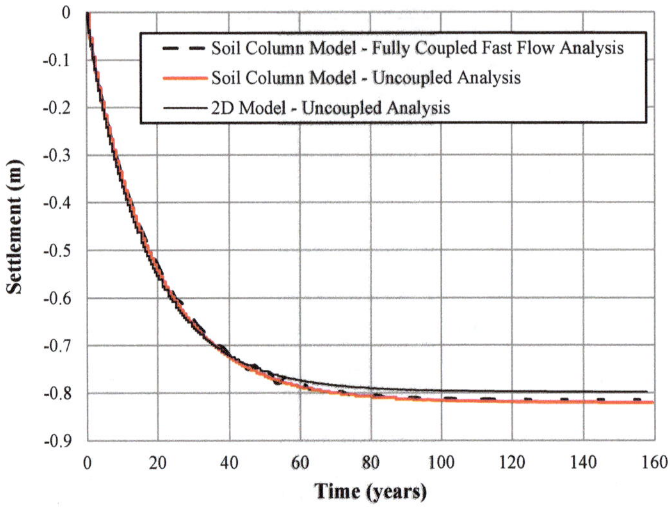

Fig. 9. Comparison of settlements from coupled and uncoupled analyses

6 Conclusions

This paper presents the FLAC modeling results for an embankment over soft soils. A fully coupled fast flow analysis and an uncoupled analysis performed on a 1D column model for the embankment centerline indicate that the uncoupled approach produces comparable results to the fully coupled model. Total settlements in the soil column models are similar to those from the 2D model at the embankment centerline.

The 2D analysis results indicate that settlements of the model are in response to dissipation of the excess pore pressures, and are greater near the embankment centerline. Deformation of the ground is mostly vertical settlement with slight lateral displacement toward the embankment centerline. The results further show that the settlement profile at the embankment centerline is close to the USACE hand calculation and the consolidation settlement may take about 100 years to complete. Structural elements may be added to the model for subsequent soil-structure interaction analyses to model the T-wall and its pile foundations.

Acknowledgements. Funding for this study is provided by the US Army Corps of Engineers (USACE). Any opinions, findings, and conclusions or recommendations expressed in this paper are those of the author and do not necessarily reflect the views of the USACE. Dr. Faiz Makdisi, Dr. Fenggang Ma, and Dr. C.Y. Chang contributed to this study. Reviewers and Agency Technical Reviewers of USACE provided review comments.

References

Itasca Consulting Group, Inc.: FLAC—Fast Lagrangian Analysis of Continua, Version 7.0. Itasca, Minneapolis, Minnesota (2011)
USACE: HSDRRS Floodwall—LPV 145, Lake Pontchartrain & Vicinity, Bayou Bienvenue to Bayou Dupre, St. Bernard Parish, Louisiana (2010)
USACE: St. Bernard Parish, Louisiana, LPV 145 T-wall, Assessment of Foundation Settlement, Prepared for New Orleans District by St. Paul District, July 2012
USACE: LPV 145 Soil Settlement Investigation. Pile Bending Calculation Summary, CEMVP-EC-D (2012)
USACE: Email and Conference Call Communications

Impact Analysis of Metro Tunnel Construction on Groundwater Flow in Nanchang, China

Youqin Zou[1,2(✉)], Baorong Huang[3], Siyu Xu[1,2], Yue Chen[1,2], Yingying Lan[4], and Ping Zhang[1,2]

[1] School of Resources, Environmental and Chemical Engineering, Nanchang University, Nanchang 330031, Jiangxi, China
zouyouqin@ncu.edu.cn
[2] The Key Laboratory of Poyang Lake Ecology and Bio-resource Utilization, Nanchang University, Ministry of Education, Nanchang 330047, Jiangxi, China
[3] Jiangxi Province Investigation, Design & Research Institute, Nanchang 330095, Jiangxi, China
[4] Nanchang Institute of Technology, Nanchang 330099, Jiangxi, China

Abstract. Five metro lines are planned to be constructed in Nanchang. Metro tunnel will go through or cut off Nanchang Quaternary aquifer, which is bound to have a significant impact on the groundwater flow. This paper aims to construct a hydrogeological conceptual model after analyzing the hydrogeological conditions of the target area. Accordingly, the numerical simulations of the groundwater flow of the target area were conducted by applying the GMS software. After the model is calibrated, the five metro lines are considered as a whole to predict the influence of metro tunnel construction on the groundwater flow in Nanchang. The study shows that due to the influence of the subway tunnel barrier, the groundwater level is lifted in upstream face of the tunnel, and the lifting effect will accumulate with the passing of time. However, the groundwater level drops in the downstream face of the tunnel, and the dropping effect will accumulate with the passing of time, too. In the area between groundwater drop funnel in the east and groundwater mound in the southwest, the lifting and dropping fluctuate with the change between upstream face and downstream face. And the other rising and declining are up to the alternation of principle influence factors. The sensitivity analysis of precipitation shows that that water level changes are very sensitive to precipitation.

Keywords: Metro tunnel construction · Impact · Groundwater in nanchang GMS (short for Groundwater Modeling System)

1 Introduction

Nanchang is the capital city of Jiangxi province, where the 13th five-year plan will promote the development of the central region. Nanchang Rail Transit (NCRT) is one of the most advanced MRT (Mass Rapid Transit) systems in Asia, China's the second batch urban construction projects of rail transit (subway, light rail), the main traffic arteries connecting East Nanchang, West Nanchang, South Nanchang, North Nanchang and four satellite cities around Nanchang. The metro system includes five transit lines.

H. Khabbaz et al. (eds.), *New Prospects in Geotechnical Engineering Aspects of Civil Infrastructures*, Sustainable Civil Infrastructures,
https://doi.org/10.1007/978-3-319-95771-5_6

Line 1 had been established since July 2009 and began to operate in December 2015, and its average daily transport of passengers was 0.2179 million during 2016. Line 2 will come into service in August 2017. NCRT plays an important role to mitigate the traffic congestion. But construction of metro needs tunnels crossing the city, as in Nanchang, the groundwater flow is crossed transversally by the tunnels. Then water table changes will be induced, thus will lead to ecological environment changes, such as contamination of the groundwater, harm to foundation of the urban structure,and so on (Kim and Lee 2003; Yang et al. 2009; Attard et al. 2016). Subway construction has an impact on the groundwater environment to some extent, which has attracted widespread attention in the world (Attard et al. 2016).

When metro is constructed, the impermeable barriers within natural aquifers must be considered. In order to assess the impact of five transit lines construction in the groundwater flow as a whole, excluding excavation, GMS (Groundwater Modeling System) software is used to develop the hydrological conceptual model based on hydrological conditions. Then, groundwater flow field is simulated under the hydrological conceptual model. After model calibration through classical trial-and-error procedure, the impact on groundwater flow field is predicted and sensitivity analysis is achieved.

2 The Study Area

The study area is located in the lower reach plain of Gan River and Fu River, is adjacent to Poyang Lake, the largest fresh water lake in China, and covers an area of 618.60 km^2 (Fig. 1). The west and north sides are bounded by Gan River.

Nanchang enjoys subtropical monsoon climate marked by four distinct seasons. For the period of 1951–2007, an average annual temperature is 17.75 °C, an average monthly temperatures peak at 29.28 °C in August and drop as low as 4.92 °C in January; an average annual precipitation is 1572.7–1656.3 mm, 45.2% of the precipitation is concentrated in the rainy months of April through June, while in the December through the following February the rainfall only accounts for 19.5%, the precipitation distributed unevenly.

3 Geological and Hydrogeological Characteristics

The study area exhibits interstream block features of Gan River and Fu River with elevation ranging from 16 to 40 m. The aquifer crossed by the metro tunnels distributed in the area is comprised of alluvium which is characterized by sand and gravel of Holocene epoch, late Pleistocene and middle Pleistocene. The aquifer top and floor elevations of Holocene epoch, late Pleistocene and middle Pleistocene aren't different obviously, and all the aquifers are laterally continuous and with close hydraulic connection (Fig. 2). So they function as an unconfined aquifer system in fact, sometime there is somewhat pressure on the edge of the study area. The aquifer has dual structure. Relative impermeable layer occurs as the upper layer and there are clay, silty sand, and silty clay lens locally. Its thickness ranges from 5 to 16 m, average 6.95 m,

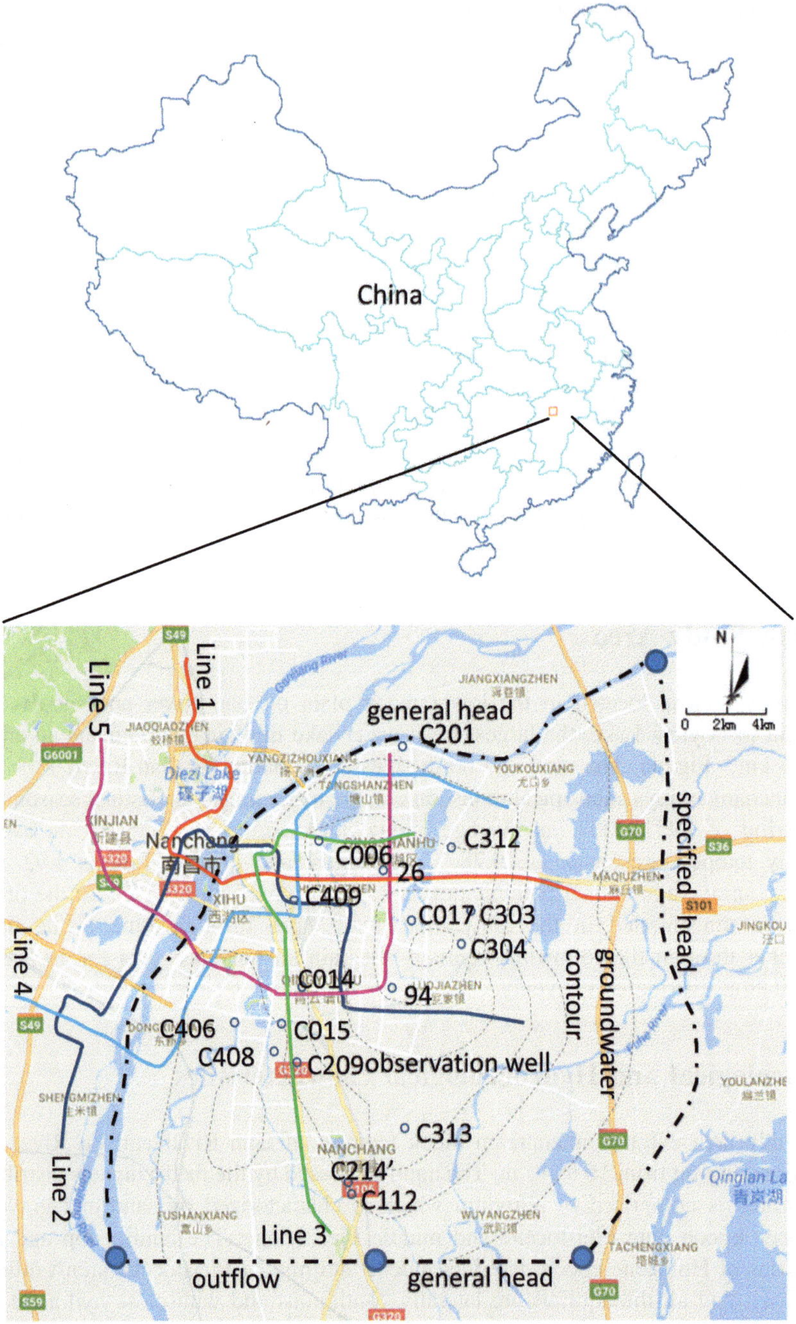

Fig. 1. The study area

no consideration of evaporation (Zou et al. 2013). Its hydraulic conductivity is 0.001–0.5 m/day. The sand, gravel layer is below the relative impermeable layer, its thickness get thicker from west to east, such as Bayi bridge 5 m, Nanchang University 17.0 m, Taizidian 28.0 m, from south to north, such as Qingyunpu 10.0 m, Longwangmiao 14.0 m, Jiangfang 20.0 m, Nanxinxiang 30.0 m. Generally, the sand, gravel layer thickness varies in the range 5–20 m, with top elevation 10–20 m (average 13.76 m) and floor elevation −10–5 m (average −5.22 m). Its hydraulic conductivity is 10.34–177.5 m/day, and specific yield value is 16–20%. Tertiary siltstone, termed red bed, forms the basement of the upper aquifer. But due to weathering, there is leakage between the confined fracture groundwater in the red bed and the pore groundwater of Quaternary within a narrow strip in Laofushan.

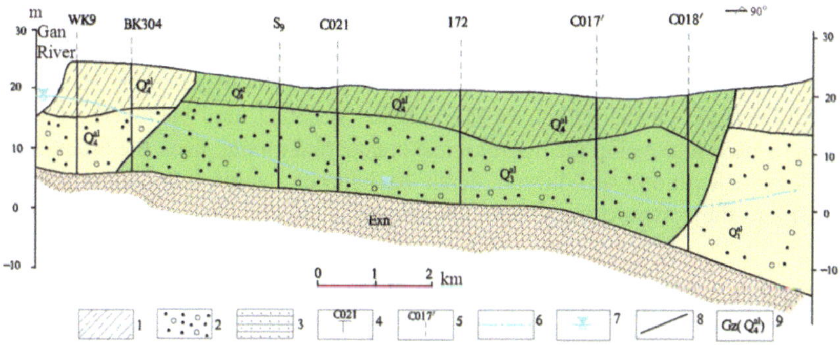

1 silty clay 2 sand gravel 3 red siltstone 4 borehole and No 5 projection borehole and No 6 groundwater level 7 surface water level 8 stratigraphic boundary 9 stratigraphic code

Fig. 2. Hydrogeological cross section of the area from Bayi Bridge to Nanchang Iron and steel Plant (modified from Zou et al. (2013))

Naturally, in the study area the Quaternary pore groundwater main direction is from southwest toward northwest. With the development of Nanchang city, the pore groundwater is substracted for industrial and domestic purpose, which induced several changes to the groundwater flow field. Until the surface water acted as the main water resource, the substraction of groundwater decreased gradually. Especially, the abstraction of groundwater declined to about 100,000 m^3/d, which is mainly for industrial purpose, when the biggest water plant, Qingyun water plant whose water resource is from Gan River, can supply 600,000 m^3/d of drinking water for the study area since 2005. According to the investigation of Jiangxi institute of geo-environment monitoring, the flow field with the depression cone in the Nanchang iron and steel plant formed and came into steady gradually until 2007. The groundwater mound located in the southwest corner of the study area is deduced by the detailed analysis of borehole lithology logs and water level fluctuations in observation wells near the edge of the groundwater dome (Fig. 1), main cause is the high floor elevation of the aquifer. And, the groundwater flow field of 2008 is similar to that of 2007. In addition, the metro was constructed since 2009. Therefore,the observation well head data of 2008 were used to calibrate the model and predict the impact of metro tunnels construction in the groundwater flow field.

4 Conceptual Model

The conceptual model for the hydrogeologic system includes the information of the potentiometric surface, recharge and discharge components, aquifer properties, and hydraulic heads (Kushwaha et al. 2009). Based on the detailed analysis of geology and hydrogeology characteristics, borehole lithology and water level fluctuations in the monitoring wells,the aquifer system influenced by the metro construction was conceptulised as a heterogeneous, isotropic, single unconfined aquifer system. GMS (Groundwater Modeling System) is a complete program for building and simulating groundwater models. It features 2D and 3D geostatistics, stratigraphic modeling and a unique conceptual model approach. Currently supported models include MODFLOW, MODPATH, MT3DMS, RT3D, FEMWATER, SEEP2D, and UTEXAS. The GMS software is used to develop the hydrological conceptual model, then, the information of water level, elevation, soil layers, depth to bedrock, aquifer basement, etc. were inputted into the software via various coverages.

5 Boundary Conditions

According to its hydrogeological condition, boundary conditions of the study area are defined (Fig. 1). The north-western part is bounded by the Gan River, where there is close hydraulic connection with the aquifer in terms of the water level fluctuations between observation wells along the riverside, BK43, BK 55 and Gan River. The Gan River is taken as a general head boundary. In the east of the study area, 18 m groundwater contour is taken as a specified head boundary. The reasons are (1) as mentioned above, the aquifer gets thicker from west to east, in the east, the thickness of the aquifer is always more than 20 m, and the groundwater is very abundant; (2) the depression cone in Nanchang and iron plant can hardly affect the 18 m groundwater contour; (3) the 18 m groundwater contour is drawn by deducing from the monitoring wells data. The south-eastern part at the edge of the depression cone is defined as general head boundary, the causes are similar with the east, whose water level is the interpolated value between the 18 m groundwater contour and the data of observation well. In case of the south-western part is inferred as flow boundary, as no outflow data are available, the discharge rates were estimated using the boreholes lithology logs initially, then were adjusted in the model when calibration.

As there is close hydraulic connection between Gan River and aquifer, Gan River will recharge the aquifer during flood season, on the other hand the aquifer will discharge to the Gan River in the dry season. Apart from Gan River, the recharge to the aquifer is still from precipitation, leakage, lakes, Fu River and irrigation return flow. The precipitation infiltration is the principle vertical recharge, but the dual structure of the aquifer and city impermeable pavements nearly prevent the infiltration from rainfall, only in the east (from Wugangwei, Luojiaji, Taizidian to Youkou) and west (Liantan town), the precipitation will recharge the aquifer. From April to November, there is irrigation return flow in rice field. As mentioned above, the confined fracture groundwater of Tertiary leaks to the unconfined groundwater via the weathered siltstone of the basement. A branching river of Fu River is located in the west of the study area, the thickness of the sludge in the riverbed is 1–3 m, the loam of Holocene epoch below the sludge is as thick as 7 m, and the depth of the branching river is less 3 m, the river water can't connect with the groundwater. Other water bodies, such as Qingshan Lake, Aixi Lake, Yao Lake, another branching river of Fu River located in the east of the study area and so on, their relationships with the groundwater are similar to the branching river in the west, according to the borehole lithology logs. Therefore, the recharge from the lakes and branching rivers is negligible. The recharge values incorporated in the model for rainfall and irrigation return flow referred to the hydrogeological report on the study area before. The leakage was estimated using Darcy's Law in terms of water level data of the two layers of aquifers.

The discharge from the aquifer includes outflow from southwestern boundary and abstraction for industries purpose except for discharging to Gan River. The outflow has been discussed above. The groundwater abstraction is about 100,000 m^3/d. The evaporation is negligible due to thickness of upper layer of the dual structure.

6 Numerical Model

The three-dimensional movement of groundwater, which is assumed to be constant density in porous media, may be described by the partial differential equation as follows (Don et al. 2005):

$$\frac{\partial}{\partial x}\left(K_{xx}\frac{\partial h}{\partial x}\right)+\frac{\partial}{\partial y}\left(K_{yy}\frac{\partial h}{\partial y}\right)+\frac{\partial}{\partial z}\left(K_{zz}\frac{\partial h}{\partial z}\right)-W-\mu\frac{\partial h}{\partial t} \tag{1}$$

where K_{xx}, K_{yy} and K_{zz} are values of hydraulic conductivityalong the x, y, and z coordinate axes, which are assumed to be parallel to the major axes of hydraulic conductivity (LT^{-1}); h is the potentiometric head (L); W is a volumetric flux per unit volume and represents sources and/or sinks of water (T^{-1}); μ is the specific yield of the porous material (L^{-1}); and t is time (T).

The MODFLOW 2000, which is the finite-difference computer code and can numerically approximate Eq. (1), was applied to simulate the groundwater flow in the study area. The groundwater modeling system (GMS), as the pre- and post- processor, was used for giving input data and processing the model output.

7 Model Calibration

Model calibration aim to calibrate and derive the hydraulic parameters or groundwater hydrological loads (such as infiltration, leakage, outflow and so on) which are inputted in the hydrogeological conceptual model, to ensure the proposed model is accurate and applicable (Yang et al. 2009). In this study, classical trial-and-error procedure was applied to calibrate the inputted hydraulic parameters and the zoned hydrogeological units. There are 17 wells distributed in the study area. The water level data observed on 10th, 20th, 30th (29th) of each month in 2008 was used to calibrate model. The calibration procedure was to vary the parameters to minimize the differences between the simulated and observed water levels, and to achieve the best overall agreement is the most important. During calibration, the upper and lower limits of the calibration indicator in GMS represent plus and minus one allowable error value, which is defined as 0.5 m. A good match was achieved through trial and error process because almost 85% absolute values of differences between simulated and measured heads were less 0.5 m, which demonstrated that the proposed model is accurate and applicable and can be applied to predict the impact of metro construction on groundwater flow field.

8 Results and Discussion

To predict the impact of the presence in the aquifer of the metro tunnels on the groundwater flow field, the model was run under the condition that the tunnels as a whole were considered as a low-permeability barrier. A very small permeability value for the tunnels was inputted in the model. The differences between the simulated water heads and observed water heads in the observation holes were analyzed to assess how the metro tunnels to affect the groundwater flow. And the sensitivity analysis of the groundwater level to the precipitation was done by varying only the precipitation over a range of ±10% under metro tunnels construction condition, the water level change percentage was used to evaluate the sensitivity.

8.1 Impact Analysis

There are 17 observation wells distributed in the study area. Their water level changes are depending on their positions relative to the subway tunnel, and their original position in the flow field. They can be divided into four categories: (1) mainly influenced by the upstream face of the tunnels, such as C408 and 26, (2) principally affected by the downstream face of the tunnels, including 94, C017, C209 and C006, (3) located between the depression cone and water dome, for example C214′, C112, C015, C406 and C409, their positions relative to the tunnels alternate between the upstream and downstream due to the natural changes of groundwater flow field, (4) influenced by multiple factors, for instant C303, C304, C312, C313, C014 and C201.

Figure 3 shows that the water levels of C408 and 26 rise due to the tunnels, and rising in 26 is more than that in C408. The situation is reflected in Fig. 7, the box plots of C408 and 26 are upon 0 m, and the box of 26 is higher than that of C408. One possible reason is that 26 is influenced by Line 1 and Line 5, C408 is affected by Line 3, another is the flow direction.

Figure 4 demonstrates that there are water level decreases in 94, C017, C209 and C006, which located in the downstream face of the tunnels. In Fig. 7, the box plots of them are under 0 m.

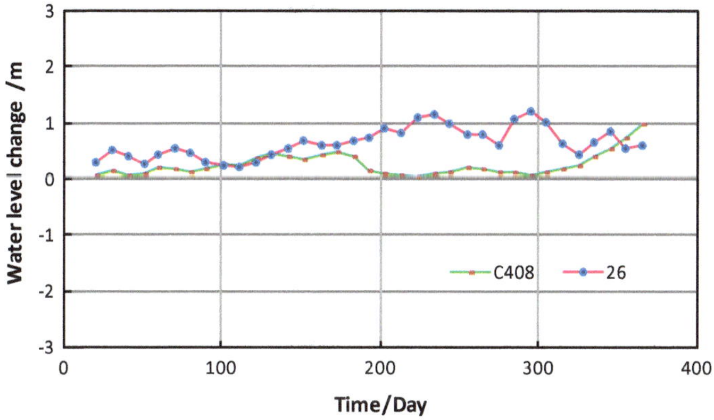

Fig. 3. Water level changes predicted in observation wells C408 and 26

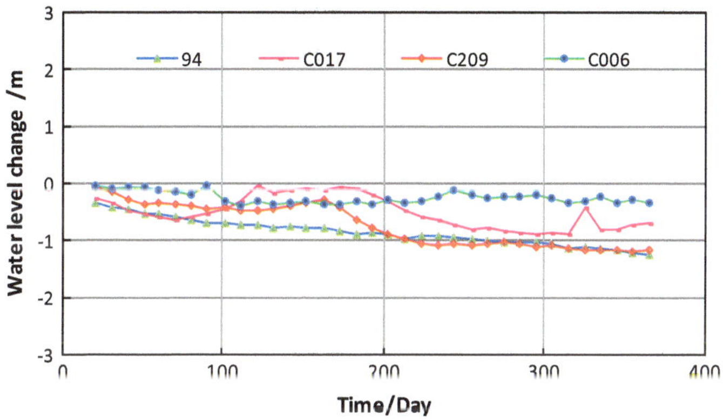

Fig. 4. Water level changes predicted in observation wells 94, C017, C209 and C006

The interesting thing is about 26 and C006 (showed in Fig. 1), they are affected by four Lines, 26 principally shows the influence of Line 1 and Line 5, as upstream face, C006 mainly reveals the influence of Line 3 and Line 4, as downstream face.

Figure 5 and Fig. 6 show that the water level rising and decreasing in wells of categories (3) and (4), which is demonstrated in Fig. 7, the 0 m is between the box plots, due to the alternation of principle influence factors. Taking C015 as an example, when the water of the depression cone recharges the water mound, C015 is the downstream face of Line 3, water head declines, if the water of the water mound recharges the depression cone, C015 is the upstream face of Line 3, water head rises. For instance, the water level change of C303 is controlled by the groundwater abstraction of Nanchang iron and steel plant, the recharge from east boundary and the Line 1.

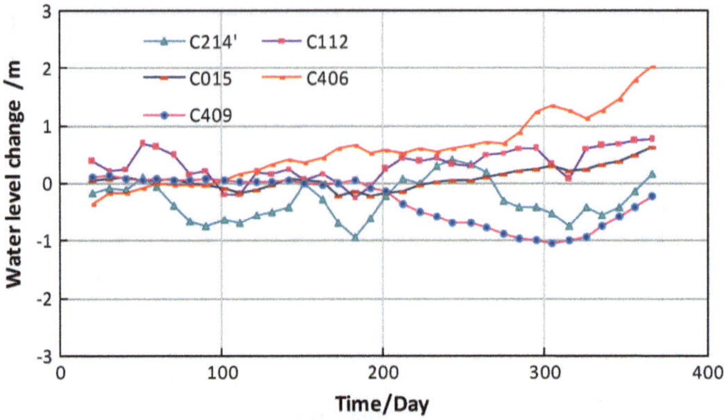

Fig. 5. Water level changes predicted in observation wells C214′, C112, C015, C406 and C409

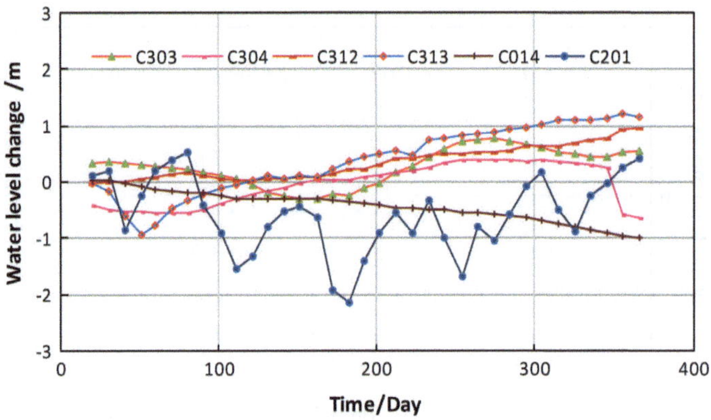

Fig. 6. Water level changes predicted in observation wells C303, C304, C312, C313, C014 and C201

Figure 3 through Fig. 6, water head changes in most of the wells show accumulation when water head changes get higher/lower over time, which is also showed in Fig. 8, the lengths between Q1 and Q3 get longer as time goes by.

8.2 Sensitivity Analysis

Rainfall is the principle vertical recharge to the groundwater, and it is abundant in the study area. The sensitivity analysis was investigated to the precipitation by rise/decline 10% of it. The result is showed in Fig. 9, which illustrates the water head changes are very sensitive to precipitation and symmetrical to the rise/decline of the rainfall.

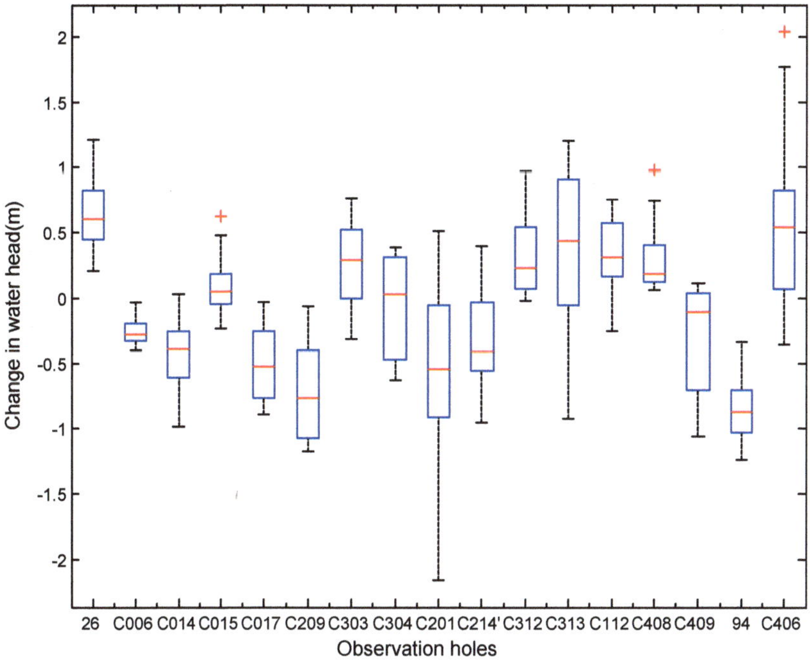

Fig. 7. Water level changes predicted temporal distribution of 2008 in 17 observation wells

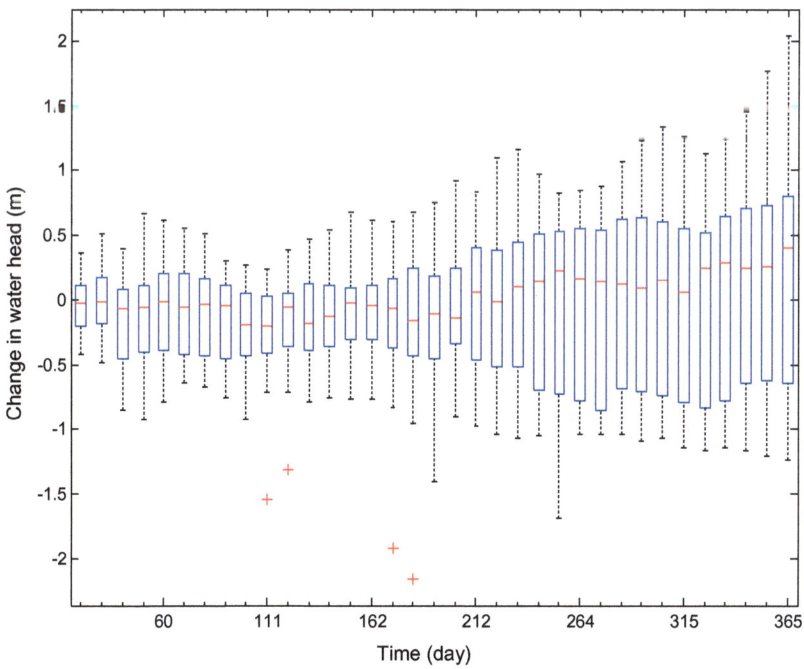

Fig. 8. Water level changes predicted spatial distribution of 17 observation wells in 2008

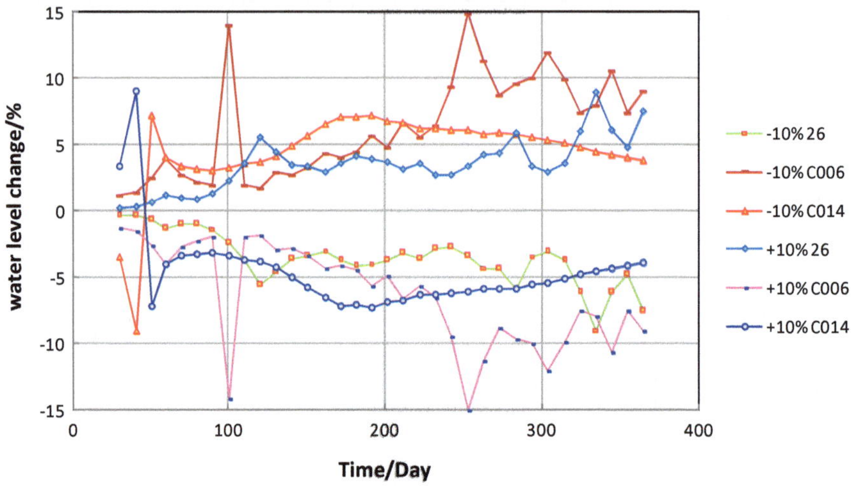

Fig. 9. Sensitive analysis to precipitation

9 Conclusions

A one-layered finite-difference flow model was applied to assess the impact of the metro construction on the groundwater flow field. The simulated results indicate that the groundwater flow field is affected by the presence of the metro tunnels due to their impeding the horizontal flow to groundwater. The one hand, the groundwater level rises in the upstream face of the tunnels, the other hand, the groundwater level declines in the downstream face, most changes of flow field show accumulate characteristic. The level changes are sensitive to the precipitation when varying only the precipitation over a range of ±10%.

Acknowledgements. The research reported in this paper was performed under Natural Science Foundation of China (project No 21467014) and Key Technology R&D Project of Jiangxi Province (project No 20151BBG70008).

References

Attard, G., Winiarski, T., Rossier, Y., Eisenlohr, L.: Review: impact of underground structures on the flow of urban groundwater. Hydrogeol. J. **24**(1), 5–19 (2016)

Don, N.C., Araki, H., Yamanishi, H., Koga, K.: Simulation of groundwater flow and environmental effects resulting from pumping. Environ. Geol. **47**, 361–374 (2005)

Kim, Y., Lee, K.: Disturbance of groundwater table by subway construction in the Seoul area, Korea. Geosci. J. **7**(1), 37–46 (2003)

Kushwaha, R.K., Pandit, M.K., Goyal, R.: MODFLOW based groundwater resource evaluation and prediction in Mendha sub-basin, NE Rajasthan. J. Geol. Soc. India **74**(4), 449–458 (2009)

Yang, F., Lee, C., Kung, W., Yeh, H.: The impact of tunneling construction on the hydrogeological environment of "Tseng-Wen reservoir transbasin diversion project" in Taiwan. Eng. Geol. **103**(1–2), 39–58 (2009)

Boundary Effects of Pile Cap on the Integrity Testing of Group Piles

Jiunnren Lai[1(✉)], Bo-Huan Yang[2], Chi-Ling Pan[1], and Chiung-Fen Cheng[3]

[1] Department of Construction Engineering, Chaoyang University of Technology, Taichung City, Taiwan
{jrlai,clpan}@cyut.edu.tw

[2] Taiwan Power Company, Taipei, Taiwan
u782146@taipower.com.tw

[3] Department of Multimedia and Game Design, Overseas Chinese University, Taichung City, Taiwan
cfcheng@ocu.edu.tw

Abstract. Low-strain surface reflection pile integrity testing (PIT) methods have been successfully used for the quality assurance of newly-built individual piles. However, for existing piles with pile cap, previous studies have shown that the stress-wave reflected from the boundaries of pile cap may cause difficulties in determining their integrity. In this paper, the boundary effects of pile cap on the signal of PITs were studied by numerical simulation and tests on model piles. Feasibility of minimizing the boundary effects by embedded accelerometers was also investigated. Results from this study indicate that signals obtained from numerical simulation are similar to those obtained from tests performed on model piles. It also shows that boundary effects of pile cap on low strain pile integrity tests can be reduced by using embedded accelerometers, thus making it easier to assess the integrity of piles with a pile cap.

1 Introduction

The ability to assess the structural integrity of piles after a major event such as earthquake or scouring of riverbed plays a important role in evaluating the safety conditions of brides. Several non-destructive testing (NDT) methods based on wave propagation theory have been used to assess the integrity of drilled shafts or cast-in-place piles. These tests are also called pile integrity test (PIT) and can be classified into two groups: direct transmission method and surface reflection method.

The most common direct transmission PIT is the Cross-hole Sonic Logging [CSL] (Stain and Williams 1991) or the Cross-hole Tomography [CT] (Han and Wang 1992). A CSL/CT test normally requires two pairs of steel or PVC access tubes installed in the shafts and tied to the rebar cage. The cage is then lowered into the borehole and the concrete is placed. A sound source and receiver are lowered into the tubes, maintaining a consistent elevation between source and sensor. A signal generator generates a sonic pulse from the emitter which is recorded by the receiver. Relative energy, waveform

H. Khabbaz et al. (eds.), *New Prospects in Geotechnical Engineering Aspects of Civil Infrastructures*, Sustainable Civil Infrastructures,
https://doi.org/10.1007/978-3-319-95771-5_7

and differential time are recorded and logged. This procedure is repeated at regular intervals throughout the pile. By comparing the graphs from the various combinations of access tubes, a qualitative idea of the structural soundness of the concrete throughout the pile can be gleaned. The CSL/CT method is considered to be more accurate than the surface reflection PITS in the determination of structural soundness of concrete within the drilled shaft (Hollema and Olson 2003). However, for existing piles with pile cap, access tubes are often not available, thus the CSL was not investigated in this study.

The surface reflection PIT methods based on one-dimensional (1D) stress wave propagation theory are often used to evaluate the integrity and determine the length of deep foundations. To perform these tests, the pile top is struck by a hand-held hammer which generates a stress wave that travels down the pile, reflects off the pile toe or other material/cross-section changes and then back to the pile top. An accelerometer or geophone is attached to the pile top by a thin coupling compound and records both the incident wave and reflective waves. Depending on how the signal were analyzed, the surface reflection PITs are further classified into Sonic Echo [SE] method (Steinbach and Vey 1975), Impulse Response [IR] method (Davis and Dunn 1974; Finno and Gassman 1998), and Ultra Seismic [US] method (Olson et al. 1998). These surface reflection PITs have been successfully applied to assess the structural integrity of newly-built individual piles.

The 1D wave propagation assumption is satisfactory in pile driving analysis since the diameter of the ram is comparable in size to that of the pile, resulting in an impact which causes a uniform response of the pile head and also at any pile cross section. However, in the case of a SE pile integrity test where the hammer is small (e.g., 40–50 mm in diameter) compared to the pile diameter (typically 1–2 m), the 1D assumption may no longer hold even if the pile is long compared to its diameter. Several investigators (e.g., Fukuhara et al. 1992; Chow et al. 2003; Liao and Roesset 1997) have shown that three-dimensional (3D) effects may have significant influence on the pile response, especially for piles with pile cap (Wu et al. 2015). Lai et al. (2017) conducted PITs on three individual model piles and found 3D effects were reduced by using embedded accelerometers. This paper extends the previous study (Lai et al. 2017) on 3D effects of pile cap through numerical simulation and tests on model piles. Feasibility of minimizing the 3D effects of pile cap by using embedded accelerometers was investigated.

2 Theoretical Background

The SE PIT method is based on the phenomena of one-dimensional (1D) stress wave propagation in solids. Information about the shape and length of a pile can be gained by a simple blow on the top of the pile with an ordinary hammer. Because of its simplicity, this method has been widely used to assess the integrity of drilled shafts or cast-in-place piles, and was adapted as a standard procedure by ASTM (2000). Since this technique was adapted in this study, a brief description about its theoretical background is presented here.

A typical surface reflection PIT setup is shown schematically in Fig. 1. This test involves impacting the top of a pile with a hammer to introduce a downward traveling transient stress wave. When the wave encounters a change of the impedance within the pile, such as a defect or pile toe, it will reflect back to the pile head and recorded by a receiver (accelerometer or geophone). The SE test requires only the particle velocity response history in time domain to perform integrity analysis. A velocity waveform of a pile containing a bulge and a neck defect is illustrated in Fig. 2, the location of the defect or the length (L) of the pile can be calculated from the travel time (dt) of stress waves reflected from the defect or the toe of the pile using the following equation:

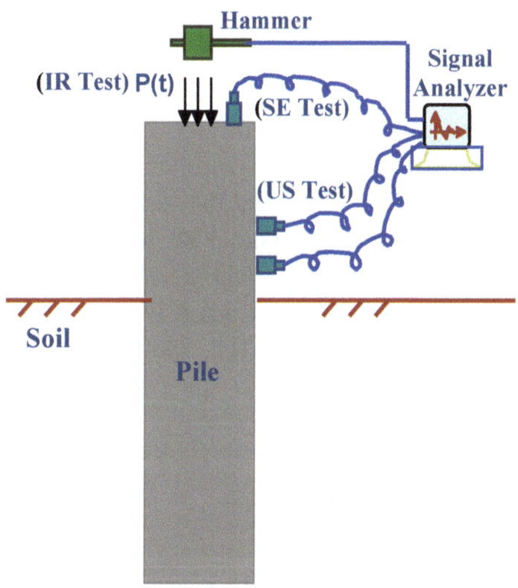

Fig. 1. Schematic drawing of a typical surface reflection PIT setup

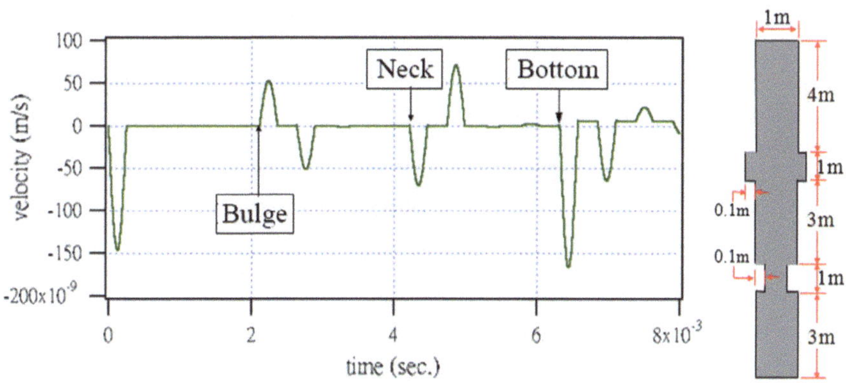

Fig. 2. Theoretical velocity profile from SE test of a pile containing defects

$$L = \frac{c \times dt}{2} \tag{1}$$

where c is the stress wave traveling speed in concrete and ranges between 3000 and 4500 m/s depends on the quality of concrete.

3 Methodology

3.1 Model Pile Test

In order to study the boundary effects of pile cap on the signal of PITs and feasibility of minimizing these boundary effects, three model piles were constructed with three embedded accelerometers for each pile, and were tested to assess their integrity. A schematic drawing of these piles is shown in Fig. 3. The diameter and length of these piles are 0.3 and 3 m, respectively. Pile P1 is an intact pile. Pile P2 contains a 100 mm by 20 mm necking defect located 1.2 m from the top. Pile P3 contains a major defect (damaged) made by 10 mm-thick Styrofoam also located 1.2 m from the top. The locations of the three embedded accelerometers are about 0.3, 1.6 and 2.8 m from the pile top, respectively. A pile cap was constructed to group these piles together. The dimensions (L/W/D) of the pile cap are 2.4, 0.5, and 0.3 m, respectively. Photographs

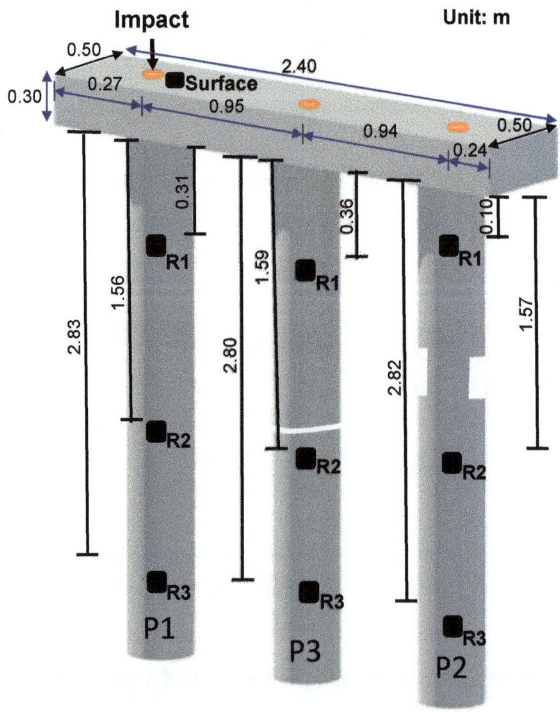

Fig. 3. Schematic drawing of the three model piles

of these three piles before and after installation of pile cap are shown in Fig. 4. Both the conventional SE test and test with embedded accelerometers were performed before and after installation of pile cap. The ability of these two methods in assessing the integrity of these piles is compared and discussed a subsequent section.

Fig. 4. Photographs taken during the construction of three model piles

3.2 Numerical Simulation

A commercial finite element program (ANSYS) is used to study 3D effects on the signal of low strain pile integrity test numerically. Under small strain condition, the behavior of concrete pile is assumed to be linear elastic and its material damping is negligible. Because the frequencies of stress wave introduced by a typical hammer are relatively low (wave lengths are relatively high) and the cross sectional area of steel rebars is much smaller than that of concrete, the effects of rebars can be simplified by just using a higher stiffness for concrete. Therefore, the following basic properties were used for concrete: density $\underline{\rho}$ = 2400 kg/m^3, Young's modulus $E = 2.81 \times 10^{10}$ Pa, and Poisson's ratio $\underline{\nu}$ = 0.2. For damaged pile P3, a 1 cm-thick Styrofoam was used to simulate a crack within the pile, and the following properties were used for Styrofoam: $\underline{\rho}$ = 40 kg/m^3, Yang's modulus $E = 5.83 \times 10^6$ Pa, and Poisson's ratio $\underline{\nu}$ = 0.45. To simplify the simulation, isotropic, axisymmetric finite elements (2D Solid 162) are used to model the circular pile. The element is defined by four nodes having six degrees of freedom at each node: translations, velocities, and accelerations in the nodal x and y directions. An example of the mesh used for Pile P2 is shown in Fig. 5. The element size is 1 cm by 1 cm. Because of symmetry, only half of the pile (0.15 m × 3 m) was modeled. The left boundary is defined as the symmetric axis. Because the piles were buried in-place, surrounding soil is much weaker than the pile, thus free boundary condition is used for the top, right, and bottom boundaries. For group plies, the top 0.15 m of the pile is buried into the pile cap and the cap is assumed to be circular with a radius of 0.25 m. The impact force is molded by a surface pressure of 1×10^8 Pa applied on the topmost element next to the symmetric axis. The surface pressure is described using a half cycle of Sin^2 function with duration of 300 μs, which is similar to an impact produced by a 0.167 kg hammer. For comparison with test signal, vertical nodal velocity for nodes corresponding to the locations of accelerometers is set to output at every 10 μs.

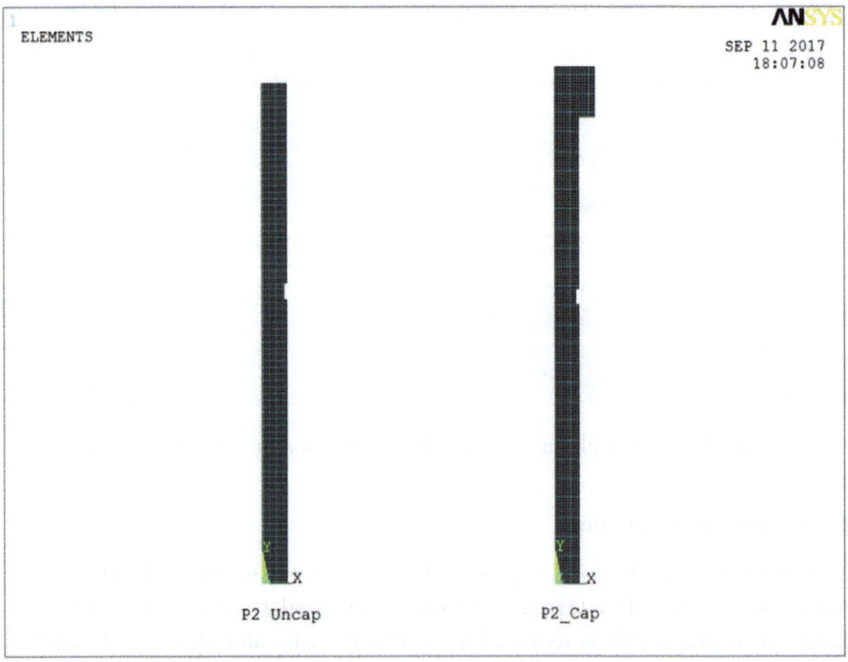

Fig. 5. Finite element mesh of pile P2

4 Results and Discussion

4.1 Intact Pile P1

Velocity profiles of intact pile P1 before installation of pile cap are shown in Fig. 6. It can be seen that the waveforms obtained from numerical simulation and test on model pile are very similar except the amplitude of toe reflection from test is smaller than from numerical simulation because the impedance of soil is neglected. The incident wave and toe reflection can be identified easily with surface receiver and embedded receivers R1 and R2. The length calculated using elapsed time between the incident wave and toe reflection of the simulated velocity waveform by Eq. (1) is 3.11 m by the surface receiver, 2.77 m by R1 and 1.44 m by R2. For receiver 3, because it is located near the toe, only the incident wave can be identified, the reflection in the waveform is from the top rather than from the toe. Bear in mind that the distance between the 1st receiver and pile top is 0.31 m while the distance between the 2nd receiver and pile top is 1.56 m. Therefore, the estimated pile length is 3.08 m by R1 and 3.00 m by R2. There is a deviation of 0.11 m in the estimated pile length by the surface receiver and a deviation of 0.08 m by R1. The deviation is resulted from 3D effects as mentioned by the investigators (Chow et al. 2003; Liao and Roesset 1997) in the literatures reviewed.

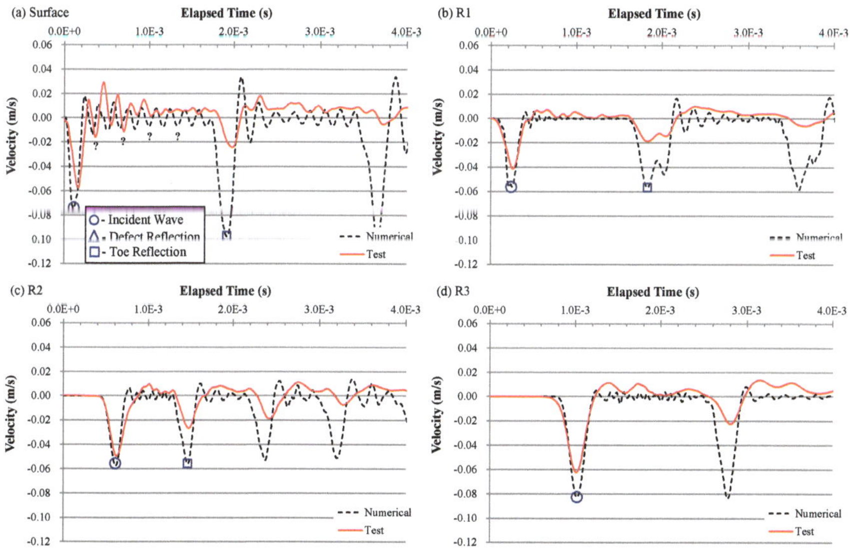

Fig. 6. Velocity profiles of intact pile P1 before installation of pile cap

As shown in Fig. 6a, the incident wave and reflected wave from pile toe can all be identified with the conventional SE test method (surface receiver). However, stress waves reflected from the side boundary of pile top (3D effects) are very likely to be misidentified as reflections from defect as shown by the "?" marks in Fig. 6a. In contrast, as shown in Fig. 6b, waveform from the 1st embedded accelerometer (R1) is more "clean" than that of surface receiver. There is less chance of misidentifying them as defects.

The velocity profiles of intact pile P1 after installation of pile cap are shown in Fig. 7. The length calculated using elapsed time between the incident wave and toe reflection from simulated velocity wave form is 3.28 m by the surface receiver, 2.72 m by R1 and 1.33 m by R2. If the distances between the receivers and top of pile cap are taken into account, the estimated total lengths are 3.28, 3.18, and 3.04 m by the three receivers. There is also a small deviation from the 3.15 m actual length which is resulted from 3D effects as mentioned earlier. By comparing Figs. 6 and 7, it can be seen that the boundary effects become more severe after installation of pile cap, thus making identification of toe reflection more difficult, especially for the case of surface receiver.

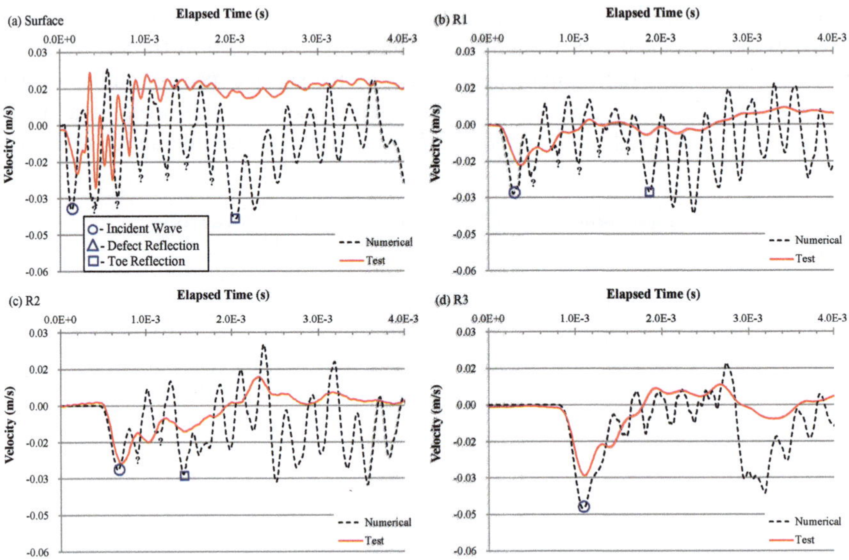

Fig. 7. Velocity profiles of intact pile P1 after installation of pile cap

4.2 Necking Pile P2

The velocity profiles of necking pile P2 before installation of pile cap are shown in Fig. 8. The simulated and tested waveforms are also very similar. Both the defect reflection and toe reflection can all be identified in the waveforms of surface receiver and the 1st receiver. However, 3D effects are more significant in the waveform of surface receiver than in the 1st receiver. The estimated defect location is 1.23 m from pile top, which is slightly higher than the actual location (1.20 m) of the defect. The estimated pile length is 3.17 m, which is also slightly higher than the actual pile length due to 3D effects. For receiver R2, only the incident wave and toe reflection can be identified. The estimated pile length by this receiver is 3.04 m and is very close to the actual pile length. For receiver R3, only the incident wave can be identified.

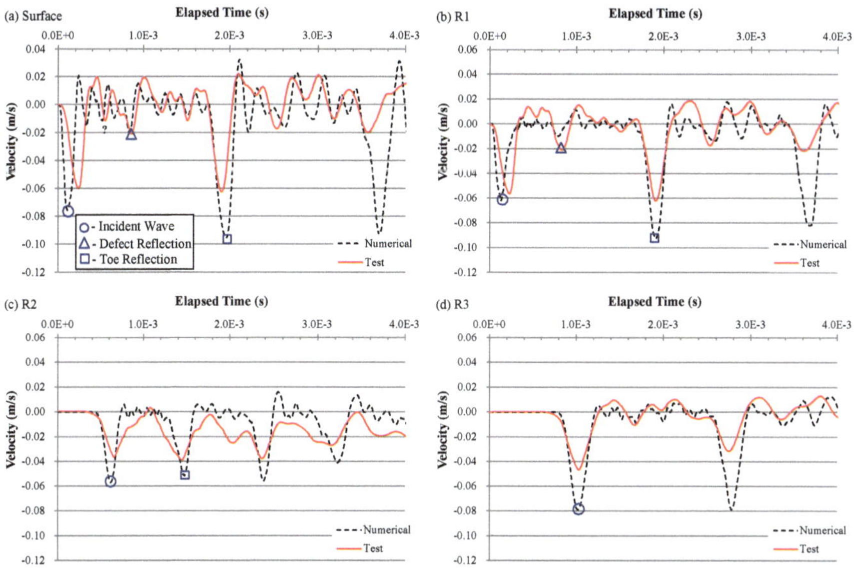

Fig. 8. Velocity profiles of necking pile P2 before installation of pile cap

The velocity profiles of necking pile P2 after installation of pile cap are shown in Fig. 9. The incident wave and toe reflection can be identified in the waveforms of surface, R1, and R2 receivers, while the defect reflection can only be identified in the waveform of receiver R1. The estimated defect location from top of pile cap and length of pile (including cap) is 1.48 and 3.34 m, which is slightly deviated from the actual of 1.35 and 3.15 m due to 3D effects.

4.3 Damaged Pile P3

The velocity profiles of damaged pile P3 before installation of pile cap are shown in Fig. 10. The shapes of simulated and tested waveforms are similar up to about 1×10^{-3} s (i.e., corresponding to the location of Styrofoam). After this point, the two curves deviate apart from each other. Furthermore, only the defect reflection can be identified in the simulated curve, while both defect reflection and toe reflection can all be identified in the tested curve. This is because that the steel rebars were neglected in the numerical simulation. When the tress wave travel downward to the defect (Styrofoam), it is reflected back to pile top. Therefore, only the defect reflection appears in the simulated waveform. However, for the case of actual test, energy induced by the impact can be transmitted through Styrofoam to the toe of pile then reflected back. Therefore, both defect and toe reflections appear in the tested curve. For the same reason, incident wave is not obvious in the simulated waveform of receivers under the defect (R2 and R3), while it is significant in the tested waveform.

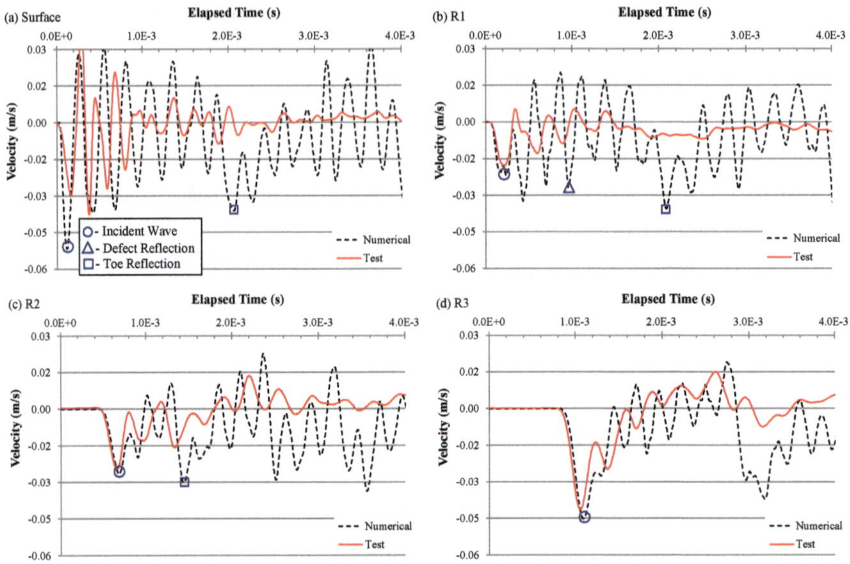

Fig. 9. Velocity profiles of necking pile P2 after installation of pile cap

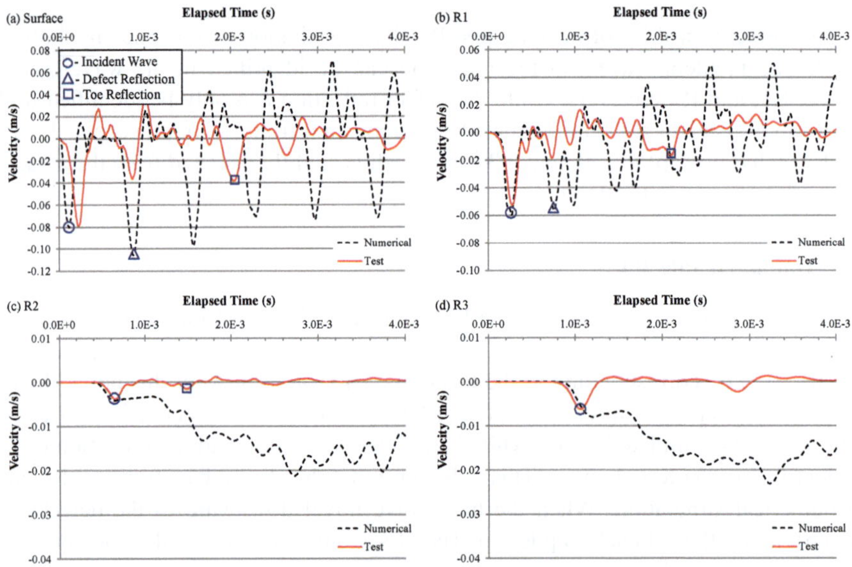

Fig. 10. Velocity profiles of damaged pile P3 before installation of pile cap

The velocity profiles of damaged pile P3 after installation of pile cap are shown in Fig. 11. As for the other two piles, with the presence of pile cap, 3D effects almost dominate the velocity profile of the integrity test. Although it is getting more difficult, the location of defect can still be identified, especially with the receiver (R1) located below the pile cap and above the defect.

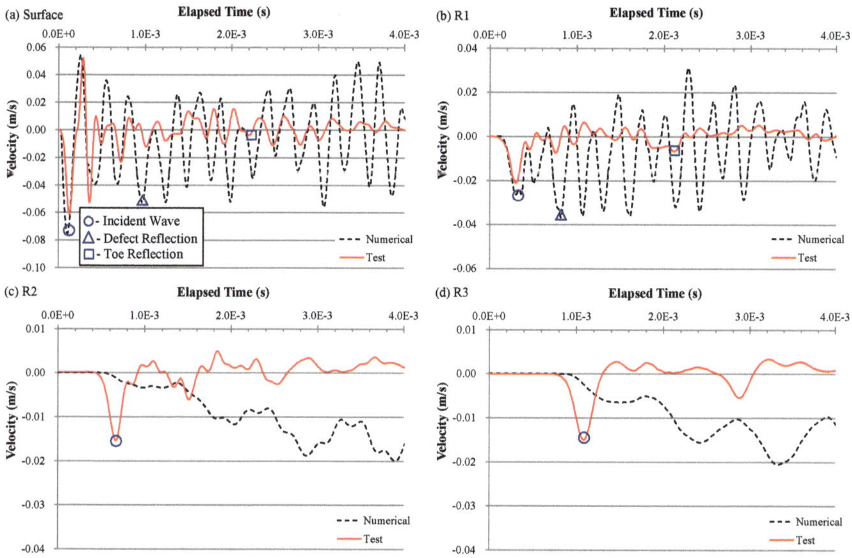

Fig. 11. Velocity profiles of damaged pile P3 after installation of pile cap

5 Conclusions

The main objective of this paper has been to investigate the boundary effects of pile cap on the integrity testing piles by numerical simulation and model pile test. From the results of this study, the following conclusions can be drawn:

(1) In general, the velocity wave forms obtained from numerical simulation are similar to those from model pile test, except the amplitude of reflective waves. The amplitude of reflective waves is higher from numerical simulation than model pile test due to soil damping was neglected in the numerical simulation.
(2) For intact piles or piles with minor defect (necking), effects of steel rebars can be neglected, numerical simulation and model pile test yield similar velocity profile of pile integrity test. For a pile with major defect, the effects of steel rebars cannot be neglected.
(3) For individual piles without pile cap, conventional PIT methods can easily identify the type of pile, whether it is intact or with minor or major defect.
(4) For a group of piles with a pile cap, boundary effects start to dominate the response of a pile integrity test. It becomes very difficult to identify a minor defect. However, a major defect can still be identified using PIT methods.
(5) Boundary effects of pile cap on the waveforms of low strain stress-wave reflection tests are reduced by using embedded accelerometer making it easier to assess the integrity of piles with a pile cap.

Acknowledgements. The work presented in this paper is sponsored by the Ministry of Science and Technology (formerly National Science Council) of Taiwan, R.O.C. under Grant No. NSC 102-2221-E-324-011-MY3. Their support is deeply appreciated.

References

ASTM: Standard Test Method for Low Strain Integrity Testing of Piles. D5882-00, Book of Standards Volume 04.09. ASTM International, West Conshohocken, PA (2000)

Chow, Y.K., Phoon, K.K., Chow, W.F., Wong, K.Y.: Low strain integrity testing of piles: three-dimensional effects. J. Geotech. Geoenviro. Eng., ASCE, **129**(11), 1057–1062 (2003)

Davis, A., Dunn, C.: From theory to field experience with the nondestructive testing of piles. Proc. Inst. Civil Eng. **57**(Part 2), 571–593 (1974)

Finno, R.J., Gassman, S.L.: Impulse response evaluation of drilled shafts. J. Geotech. Geoenviro. Eng., ASCE, **124**(10), 965–975

Fukuhara, T., Kakurai, M., Sugimoto, M.: Analytical evaluation of defective piles. In: Proceedings 4th International Conference on Application of Stress-Wave Theory to Piles, Hague, The Netherlands, 1992, pp. 563–569

Han, B.W., Wang, D.G.: Application of sonic tomography in the integrity testing of concrete piles. In: Proceedings, the 4th International Conference on the Application of Stress-Wave Theory to piles, Hauge, Netherlands, 1992, pp. 231–234

Hollema, D.A., Olson, L.D.: Crosshole sonic logging and velocity tomography imaging of drilled shaft foundations. In: Proceedings of NDT-CE 2003, Berlin, Germany, 2003

Lai, J., Yang, P.C, Yang, B.H., Chang, D.W.: Integrity testing of model pile with embedded accelerometers. In: Proceedings 2017 International Conference on Transportation Infrastructure and Materials, Qingdao, China, 2017, pp. 801–808

Liao, S.T., Roesset, J.M.: Dynamic response of intact piles to impulse loads. Int. J. Numer. Analyt. Meth. Geomech. **21**, 255–275 (1997)

Olson, L.D., Jalinoos, F., Aouad, M.F.: Determination of Unknown Subsurface Bridge Foundations. NCHRP 21-5 Project Report, Transportation Research Board, 1998, p. 71

Stain, R.T., Williams, H.T.: Interpretation of sonic coring results: a research project. In: Proceedings, the 4th International DFI Conference, Balkema, 1991, pp. 633–640

Steinbach, J., Vey, E.: Caisson evaluation by stress wave propagation method. J. Geotech. Eng. Div., ASCE **101**(GT4), 361–378 (1975)

Wu, S., Lai, J., Cheng, C.F., Yang, B.H.: Integrity testing of model piles with pile cap. In: Proceedings of NDT-CE 2015, Berlin, Germany, 2015, pp. 920–927

Soft Soil Improvement Using Rigid Inclusions: Toward an Application for Transport Infrastructure Construction in Vietnam

Van Duy Tran[1(✉)], Jean-Jacques Richard[1], and Tung Hoang[2]

[1] SEFI-INTRAFOR, FAYAT Fondations, Paris, France
vd.tran@sefi-intrafor.fayat.com
[2] National University of Civil Engineering, Hanoi, Vietnam

Abstract. Soft soil improvement by vertical rigid inclusions is a widespread technique in France for reducing and homogenizing surface settlements. This technique consists of a rigid inclusion network associated with a granular earth platform (so-called mattress), intercalated between the reinforced soil and the upper structure. The overload is transferred partially onto the inclusions through arching effect occurring in the platform due to shearing mechanisms in the fill. Such a process exhibits an economic and effective solution for large-scale civil works (roadways, railways and industrial building foundations) constructed over soft soils, especially in cases of requiring rapid construction. This paper presents, in the first part, a concise introduction to the soil improvement including general principles, operating mechanisms, calculation rules for the design and execution methods. In the second part, a parametric analysis based on a highway construction project in Dong Nai, Vietnam is rigorously performed to show the influence of different parameters of the technique such as the thickness of the granular earth platform and inclusion diameter on the behavior of a rigid inclusion supported embankment in the aim of highlighting the functioning mode of the new improvement solution. This leads to a preliminary design solution of soft soil improvement using rigid inclusions for the existing road construction project in Vietnam.

1 Introduction

The construction of civil works on soft soils necessitates the use of soil improvement techniques aimed at reducing absolute and differential settlements. Rigid inclusions method has showed its advantages for the construction of transport infrastructures and large-scale industrial facilities such as homogenization and reduction of surface settlements, rapid installation and no impact to the environment. Therefore, this technique has strongly grown in France especially after the issue of the design and construction guideline—ASIRI national project [5].

In Vietnam, the enormous increase of transport infrastructure construction projects in recent times presents a challenge for ground reinforcement techniques in constructing embankments on soft soil areas. Among the ground improvement methods currently used in Vietnam, Deep Cement Mixing (DCM) method occupies the majority of market shares due to its relatively good performance and competitive cost [6].

H. Khabbaz et al. (eds.), *New Prospects in Geotechnical Engineering Aspects of Civil Infrastructures*, Sustainable Civil Infrastructures,
https://doi.org/10.1007/978-3-319-95771-5_8

However, the quality of DCM columns is difficult to control and dependent significantly of the surrounded soils. In addition, DCM often fails in improving soft soils rich in organic matters which are widely encountered in the southern Vietnam. This problem could be totally remedied by rigid inclusions which differentiate to DCM columns by using grout or concrete inclusions.

This paper includes a brief description of the principles of the rigid inclusions technique for soft soil improvement and the specific calculation approach used in practice design in France. A parametric study of the behavior of an embankment on reinforced foundation based on an existing highway project in the southern Vietnam is then investigated, which allows to propose an application of the rigid inclusions technique for this project.

2 Principle of the Reinforcement Technique

2.1 Technique Description

The principle of soft soil improvement by rigid inclusions is shown in Fig. 1, which is based on the use of a vertical rigid inclusion network inserted into the soft soil, a granular earth mattress located between the inclusions and the upper structure. The structure load is partially transferred onto the rigid inclusions and then down to the bearing layer through arching effects within the mattress and side friction below the top of the inclusions. The load transfer capacity of the mattress depends on its thickness and frictional characteristics [1]. The coverage rate that is defined by the proportion of the surface covered by the inclusions can be increased by adding pile caps. This ratio currently lies in a range of 2–10% [5].

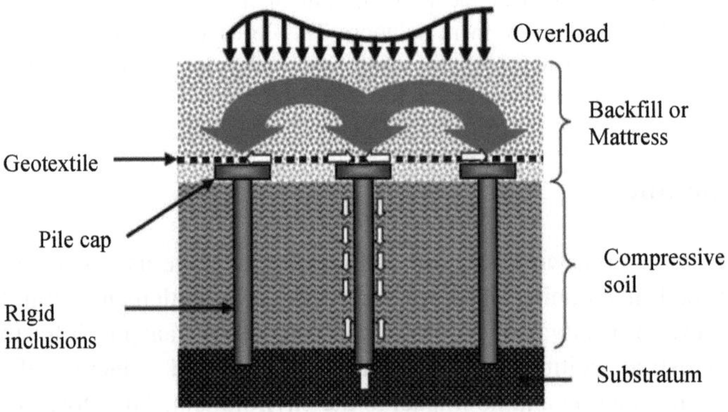

Fig. 1. Rigid inclusion reinforcement principle

Geosynthetic reinforcement can be added directly on the heads of inclusions or between the mattress and the inclusions. The differential settlement between the inclusions and the soft soil induces the tension in the geosynthetic sheet, which is involved in the load transfer on the inclusions.

Skin friction along the rigid inclusions is also implicated in the load transfer mechanism. Under the structure load, the soft soil settlement is greater than that of the inclusions itself; the inclusion tops punch in the mattress. This allows to develop a negative friction. In contrast, in the lower part, the inclusions settle more than the surrounding soil and move into the bearing layer, leading to positive friction. A toe resistance is developed in the lower end anchored in the bearing layer.

Rigid inclusions are generally made of sand/cement grout or poor concrete, which can be included eventually steel reinforcement for supporting tension or lateral loads. Rigid inclusions are possibly preformed or manufactured in situ. The design parameters such as inclusion dimensions, spacing, and composition are based upon the development of an optimal support distribution between the inclusions and the surrounding soil to reduce and homogenize surface settlements to within the admissible range, and to achieve the desired equivalent deformation modulus of the soft soil-inclusion composite. In construction reality, the inclusion diameter varies from 250 to 500 mm and the inclusion depth from 3 up to 30 m.

2.2 Installation Methods

As mentioned below, there are two types of rigid inclusions, preformed or manufactured in situ by pilling techniques. Preformed inclusions are usually considered in the case of very soft soil to avoid the overconsumption of grout or concrete and ensure the inclusion integrity. While manufactured in situ rigid inclusions are installed into soft ground by rigs with specially designed augers, which are displacement and non-displacement rigid inclusions.

The drilling auger for the construction of displacement rigid inclusions is hollow and has a special geometry (Fig. 2) allowing to displace soil laterally when screwing into the soil. On reaching the design depth, the displacement auger is withdrawn and grout/concrete is simultaneously pumped under pressure until the platform level. Rebar cage can be eventually installed into the fresh grout/concrete. During construction, displacement inclusion strength is greatly enhanced by the displacement of soil laterally and the pressure grout/concrete effect [2].

The non-displacement rigid inclusion is usually constructed by continuous flight auger (CFA) method whose drilling tool is composed of a hollow central shaft and a surrounding helical flight system. The constructive sequence of non-displacement inclusions is similar to that of displacement inclusions except that the soil spoil is not displaced laterally but is loaded onto the flights to the surface level. The grout/concrete pump is also carried out simultaneously with the auger extraction (Fig. 3).

In case of enlarged head inclusions, squared concrete slabs are usually added on the heads of the inclusions, which necessitates two separate interventions (completion of the inclusion and installing the precast slab) and is difficult to set up. In order to reduce the construction time, a particular construction method for producing enlarged head inclusions in a single operation has been developed to achieve the rigid circular inclusions associated with a cone shaped head (Rigid Inclusion with Variable Inertia—INIV inclusion). Such a method has been successfully applied in a pile-supported road project near Bourgoin-Jallieu in France [3].

Fig. 2. An displacement drilling tool

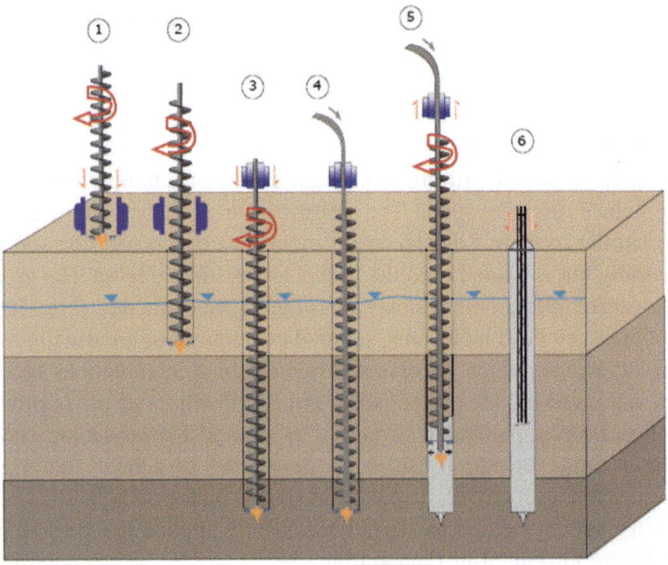

Fig. 3. Execution sequence of a non-displacement rigid inclusion

During rigid inclusion construction, real-time monitoring and recording of drilling and extraction parameters such as drilling depth, pressure, grout/concrete volumes, advancement speed, auger rotation and downward force of the rotor torque of the auger are also performed, which ensures a high quality control of the construction.

3 Calculation Methods

The rigid inclusion reinforcement behavior is commonly studied in a complete manner by numerical models in continuous medium such as finite element and different element methods which allow to account for the entire configuration of the reinforced soil, rigid inclusions and the interfaces between the different components as well as the structure construction history. However, numerical methods require a dedicated implementation of soil-structure simulation including the relevance of the adopted soil constitutive models and the quality of soil parameter input data. This causes many difficulties in practice design for geotechnical engineers. Therefore, the French guideline ARISI proposes an analytical approach to model the load transfer in soil reinforced by rigid inclusions which is in good agreement with the numerical results.

The principle is to consider an elementary reinforced cell centered by an inclusion, which is subjected to a vertical uniform load. This cell is decomposed in two sub-domains: the inclusion and the soil parts extended up to the top or bottom boundaries, and the all remaining soil volume (Fig. 4). The interaction between the two sub-domains is assumed to be fully described by the shear stress τ which is developed on their interface. The elementary cell is assimilated to an equivalent cylinder. The model is assumed to be periodic which implies that both lateral deformations and shear stresses are vanished at the exterior boundaries.

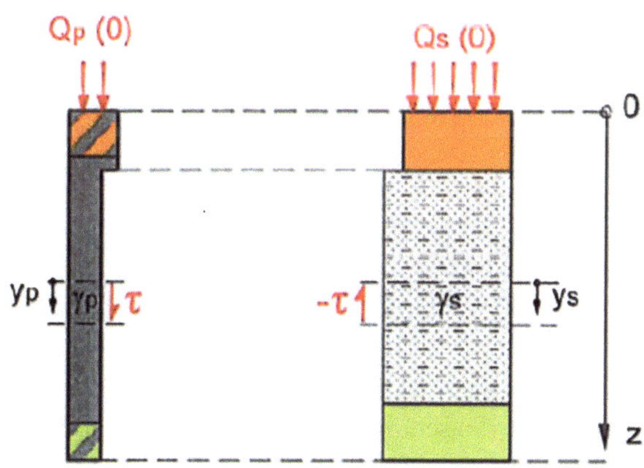

Fig. 4. Model of the elementary cell around the inclusion

The equilibrium state is described by the variation of forces within the inclusion (Eq. 1) and the soil (Eq. 2)

$$dQ_P(z) = (P\tau + \gamma_P A_P)dz \tag{1}$$

$$dQ_s(z) = (-P\tau + \gamma_s A_s)dz \tag{2}$$

where τ is the shear stress developed at the soil-inclusion interface, γ_s the soil unit weight, γ_P the inclusion unit weight, P the inclusion perimeter, and A_p and A_s the respective cross section areas of the inclusion and the soil.

The inclusion and average soil settlements are calculated by the following equations

$$dy_p(z) = \frac{Q_P(z)}{A_p E_p} dz \tag{3}$$

$$dy_s(z) = \frac{Q_s(z)}{A_s E_s} dz \tag{4}$$

where E_p is the Young's modulus of the inclusion and E_s the soil oedometric modulus of deformation.

The shear stress represents for the soil-inclusion interaction which is described by a transfer function expressing the dependence of the shear stress τ with the relative displacement between the two sub-domains $\gamma_p(z) - \gamma_s(z)$. In France, one uses widely the transfer functions proposed by Frank and Zhao 1982 [4] which are based on Menard pressuremeter modulus E_M, the limit unit shaft friction q_s and end bearing q_p. Such an analytical approach has been incorporated in a specific module "Taspie+" of the oriented foundation design program "FOXTA".

4 Case Study

4.1 Geotechnical Context

The numerical modeling is based on the geotechnical context of the highway construction project Ben Luc—Long Thanh in Dong Nai, Vietnam. The considered section in the package A-1 from km 2 + 090.0 to km 2 + 130.0 is composed two layers:

- Layer 1a: this layer is fluvial sediments of Alluvial Epoch, which consists of very soft clay and contains organic matters sparsely.
- Layer 2b: this layer is fluvial sediments of the latter Diluvial Epoch, which consists of stiff clay with some sandy considering the bearing layer.

A surcharge of 16 kN/m^2 is setup on the embankment of 6 m of height which is supported by the soft layer 1a. This causes probably important surface settlements and stability problems. According to the Vietnamese Standards (22TCN262-2000), the settlement criteria for the considered section shall be less than 100 mm. Hence, this section has been reinforced by DCM columns anchored into the Layer 2b. However the soil reinforcement quality would be not good as expected because of the presence of 10% organic matters in the Layer 1a. The authors propose therefore an alternative solution using the rigid inclusions technique. The project parameters are shown in Fig. 5. The effectiveness of reinforcement can be determined in terms of stress efficacy which is defined by the ratio between the load transmitted to the head of the inclusion and the total load on the unit grid.

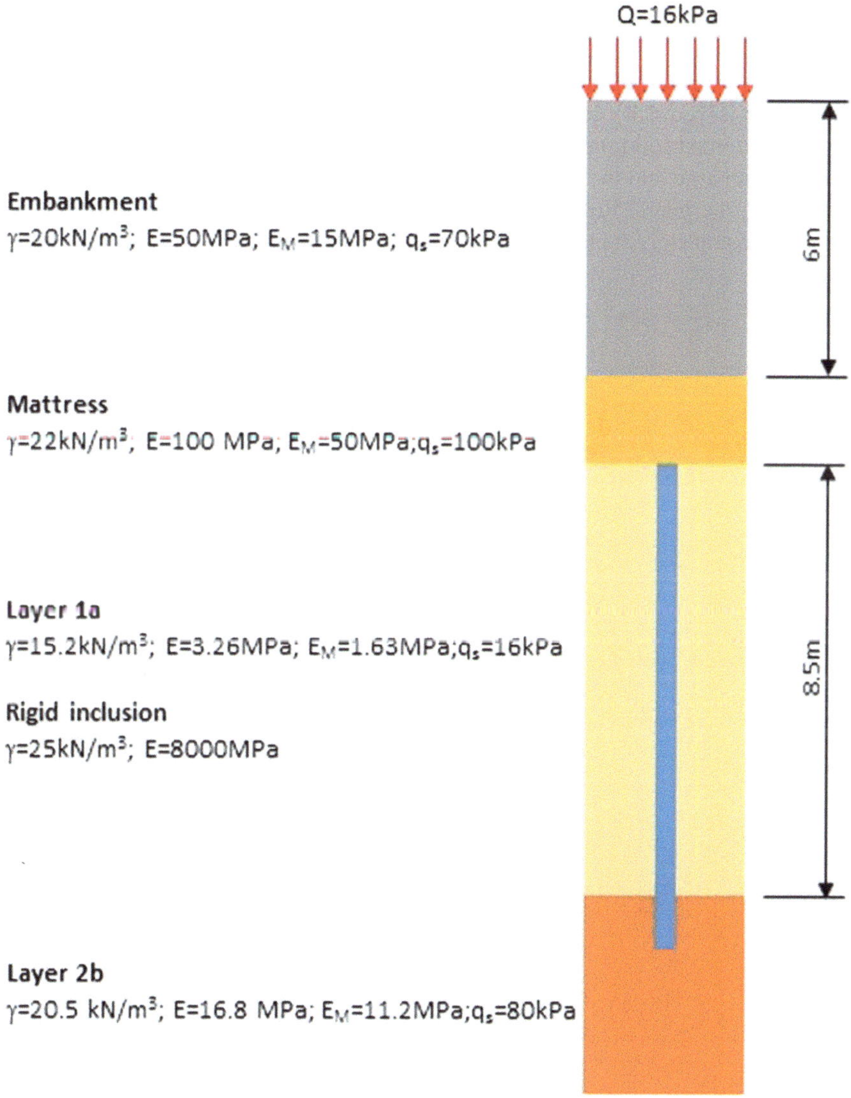

Fig. 5. Model parameters for soil reinforced by rigid inclusions

5 Results and Discussion

5.1 Piled Embankment Behavior

Firstly we consider the reference case consisting of 8.5 m compressible soil reinforced by 0.42 m of diameter inclusions separated by a distance of 1.5 m. The inclusions are 10 m of depth with 1.5 m of anchor in the bearing layer—layer 2b. The mattress of 1 m height is placed between the embankment and the inclusions. Figure 6 shows the settlement and force distribution profiles in the inclusion and in the soil. The structure settlement is 60 mm at the top embankment surface. The neutral point is located at 6 m of depth from the natural level where soil and inclusion settlements are equal and where the inclusion is vertically loaded at its maximum. Obviously, at the inclusion head level, the reinforced soil foundation is subjected to a total force of 310.5 kN (consisting of vehicle surcharge and the embankment and earth mattress weights) which is mainly transmitted to the pile (250 kN), the remaining force is received by the soil. The total load is then completely transmitted to the bearing layer under the inclusion.

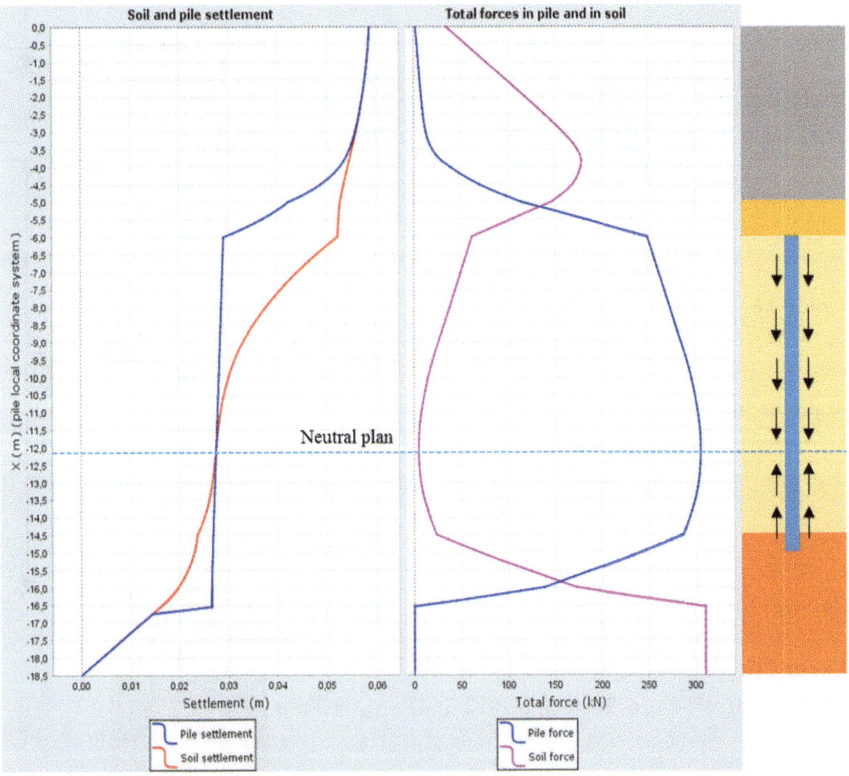

Fig. 6. Settlement and force distributions in the elementary cell

The load transfer role of the granular earth mattress is considered by studying the influence of the mattress thickness in terms of the structure settlement and the stress efficacy. Figure 7 illustrates the evolutions of the two parameters versus the variation of the mattress thickness between 0 and 3 m, it is obvious that in range of 0–1 m of the mattress thickness (reference case) the structure settlement rapidly decreases and the efficacy strongly increases. After this value, they are insignificantly improved and tends to the constant values when the mattress thickness approaching to 2 m. This can be explained by the arching effect occurring within the mattress. In this case, this effect requires a necessary thickness of 2 m to be completely developed.

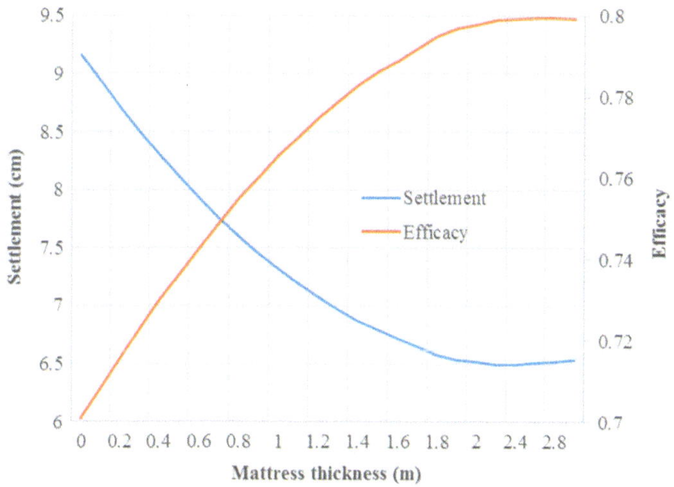

Fig. 7. Settlement and stress efficacy evolutions with the mattress thickness

The influence of the inclusion diameter is then investigated, which is also based on the reference case. The inclusion diameter is varied from 0.25 to 1.5 m. Figure 8 presents the structure settlement evolution as a function of the inclusion diameter. The decrease of this parameter is important till a diameter of 0.5 m. After this diameter value, the structure settlement reaches a threshold. A possible explication is that when the inclusion diameter is large enough, the total charge is mostly transmitted to the inclusion, which leads to the fact that the inclusion and soil settlements are equal and uniform over the entire depth of the inclusion.

5.2 Preliminary Design

According to the ASIRI design guideline, the inclusion design should be verified by the following criteria:

- The granular earth mattress thickness $H_M > 0.7(s - a)$ with s is the grid size and a the inclusion diameter.

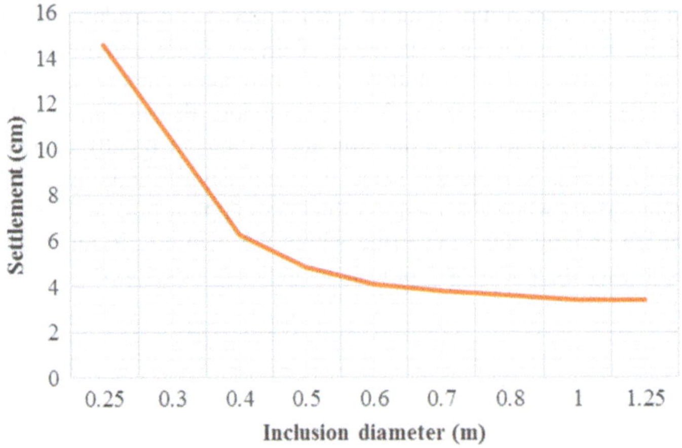

Fig. 8. Structure settlement as a function of the inclusion diameter

- Prandtl's punching failure for the mattress (Fig. 9): it is convenient to verify that the stress on the inclusion head remains below

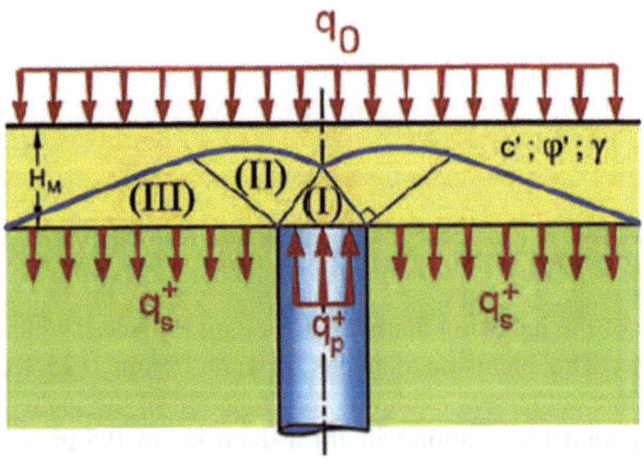

Fig. 9. Prandtl's punching failure principle of the mattress

$$q_p^+ = \frac{1}{\alpha(N_q - 1) + 1}\left[N_q(q_0 + \gamma \cdot H_M)\right] \tag{5}$$

with α is the coverage ratio and $N_q = 64.2$ (for $\varphi' = 40°$).

Obviously, with the reference case study $H_M = 1$ m > 0.7 × (1.5–0.36) = 0.8 m and the stress at the inclusion head is 1804 kPa less than q_p^+ = 1813 kPa, combined with the estimated settlement of 60 mm less than 100 mm required by the Vietnamese standard. Therefore, the authors propose this case study as a preliminary design of soil reinforcement by using rigid inclusions for this highway construction project.

6 Conclusions

Soft soil reinforced by rigid inclusions is widely used in France. However this technique is not well-known in Vietnam. This work contributes to clarify the reinforcement principles by a study of the influence of the different parameters such as the thickness of mattress layer and the cover rate on the pied-embankment behavior based on an existing highway construction project in Vietnam.

From the outcome of this work, the author proposed a design solution of soil reinforced by rigid inclusions which is justified in accordance with the French guideline ASIRI for this construction project. This would be an interesting scientific base for the application of rigid inclusions to future infrastructure construction projects in Vietnam.

References

1. Boussetta, S., Bouassida, M., Zouabi, M.: Assessment of observed behavior of soil reinforced by rigid inclusions. Innov. Infrastruct. Solut. **1**, 27 (2016). https://doi.org/10.1007/s41062-016-0027-6
2. Cirión, A., Paulín, J., Racinais, J., Glandy, M.: Displacement rigid inclusions. In: Proceedings of the 18th ICSMGE, Paris, France, 2013
3. Dias, D., Grippon, J., Nunez, M.: Behavior of a pile-supported embankment using rigid piles with variable inertia. ISSMGE—TC 211, Brussels, Belgium (2012)
4. Frank, R., Zhao, S.R.: Estimation par les paramètres pressionmétriques de l'enforcement sous charge axiale des pieux forés dans les sols fins. Bulletin Liaison Laboratoire des Ponts et Chaussées **119**, 17–24 (1982)
5. IREX: Recommandations pour la conception, le dimensionnement, l'exécution et le contrôle de l'amélioration des sols de fondation par inclusions rigides (for the French version). Recommendations for the design, construction and control of rigid inclusion ground improvements (for the English version). Projet National ASIRI. Presses des Ponts, ISBN 978-2-85978-462-1 (available in French and English) (2012)
6. Le, T.T., Dao, T.K.C., Nguyen, T.K.: Development and application of cement deep mixing method in Vietnam. In: Geotechnics for Sustainable Infrastructure Development—Geotec Hanoi 2016, Phung (edt). ISBN 978-604-82-1821-8 (2016)

Mechanical and Deformation Properties of Deep Foundation Pit Supporting System Subjected to Asymmetric Loadings

Shouhua Liu[1(✉)], Junsheng Yang[1], Wuyi Liu[2], and Jianhong Wang[3]

[1] School of Civil Engineering, Central South University, Changsha 410075, Hunan, China
{shouhualiu, jsyang}@csu.edu.cn, 18945087415@163.com

[2] SINOHYDRO BUREAU 8 CO., LTD., Changsha 410075, Hunan, China
18945087415@163.com

[3] M.C.E. School of Civil Engineering, Central South University, Changsha 410075, Hunan, China
154811075@csu.edu.cn

Abstract. Due to the limitation of the terrain and environment condition, two sides of deep foundation pit often appear in different elevation level, which makes foundation pit subjected to asymmetric loadings. The mechanical characteristics of retaining structure with the influence of asymmetric loadings are quite different from that with the influence of symmetric ones. Currently, the relevant code doesn't present a detailed method for supporting system design of foundation pit under asymmetric loading. Based on Shazitang Station of Changsha Metro Line 4, two sides of them are at different elevation level, the supporting system design scheme and the excavation process was then introduced. Especially, to ensure the safety of foundation pit as well as surrounding buildings, the first supporting was designed as concrete diagonal brace. The horizontal displacement of diaphragm walls and the settlement of surrounding buildings were monitored. Finite element models, which aimed at analyzing the stress and deformation of supporting structures, were then established. With respect to the action of asymmetric load, the horizontal displacement of diaphragm wall on the side of higher elevation level is bigger than the other side. At the top of the wall, the horizontal displacement appears to be a "drift" towards the side of lower elevation level. The horizontal displacement of diaphragm wall and settlement of surrounding ground is relative small and meets requirements of relevant code. In all stages of construction, the deep foundation pit and surrounding buildings remain safety. It can be concluded that the design of supporting system is highly effective in controlling deformation during the foundation pit excavation, which can provide a reference for the design and construction of similar projects.

Keywords: Deep foundation pit · Supporting system · Asymmetric loadings
Mechanical characteristics

H. Khabbaz et al. (eds.), *New Prospects in Geotechnical Engineering Aspects of Civil Infrastructures*, Sustainable Civil Infrastructures,
https://doi.org/10.1007/978-3-319-95771-5_9

1 Introduction

At present, the construction methods of subway stations in China mainly include cut and cover method, undercutting method, semi-cover excavation method and cover and cut method. Cut and cover method is the main construction method which has the advantages of convenient construction, rapid construction speed and low cost. However the large land occupation brings traffic congestion. It is not suitable in heavy traffic areas. Either undercutting method or cover and cut method has no effect on traffic with high cost and long construction period. On the contrary, the semi-cover excavation method has the advantages of low cost, short construction period and no influence on traffic. Metro stations are generally located in urban trunk roads. Therefore, semi-cover excavation method has been widely used in subway stations. Due to the limitation of the terrain and environment condition, two sides of deep foundation pit often appear in different elevation level, which makes foundation pit subjected to asymmetric loadings. Under the asymmetrical geotechnical and structural loads, the deformation law and stress characteristics of foundation pit have its particularity.

Currently, the relevant study doesn't present a detailed method for supporting system design of foundation pit under asymmetric loading. At present, the research is mainly about the deformation characteristics of symmetrical foundation pit (Wang et al. 2010; Tan and Wang 2013a, b). There are few researches on the deformation characteristics of foundation pit under asymmetric loading (Shi et al. 2011), especially the ground elevation of both sides of the foundation pit is different. Therefore, it is necessary to study the deformation characteristics of excavation under asymmetric loading.

In this paper, taking Shazitang Metro Station of Changsha Metro Line 4 as the research background, field monitoring is carried out during construction. Firstly the design of retaining structure of foundation pit is introduced. Secondly sorting and analyzing the measured data. And then a numerical model is presented to simulate the excavation of the deep foundation pit. Finally based on the construction monitoring data and numerical simulation results, under asymmetric loading the characteristics of foundation pit is obtained.

2 Project Overview

Shazitang Metro Station of Changsha Metro Line 4 is located at the northwest quadrant of Huangtuling Road and Shaoshan Road. One part of the pit is located in the original brewery block, and the other part is under the Huangtuling Road. To ensure the normal passage of Huangtuling Road, both cut and cover method and semi-cover method are adopted to construct the station (Fig. 1). The length of foundation pit is 169.6 m, the width of standard section is 22.9 m, and the excavation depth is 26.8–29.5 m. The depth of excavation is mainly silty clay, sandy gravel and argillaceous siltstone, the pit section as shown in Fig. 2. The schematic diagram of the plane is shown in Fig. 1.

The designed system consisted of diaphragm wall with four levels of internal brace and center post. The dimensions of the supporting structures for typical sections are shown in Figs. 2, 3 and 4. The pre-stress of both steel supports are 500.0 kN. Due to

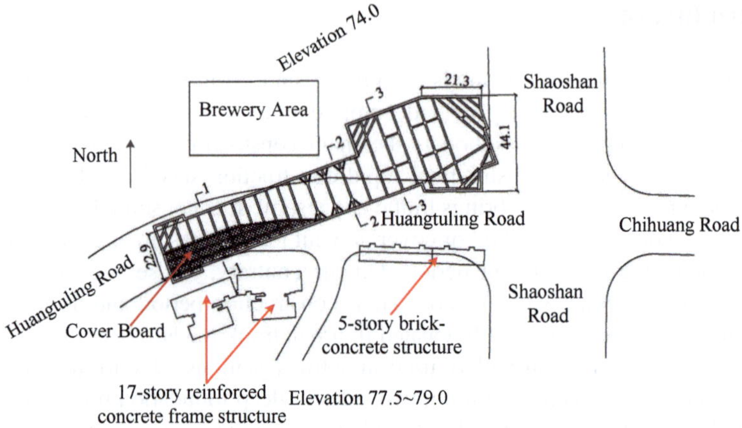

Fig. 1. Layout of foundation pit (unit: m)

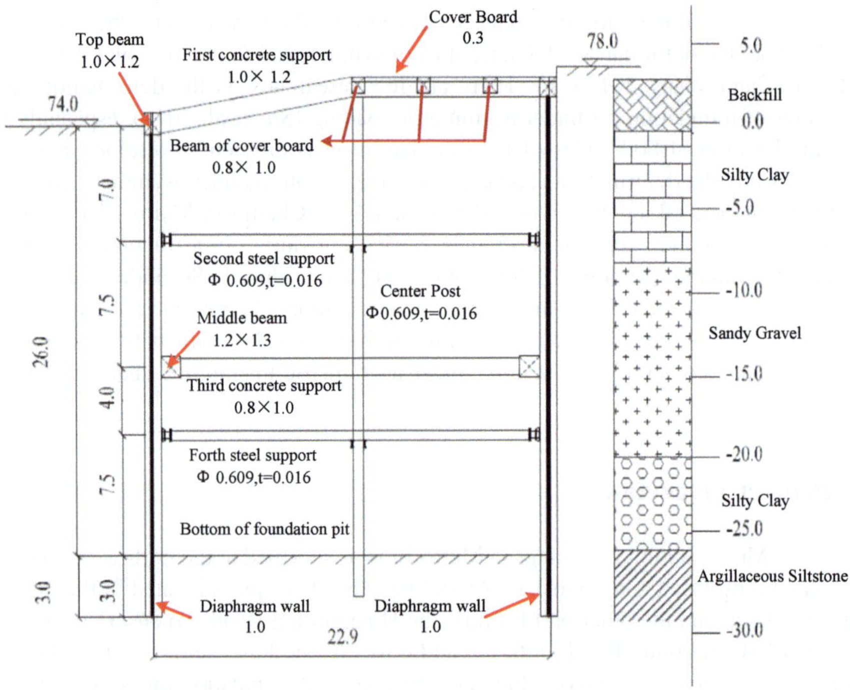

Fig. 2. 1-1 section of foundation pit (unit: m)

the lower elevation of brewery area and the cover board, the lower elevation side of first brace is supported by a inclined brace in 1-1 and 2-2 cross-section (Figs. 2, 3 and 5).

The foundation pit in semi-cover area (Fig. 1) is excavated and built with the following construction flow: step 1: building diaphragm wall, uplift pile (south side),

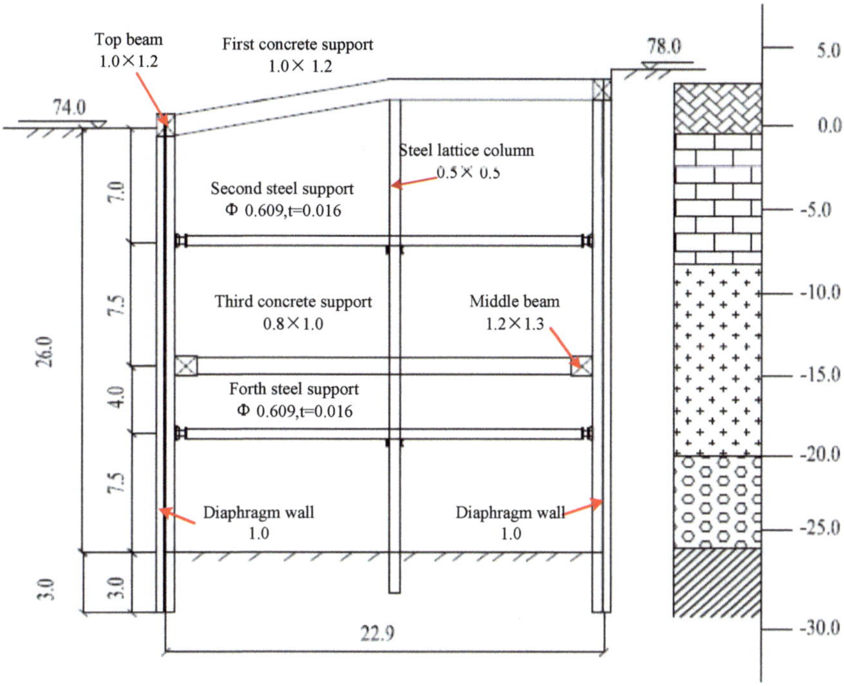

Fig. 3. 2-2 section of foundation pit (unit: m)

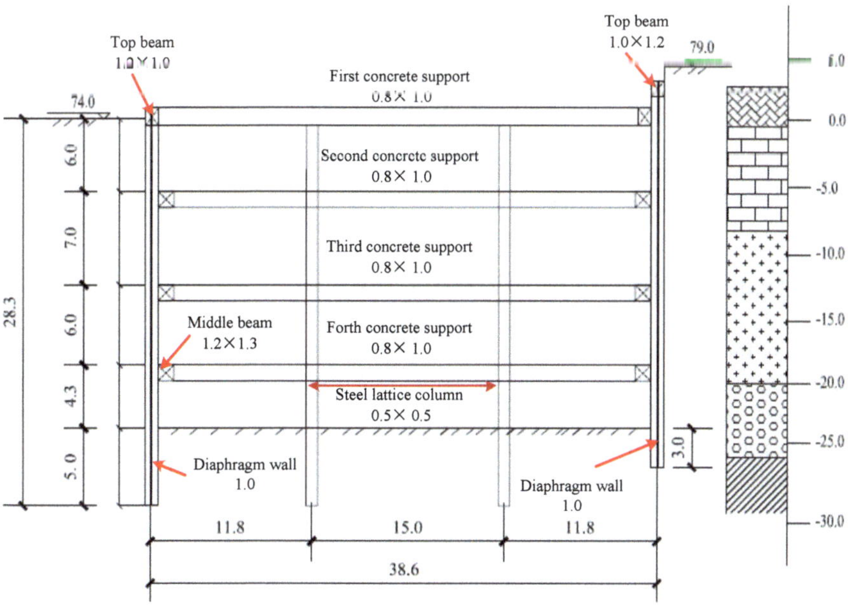

Fig. 4. 3-3 section of foundation pit (unit: m)

Fig. 5. Spot map of first inclined brace

erect column pile → step 2: building cover board (south side) → step 3: building diaphragm wall, uplift pile, first concrete support (north side) and removing cover soil → step 4: excavating first floor space → step 5: building second steel support → step 6: excavating second floor space → step 7: building third concrete support → step 8: excavating third floor space → step 9: building forth steel support → step 10: excavating fourth floor space and building bottom plate.

3 Field Measurement

There is system test for the foundation pit which aims to ensure the safety of the foundation pit, verify the rationality of the design. The monitoring items include: the settlement and horizontal displacement of the center post, the settlement and horizontal displacement of the diaphragm wall, the settlement of the ground, the cross-section force of the internal brace struts.

4 Vertical Displacement of Center Post

Figure 6 is the vertical displacement curve at the top of the erect column pile under different excavation stages. As can be seen from Fig. 6, the erect column pile sunk during construction. The rule of erect column pile settlement of bias foundation pit is contrary to the symmetrical foundation pit (Tan and Li 2011; Liao et al. 2015). With the proceeding of excavation, the settlement increases gradually and tends to be stable at the end of excavation. Due to the bias loads, the vertical character of the center post is settlement.

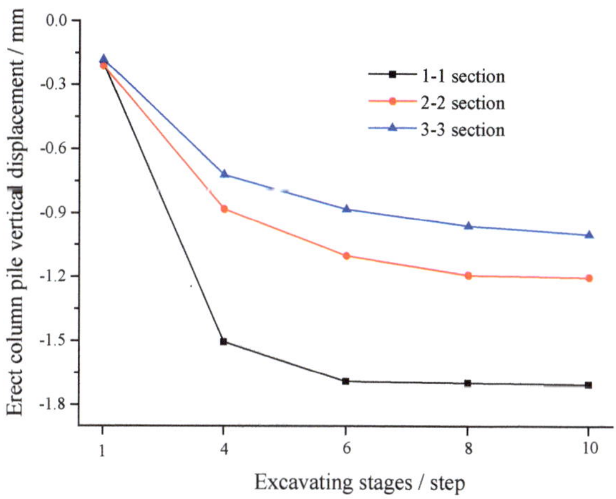

Fig. 6. Vertical displacement of center post

5 Horizontal Displacement of Diaphragm Wall

Figure 7 is horizontal displacement of diaphragm wall in 1-1 cross-section. When the diaphragm wall deform to the inside of the foundation pit, the horizontal displacement of the diaphragm wall is positive. As shown in Fig. 7, the horizontal displacement of diaphragm wall develops continuously with increasing of excavation depth.

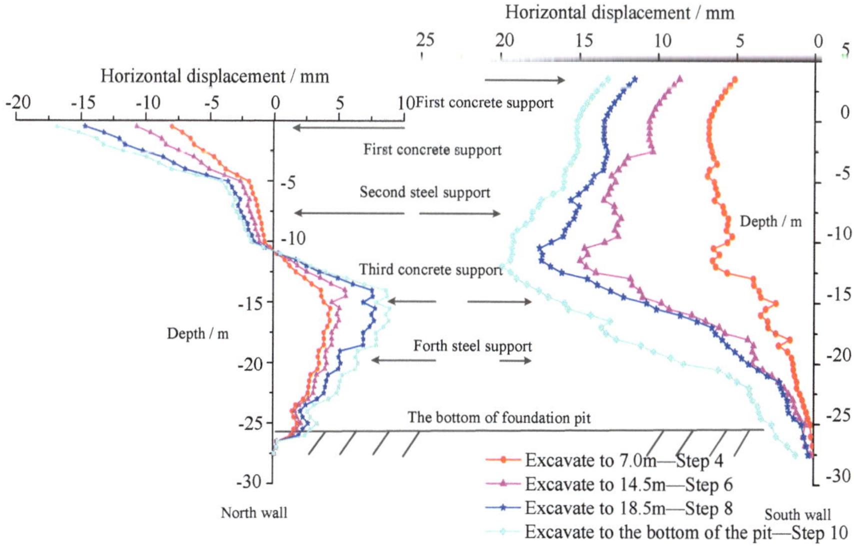

Fig. 7. Monitored horizontal displacement of diaphragm wall for 1-1 section

At higher elevation level (South), the diaphragm wall towards to the inside of the excavation. The maximum horizontal displacement of the diaphragm wall on the higher side is 19.8 mm. At lower elevation level (North), the diaphragm wall is bounded by 1/2 H (H is the depth of excavation). The upper part of the diaphragm wall moves to the outside of the foundation pit, maximum value −16.8 mm located at the top. Conversely, the lower part moves to the inside, maximum value 8.9 mm which is located at 15.5 m depth. At same condition, the horizontal displacement of the diaphragm wall on the higher side is bigger than the offside because of the asymmetric loads. The horizontal displacement of the diaphragm wall is smaller than the precaution value (25.0 mm). The design of retaining structure with Semi-covered method under asymmetric loads is safe and reasonable.

The test results show that the deformation mode of the diaphragm wall of the pit which bearing asymmetric load is different from that of symmetrical foundation pit, as shown in Fig. 8. At lower elevation level (North), the upper wall appears to be a "drift" towards the side of lower elevation level, the lower wall appears to towards the inside of the foundation pit. At higher elevation level (South), the diaphragm wall shows a compound deformation law (Long 2001).

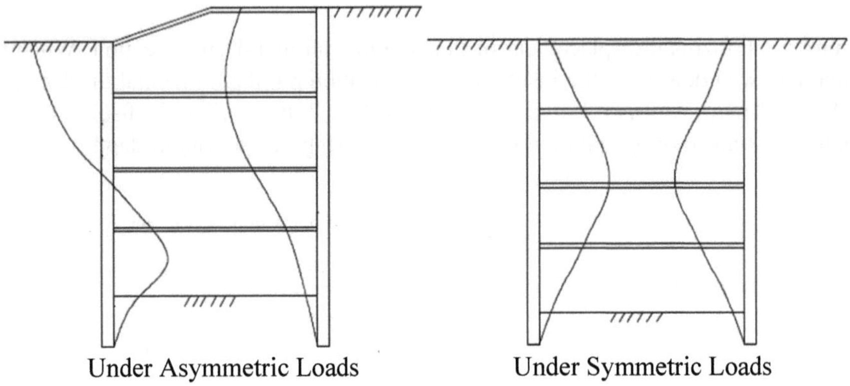

Fig. 8. Comparison between deformation modes of diaphragm walls for unsymmetrical loaded and symmetrical loaded foundation pit

6 Finite Element Analysis

According to the measured data, the deformation characteristics of the diaphragm wall and erect column pile are different from those of the traditional symmetrical foundation pit. Further research on the deformation and stress characteristics is necessary. Therefore, numerical method is adopted to study the characteristics of bias foundation pit.

7 FEA Model and Boundary Conditions

ABAQUS (2013) was used to investigate the excavation process. This model is comprised of 7984 nodes and 6803 elements. In order to reduce boundary effect, a model dimension of 200.0 m (length) × 100 m (depth) (Fig. 9) was chosen. A total two-dimensional (2D) FE model was generated considering diaphragm wall, all support and center post.

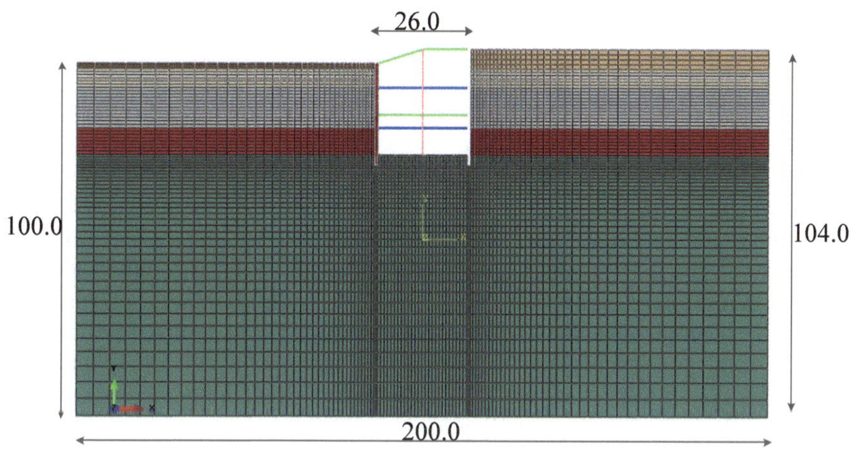

Fig. 9. 2D FE mesh employed in simulation of excavation (All units are in m)

Modified Cam-clay (MCC) model is one of the most commonly used constitutive models for clays (Huang et al. 2011), and can distinguish stiffness difference of the soil between the loading and unloading. Then the constitutive behavior of the silty soil is modeled with a Modified Cam-Clay plasticity model (Table 1). The gravel and gravel layer soil, is modeled with a Mohr–Coulomb elasto-plastic model. The diaphragm wall, internal brace struts and center post are treated as linear isotropic elastic materials.

Table 1. Geotechnical properties involved in the study

Soil layer	γ	M	λ	K	μ	e_o	c′ (kPa)	Φ′ (°)	E (kPa)
Backfill	19.0	1.20	0.0293	0.0036	0.35	1.12			
Silty clay	19.8	1.29	0.0312	0.0065	0.35	0.71			
Sandy gravel	23.0				0.30	0.82	0	35	150,000
Silty clay	19.8	1.38	0.0445	0.0098	0.35	0.76			
Argillaceous siltstone	22.8				0.28	0.64	200	32	500,000

M—Slope of critical state line ln q–p′ space for MCC; λ and K—compression and re-compression values for MCC; μ—Poisson's ratio; e_0—initial void ratio; c′—effective cohesion; Φ′—effective friction angle; E—Young's modulus

The horizontal displacement of the left and right sides and the bottom side of the model are restricted. Soil mass, wall and cover board in this model were considered as continuum solids. Internal brace struts and center post in this model were considered as beam elements (2013). The ground profile is shown in Fig. 3. The excavation sequence simulated by ABAQUS of is consistent with the practical construction process. The excavation process was simulated using the "element death" technique.

8 FEA Results Analysis

8.1 Deformation of the Diaphragm Wall

Figure 10 is the horizontal displacement curve of the diaphragm wall. The maximum horizontal displacement of the diaphragm wall on the higher side is 16.0 mm, which is located at the top of the wall. On the lower side the diaphragm wall is bounded by 2/5 H (H is the depth of foundation pit). The upper part of the diaphragm wall is deformed to the outside of the foundation pit, maximum value −16.5 mm is located at the top. The lower part is deformed to the inside, maximum value 9.5 mm is located at 20.0 m below the ground. The numerical simulation results show that the deformation mode of the diaphragm wall is the same as the measured results. The horizontal displacement values of diaphragm wall which is conducted by numerical simulation is slightly larger than measured values. The main reason is that the numerical simulation ignores the influence of the surrounding load.

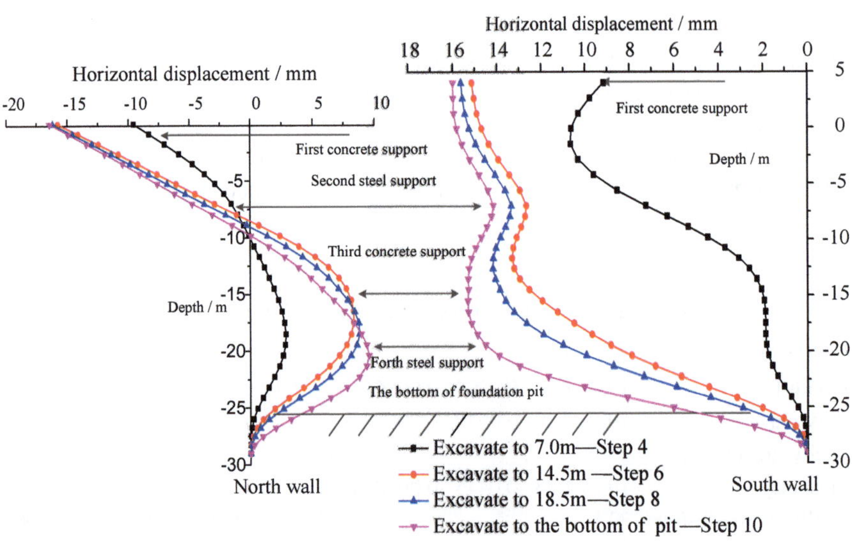

Fig. 10. Horizontal displacement of diaphragm wall for 1-1 section conducted by FEA

8.2 Bending Moment of the Diaphragm Wall

The bending moment of diaphragm wall of 1-1 section is shown in Fig. 11. When the inside of the wall is pulled, the bending moment is negative. With the excavation of the foundation pit, the maximum bending moment position moves downward. On same condition, the bending moment of the south wall at the same side is bigger than offside. The maximum bending moment of the north wall is 1.03 MN m, the south is 1.45 MN m. The maximum bending moment of both side walls are located in 3.0 m above the bottom. Based on the reinforcement of diaphragm wall, the ultimate bending moment of the wall is 4.7 MN m. The minimum safety factor of the south wall is 3.2, the north wall is 4.6.

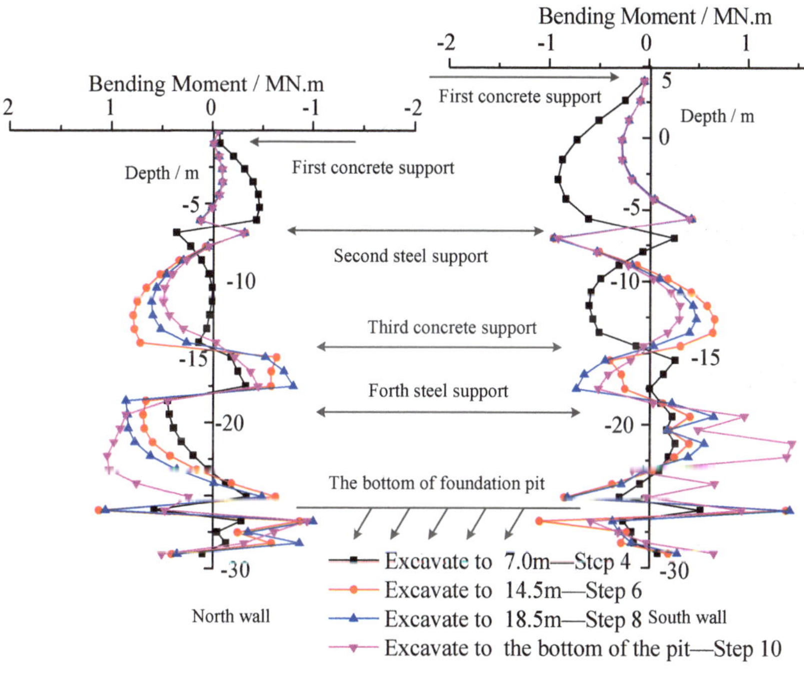

Fig. 11. Bending moment of diaphragm wall in 1-1 section

8.3 The Axial Force of the Internal Brace

Table 2 is the axial force of the support in different excavation steps. With the excavation and support erection, the axial force of the upper support decreases gradually. It can be seen from the Table 2 that the maximum of the axial force is located at the second support. The minimum is the first support. The axial force values are far less than the warning values. This shows that the design of the internal brace in the excavation still have greater optimization space.

Table 2. Axial force of the support

Supporting axis force (kN)	First support	Second support	Third support	Forth support
Step 4	217.0	–	–	–
Step 6	124.9	958.1	–	–
Step 8	113.0	917.4	629.0	–
Step 10	102.6	872.4	727.3	747.1
Warning value (kN)	2191.0	2327.5	8603.0	2327.5

8.4 Deformation of the Center Post

Figure 12 is the horizontal displacement along the depth direction of the center post. The maximum horizontal displacement 15.9 mm is at the top. The horizontal displacement gradually decreases along the depth direction. The deformation mode of center post is different from common foundation pit. Under asymmetric loading, the center post is not only a structural structure, but a slender stress bar structure. Figure 13 is the vertical displacement at the top of the temporary column. In the whole process of excavation, the center post under vertical is expressed as settlement and gradually increases with the excavation. The deformation tends to be stable at the end of excavation. The numerical simulation results are in good agreement with the measured results.

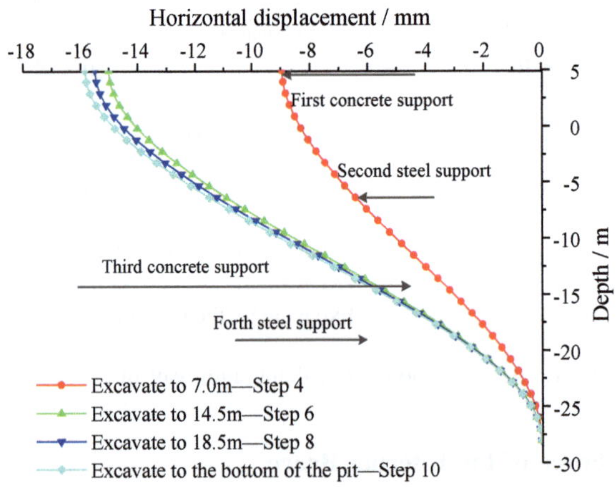

Fig. 12. Horizontal displacement of center post

8.5 Surface Subsidence

The surface subsidence curves of the higher and lower side are shown in Figs. 14 and 15 respectively. The settlement of the ground surface increases gradually with the excavation (Hsieh and Ou 1998; Ou et al. 1993). In the higher side the maximum

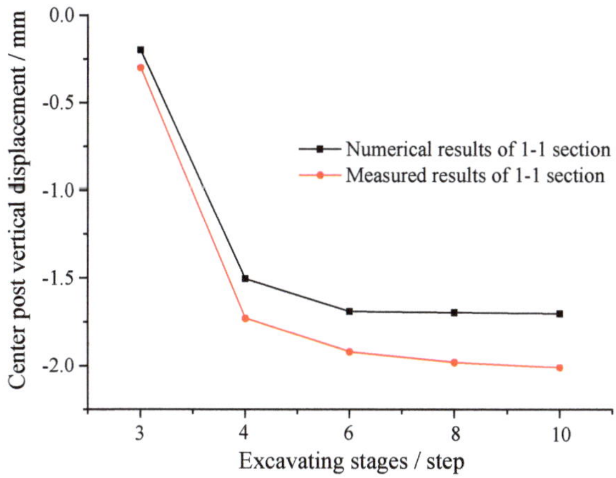

Fig. 13. Vertical displacement at the top of the center post

settlement is 11.0 mm, which is located at 7.7 m behind the wall. In the lower side the maximum settlement of the ground is 3.7 mm, which is located at 14.4 m behind the wall. The maximum settlement of the surface subsidence in the higher side is far bigger than offside. Due to asymmetric loading, the surface settlement trough on both sides of the foundation pit is asymmetrically distributed.

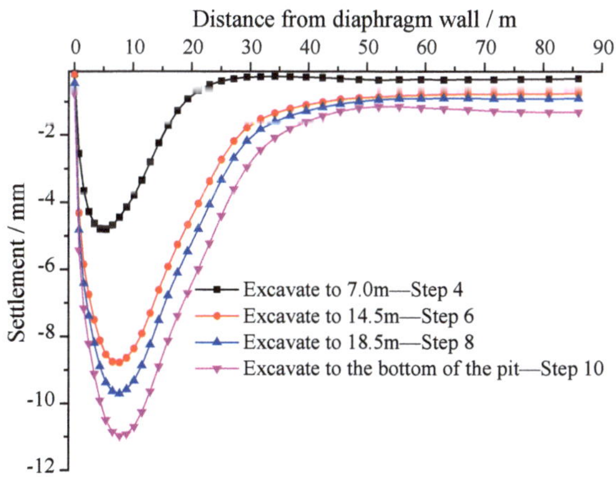

Fig. 14. Surface subsidence curves in the higher side

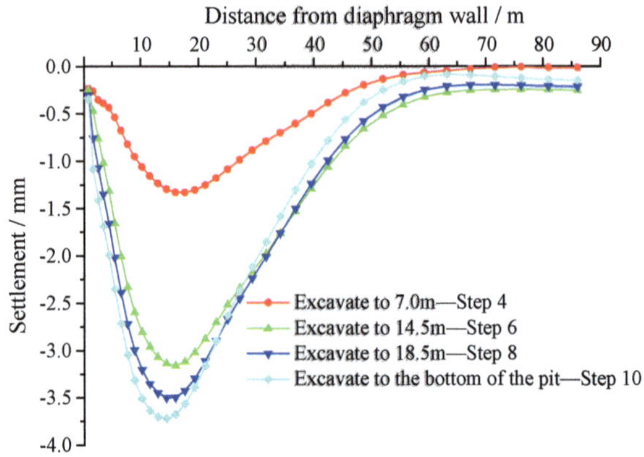

Fig. 15. Surface subsidence curves in the lower side

9 Conclusions

A case study of the stress and deformation characteristics of foundation pit under asymmetric loading was conducted by both field observation and numerical analysis by ABAQUS. Measured and numerical results analysis revealed that the stress and deformation mode of the semi-cover excavation which bearing asymmetric load is different from that of symmetrical foundation pit:

1. The deformation and force of the foundation pit is less than the warning value. The axial force values are far less than the warning values. The design of retaining structure of deep excavation with semi-covered method under asymmetric load is safe.
2. The axial force of the first support is far less than others. The axial force values are far less than the warning values which still have greater optimization space.
3. Vertically the center post shows settlement, horizontally the deformation of the center post toward to the lower elevation level of the foundation pit.
4. The surface subsidence on both sides is asymmetrically distributed. The shape of subsidence trough at the higher side is leptosome, on the offside pyknic.
5. At the top of the wall, the horizontal displacement appears to be a "drift" towards the side of lower elevation level. When the working procedure and depth is the same, the stress and deformation of the wall at the lower elevation level is smaller the offside.

In view of the semi-cover excavation which bearing asymmetric load, the characteristics of the pit should be considered in the follow-up design.

Acknowledgements. The research was conducted with funding provided by the transverse project of SINOHYDRO BUREAU 8 CO., LTD. (Grant NO.SDBJ-CSDT-JS-2017-001). The first and second authors would like to acknowledge the support provided by SINOHYDRO BUREAU 8 CO., LTD.

References

Abaqus: Abaqus User's Manual, Version 6.13. Dassault Simulia Corp., RI (2013)

Hsieh, P.-G., Ou, C.-Y.: Shape of ground surface settlement profiles caused by excavation. Can. Geotech. J. **35**(6), 1004–1017 (1998)

Huang, M.S., Liu, Y.H., Sheng, D.C.: Simulation of yielding and stress-strain behaviour of Shanghai soft clay. Comput. Geotech. **38**(3), 341–353 (2011)

Liao, S.-M., Wei, S.-F., Tan, Y., et al.: Field performance of large-scale deep excavation in Suzhou. Chinese Journal of Geotechnical Engineering **37**(3), 458–469 (2015). (in Chinese)

Long, M.: Database for retaining wall and ground movements due to deep excavation. J. Geotech. Geoenviron. Eng. **127**(3), 203–224 (2001)

Ou, C.-Y., Hsieh, P.-G., Chiou, D.C.: Characteristics of ground surface settlement during excavation. Can. Geotech. J. **30**(5), 758–767 (1993)

Shi, Y.-F., Yang, J.-S., Bai, W., et al.: Analysis of field testing for deformation and internal force of unsymmetrical loaded foundation pit's enclosure structure close to railway. Chinese J. Rock Mech. Eng. **30**(4), 826–833 (2011) (in Chinese)

Tan, Y., Li, M.: Measured performance of a 26 m deep top-down excavation in downtown Shanghai. Can. Geotech. J. **48**(5), 704–719 (2011)

Tan, Y., Wang, D.L.: Characteristics of a large-scale deep foundation pit excavation by the central-island technique in Shanghai soft clay.I: bottom-up construction of the central cylindrical shaft. J. Geotech. Geoenviron. Eng. **139**(11), 1875–1893 (2013a)

Tan, Y., Wang, D.L.: Characteristics of a large-scale deep foundation pit excavation by the central-island technique in Shanghai soft clay.II: top-down construction of the peripheral rectangular pit. J. Geotech. Geoenviron. Eng. **139**(11), 1894–1910 (2013b)

Wang, J.H., Xu, Z.H., Wang, W.D.: Wall and ground movements due to deep excavations in Shanghai soft soils. J. Geotech. Geoenviron. Eng. **136**(7), 985–994 (2010)

Modelling Linear Viscoelastic Behaviour of Kanpur Local Soil Using Prony Series, Parameter Fitting

Abhijeet Swain(✉) and Priyanka Ghosh

Department of Civil Engineering, Indian Institute of Technology (IIT) Kanpur, Kanpur, Uttar Pradesh, India
abswain@iitk.ac.in

Abstract. This paper aims at presenting a more versatile and general method to determine the rheological (time-dependent) properties of soil, using prony series parameter fitting. The time dependent stress-strain test data (creep data) of soil is used to evaluate the prony-series parameters adopting the method of nonlinear least squares fit. This paper presents the extraction of prony-series parameters of local soil collected from the geotechnical in-situ lab site, IIT Kanpur. Creep test has been performed on the soil sample and the prony-parameters are calculated. Two finite element models are used to simulate and validate the time-dependent constitutive relationship of the soil in both static and dynamic loading conditions.

1 Introduction

In geomechanics, in several occasions the soil is assumed to show a linear elastic or non-linear elastic stress-strain behavior (Winkler's model), neglecting the effect of time dependency in the constitutive relationship. This assumption simplifies the theoretical and the numerical modeling of soil and yields an acceptable prediction of the bearing capacity and the settlement characteristics of soil in various soil-structure interaction problems. However, it is very important and significant to understand the behavior of soil at different time intervals, most critically in case of expansive soils subjected to static and dynamic loading. To capture such time-dependent behavior of soils, various researchers (Tan 1957; Kerr 1961; Schiffman et al. 1964; Lyakhov 1968) have applied the concept of viscoelasticity. This application has been limited to Maxwell model (2-parameter), Kelvin-Voigt model (2-parameter), Poynting-Thompson model (3-parameter) and Burger model (4-parameter). These time-dependent constitutive models have been limited to the study of stress-strain behavior of soil subjected to static loads (soil consolidation problems), as discussed by Dey and Basudhar (2010), whereas the problems under dynamic condition are mostly modelled using simple 2-parameter Kelvin-Voigt model. The difficulty involved in the lumped parameter models is the estimation of correct values of different component parameters to replicate the exact material behavior.

The present study focuses on a more versatile and general method to determine the rheological properties of soil, using prony series fitting. Prony series is a generalized

H. Khabbaz et al. (eds.), *New Prospects in Geotechnical Engineering Aspects of Civil Infrastructures*, Sustainable Civil Infrastructures,
https://doi.org/10.1007/978-3-319-95771-5_10

mathematical representation of a bunch of Maxwell elements (spring and dashpot in series) in series and a spring in parallel to the whole array. The time dependent stress-strain test data (creep data) of soil is used to evaluate the prony-series parameters by the method of fitting. The creep data for the local soil collected from the geotechnical in-situ lab site, IIT Kanpur is recorded experimentally and is used to evaluate the prony-series parameters. Commercially available finite element (FE) software ABAQUS 3.14, is used to obtain 2D and 3D FE model for static and steady state dynamic analysis of the soil-structure interaction problems, respectively. Finally, the time-dependent constitutive relationship obtained for the soil is validated with the experimental results obtained from consolidation and block vibration tests (Bhowmik 1989; Swain and Ghosh 2016).

2 Modelling Rheology of Soil

Rheology is a science devoted to study the changes in stress-strain states of real materials over time. Soil generally exhibits viscosity, which is responsible for time-dependent stress-strain characteristics (Graham et al. 1983; Mitchel 1993). In creep test, the soil experiences strain increment with time at a constant stress (σ_0) and the ratio of the output strain ($\epsilon(t)$) to the input constant stress (σ_0) is defined as the creep compliance (C_c). On the contrary, the reduction in stress over time due to an application of a constant strain (ϵ_0), defines a relaxation test and the ratio of the output stress ($\sigma(t)$) to the input constant strain (ϵ_0) is defined as the relaxation modulus (E_R). Linear viscoelastic system may consist of simple models composed of springs and dashpots and the constitutive equation for a linear-elastic component spring is

$$\sigma = E\epsilon \tag{1}$$

where, σ, E and ϵ are the stress, modulus of elasticity and strain, respectively.

The constitutive equation for viscous dashpot can be expressed as,

$$\sigma = \eta \frac{\partial \epsilon}{\partial t} \tag{2}$$

where, η is the coefficient of viscosity. The basic arrangement of these components in Maxwell model, Kelvin-Voigt model and 3-parameter model has been already reported in literature and the constitutive stress-strain relationship for these models can be seen from Fig. 1.

2.1 Prony Series

It can be seen from Fig. 1 that the 3-parameter model has a Maxwell element attached in parallel with a spring element. Prony series is defined as a mathematical expression representing ‘n’ Maxwell elements arranged in series as shown in Fig. 2. This arrangement of Maxwell elements in parallel with a spring element represents a general

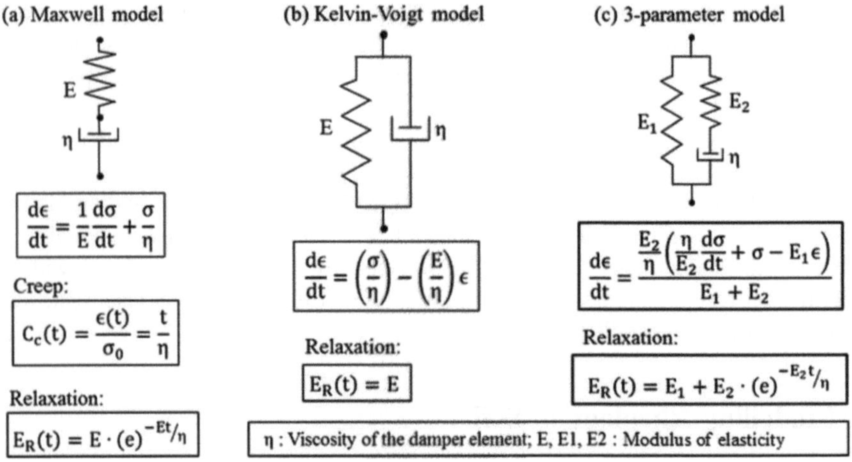

Fig. 1. Three basic lumped parameter models for soil

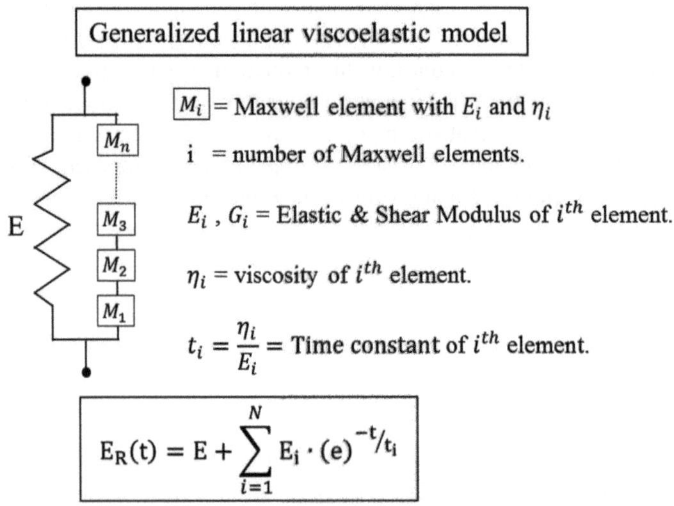

Fig. 2. Generalized linear viscoelastic lumped parameter model

model for linear viscoelastic stress-strain behaviour for non-homogenous solids. From Fig. 2, the relaxation modulus and Prony series can be expressed as

$$E_R(t) = E + \sum_{i=1}^{n} E_i \cdot (e)^{-t/t_i} \tag{3a}$$

$$\text{Prony Series} = \sum_{i=1}^{n} E_i \cdot (e)^{-t/t_i} \tag{3b}$$

In terms of shear modulus of soil, the shear relaxation modulus (G_R) can be expressed as,

$$G_R(t) = G + \sum_{i=1}^{n} G_i \cdot (e)^{-t/t_i} \tag{4}$$

$$\Rightarrow \frac{G_R(t)}{G_0} = \frac{G_\infty}{G_0} + \sum_{i=1}^{n} \frac{G_i}{G_0} \cdot (e)^{-t/t_i} \quad \begin{cases} G_\infty = \lim_{t\to\infty} G_R(t) \\ G_0 = G_R(t=0) \end{cases}$$

$$\Rightarrow g_R(t) = \left(1 - \sum_{i=1}^{n} g_i\right) + \sum_{i=1}^{N} g_i \cdot (e)^{-t/t_i}$$

$$\Rightarrow g_R(t) = 1 - \sum_{i=1}^{n} g_i \cdot \left(1 - (e)^{-t/t_i}\right) \tag{5}$$

The shear modulus terms $G_R(t)$ and G_i are normalized with respect to G_0 and represented by $g_R(t)$ and g_i, where $g_R(t)$ is defined as the normalized relaxation modulus. It can be clearly conceived that a general model with 'n' number of Maxwell elements will have (2n + 1) unknown parameters (spring elements = n + 1; damping elements = n). After normalization, g_i and t_i are the material parameters that need to be determined. Similarly, the normalized creep compliance (j_c) can be computed as a function of g_i and t_i. The parameters g_i and t_i can be obtained by performing nonlinear least square fit for the creep/relaxation test data with Eq. 5. The relationship between creep compliance and relaxation modulus can be established by a convolution integral (Ferry 1980) given in Eq. 6

$$\int_0^t E_R(t-\tau) C_c(\tau) \cdot d\tau = t; \quad t > 0 \tag{6}$$

Performing Laplace transform of Eq. 6, the following relation can be instituted

$$\overline{E}_R(s) \times \overline{C}_c(s) = \frac{1}{s^2} \tag{7}$$

where, $\overline{f}(s)$ is the Laplace transform of $f(t)$; and s = transform parameter. There are various approximate methods for the interconversion of the relaxation modulus and the creep compliance in time domain (Park and Kim 1999; Park and Schapery 1999; Schapery and Park 1999). However, for materials exhibiting very weak viscoelastic behaviour, $E_R(t) \times C_c(t) \cong 1$ provides a good approximation.

3 Creep Test on Local Soil

The local soil explored from the geotechnical in-situ lab site of IIT Kanpur, Uttar Pradesh, India is considered in this study. The basic index properties of the soil have been obtained from routine laboratory experiments and are reported in Table 1.

Table 1. Index properties of Kanpur local soil at depth = 0.8 m

Properties	Values
Liquid limit (%)	28.88
Plastic limit (%)	18.51
Shrinkage limit (%)	13.48
Specific gravity	2.62
Field density (kN/m^3)	18
Dry density (kN/m^3)	15.5
Sand fraction (%)	15
Silt fraction (%)	71
Clay fraction (%)	13
USCS classification	CL

Since the creep test involves a constant load application, consolidation test (fixed-ring type oedometer test) has been performed on the collected soil sample measuring 6 cm in diameter and 2 cm in height. A load equivalent to 10 kPa has been applied on the soil sample and the corresponding strain at different time intervals has been recorded as shown in Fig. 3. The strain data obtained from the consolidation test

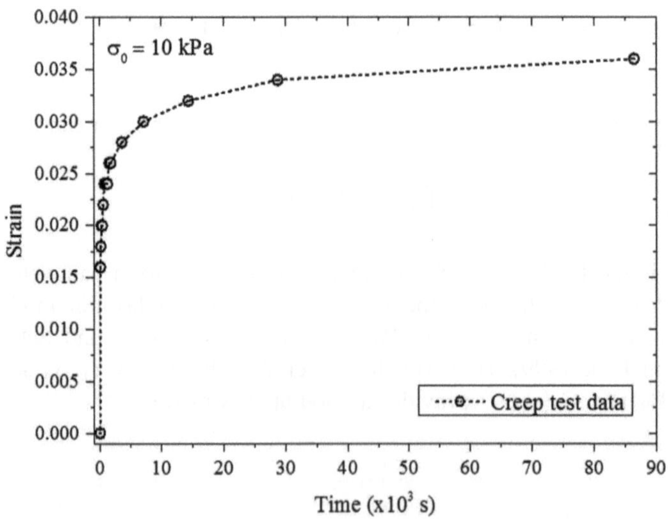

Fig. 3. Variation of strain output with time obtained from consolidation test

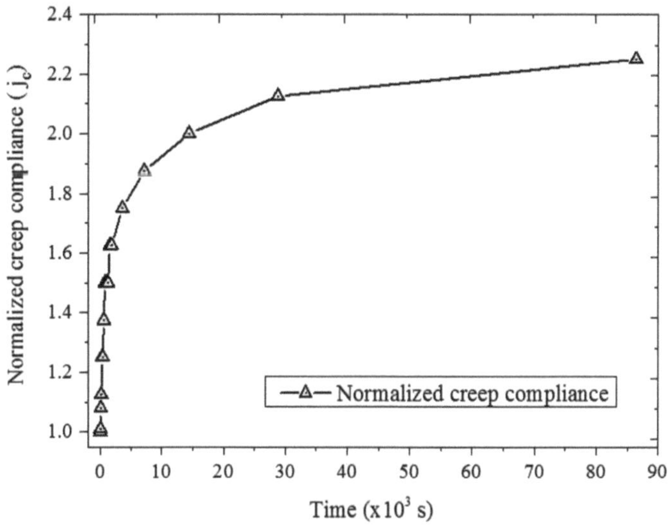

Fig. 4. Variation of normalized creep compliance (j_c) with time

is then converted into creep compliance (C_c) and is further normalized to j_c (normalized creep compliance). Figure 4 shows the variation of j_c with time.

Since the soil is mostly silty with relatively less fraction of clay, the viscosity in this material is considered very subtle. Hence, assuming the quasi-elastic interrelationship for this soil the normalized relaxation modulus ($g_R(t)$) is calculated and a nonlinear least-square fitting is performed with Eq. 5. Table 2 shows the fitting parameters obtained with an average root mean square error of 0.05. The stress-strain behavior of the soil has been modeled with 3 parameters; 2 Maxwell's elements in series along with a spring in parallel.

Table 2. Prony series fitting parameters

i	Maxwell element	g_i	t_i (s)
1	M_1	0.37409	380.05
2	M_2	0.17948	9487.7

4 Finite Element Modelling of Creep Test

In order to verify the model parameters, a 2D axisymmetric finite element model has been developed in ABAQUS 3.14 for the creep test similar to the experimental test setup (Fig. 5).

The modelling space chosen for soil is a 2D planar deformable shell. Since the oedometer used is a fixed type and the soil only moves in downward direction, the base

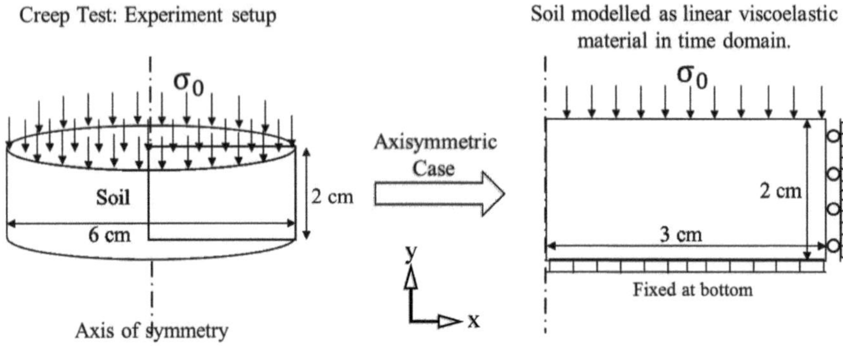

Fig. 5. 2D axisymmetric modelling of creep test in ABAQUS

of the soil in the FE model is kept fixed in vertical ($U_y = 0$) as well as horizontal ($U_x = 0$) directions, whereas the two vertical sides are provided with roller supports ($U_x = 0$). The top surface of the model is free and provided with a constant vertical stress ($\sigma_0 = 10$ kPa). Table 3 shows the input material properties for the soil. The prony series parameters given in Table 2 act as the input to define the viscosity of the material in ABAQUS. The model has been discretized with 840 number of 8-noded biquadratic axisymmetric quadrilateral (CAX8R) elements. The results of the FE analysis are plotted and compared with the experimental data in Fig. 6.

Table 3. Input soil properties in ABAQUS

Properties	Values
Mass density (kg/m^3)	1832
Young's modulus—E_0 (MPa)	3.07 (creep experiment)
Poisson's ratio	0.33 (assumed)

The strain output determined from the FE analysis shows a very good agreement with the strain values obtained experimentally and hence, ensures the calibration of the 5-parameter model with the rheology of the tested soil. The creep experiment has been continued with further stress increment of 15 kPa and the strain output has been recorded for another 24 h. To examine the efficiency of the FE analysis, another load step is added to the model and the simulation has been performed. The output obtained from the experiment and the simulation is plotted in Fig. 7. It is evident from Fig. 7 that the 5-parameter linear-viscoelastic model captures the rheology of Kanpur local soil very effectively for static load response.

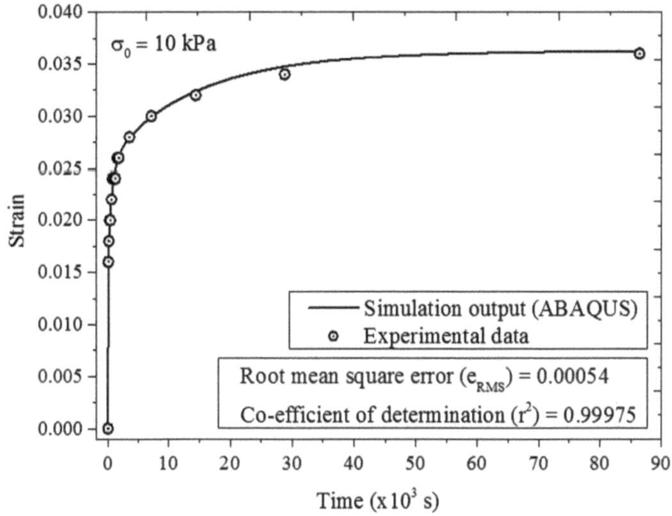

Fig. 6. Camparison of creep output from FE analysis with experimental data for $\sigma_0 = 10$ kPa

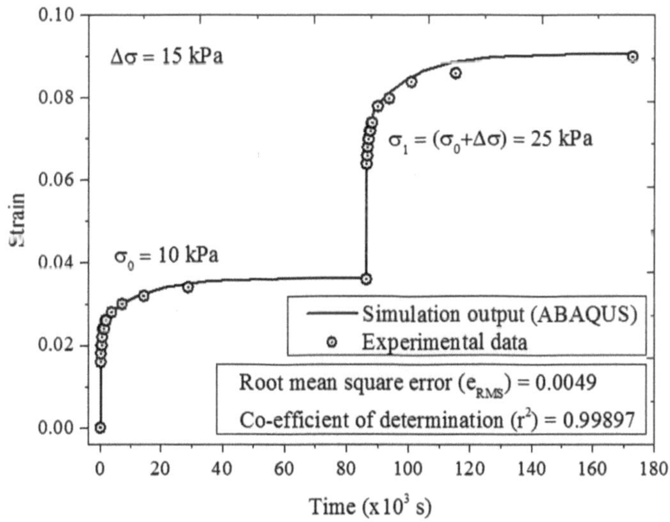

Fig. 7. Camparison of creep output from FE analysis with experimental data for $\sigma_0 = 10$ kPa and $\sigma_1 = 25$ kPa

5 Finite Element Analysis in Frequency Domain

The 5-parameter viscoelastic model obtained from the creep test data is found to provide a good approximation for the static response of the selected soil. The same constitutive stress-strain relation is implemented in a 3D FE model to capture the response of the soil subjected to dynamic loading. Swain and Ghosh (2016) have presented various block vibration test results. One of such block vibration tests with a square RCC footing (0.65 m × 0.65 m × 0.2 m) has been modeled by considering the soil domain as a 3D hemisphere.

A sensitivity analysis performed on the response of the footing with different diameters of the hemispherical soil domain results in an optimum domain size of 5.2 m in diameter. The footing is placed exactly at the center of the circular surface of the soil domain as shown in Fig. 8. The material properties considered for the dynamic analysis are given in Table 4. The footing and the soil are assumed to be elastic and viscoelastic in nature respectively, where the viscoelastic properties of the soil are taken from Table 2. The value of E_0 is chosen from Swain and Ghosh (2016). The contact between the soil and the footing is completely constrained using tie constraint, which ensures no separation and sliding of the footing relative to the soil surface. The soil domain discretization has been executed into finite (inner) and infinite (outer) zones as shown in Fig. 8. The footing and the finite zone are discretized with 8-noded linear brick elements of smaller size with reduced integration and hourglass control (C3D8R). However, the infinite zone is discretized by using 3D 8-noded linear continuum one-way infinite elements (CIN3D8). It is worth mentioning that the infinite elements act as an absorbing boundary condition to represent the semi-infinite soil region.

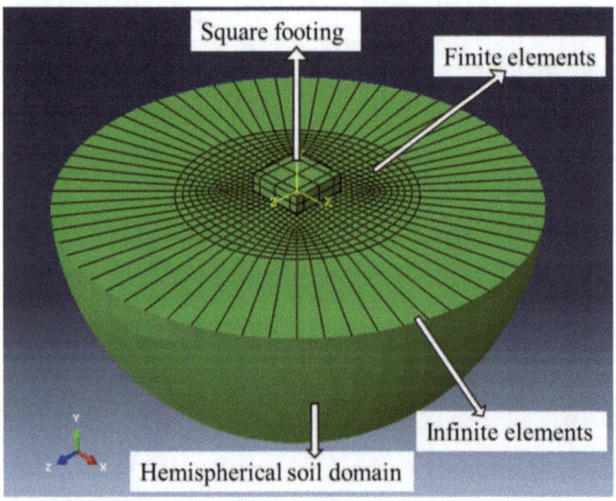

Fig. 8. 3D FE model for footing placed on soil surface

Table 4. Input parameters for 3D FE analysis in ABAQUS

Material properties	Soil	Footing
Mass density (kg/m^3)	1832	2500
Young's modulus—E_0 (MPa)	12	2500
Poisson's ratio	0.33	0.2

In the block vibration test, the basic assumption is that the foundation resting on soil is relatively very rigid and thus anticipating rigid body motion. Therefore, a rigid body constraint is defined and applied to the footing in the analysis. This allows the constituent elements not to deform but undergo large rigid body motions. The footing has been discretized with 16 elements. The maximum grid spacing (Δh), as reported by Kuhlemeyer and Lysmer (1973) can be expressed as

$$\Delta h \leq \frac{\mathrm{v}}{8\mathrm{f}_{\max}} \tag{8}$$

where, v is the lowest wave velocity, which is generally considered as the shear wave velocity versus f_{max} is the maximum operating frequency and for most of the industrial machines, it can be taken as 50 Hz.

In this study, the maximum value of Δh is calculated to be 0.125 m. Therefore, the mesh size of the soil domain is varied from 0.05 m to 0.12 m, generating 8960 elements. As proposed by Swain and Ghosh (2016), the footing is excited with a harmonic force as given in Eq. 9.

$$\mathrm{F} = \mathrm{F}_0 \times \sin\omega t = (\mathrm{m}_e \cdot e \cdot \omega^2) \times \sin \omega t = (0.11904 \times \omega^2 \times \sin\frac{\theta}{2}) \times \sin \omega t \tag{9}$$

where, m_e is the unbalanced mass in the rotating mass type excitation system, e is the radius of rotation of the eccentric mass and $\theta = 24^\circ$ is the eccentric angle setting between the unbalanced rotating masses. To evaluate the dynamic response of the footing-soil system in the frequency domain, the direct solution—steady state dynamic procedure is chosen in the analysis, which performs a frequency sweep in the desired frequency range to obtain the steady state amplitude response of the system at every frequency increment due to the harmonic excitation. A frequency range of 0–45 Hz has been chosen with increments of 1 Hz to capture the response of the foundation-soil system. Figure 9 shows the variation of vertical displacement amplitude response with frequency as obtained from the FE analysis. The resonant frequency (f_{mr}) as obtained from the response curve is found to be 35 Hz and the natural frequency (f_n) is estimated as 33.8 Hz. Figure 10 and Table 5 present a comparison of the steady state dynamic response output of the present analysis with Swain and Ghosh (2016), and Bhowmik (1989) for the same investigation site. Bhoumik (1989) has conducted block vibration tests at the same location chosen for the present study and the response curves have been presented for a concrete block (1 m × 0.75 m × 0.7 m) vibrating on the soil surface.

In Fig. 10, the variation of the magnification factor (M) with the frequency ratio (f_r) is plotted in non-dimensional form, where the frequency ratio and magnification factor can be expressed as:

$$\mathrm{f}_r = \frac{\mathrm{f}}{\mathrm{f}_n} \tag{10}$$

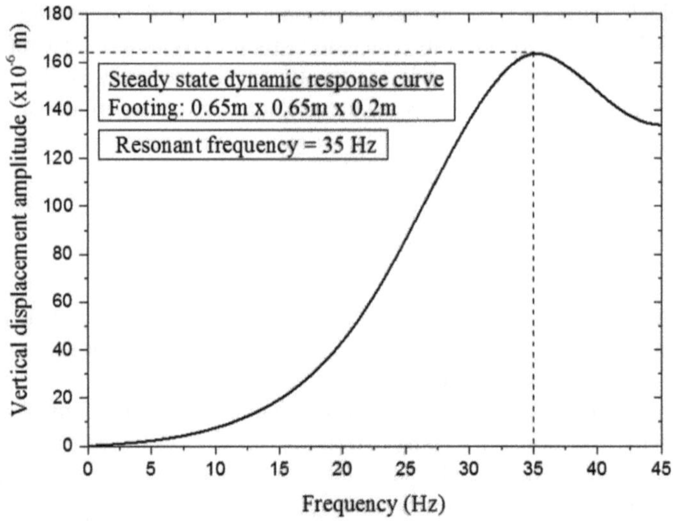

Fig. 9. Dynamic response of foundation-soil system obtained from FE analysis

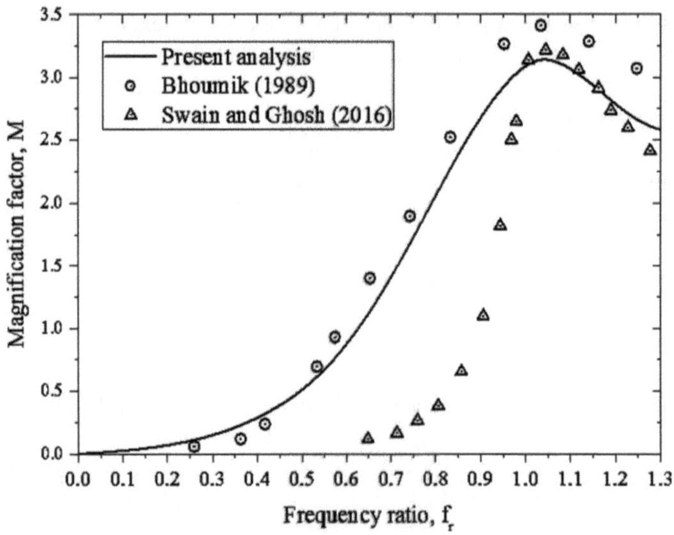

Fig. 10. Comparison of isolated footing response with available experimental results

Table 5. Output comparison with available experimental results

Output parameters	Present study	Swain and Ghosh (2016)	Bhoumik (1989)
Resonant displacement ($\times 10^{-6}$)	164	172	127
Resonant frequency, f_{mr} (Hz)	35	36.28	40.98
Natural frequency, f_n (Hz)	33.80	35.40	39.13
Damping (%)	17.30	15.50	14.50

$$M = \frac{A}{\left(\frac{m_e \cdot e}{m}\right)} \tag{11}$$

where, A is the vertical displacement amplitude of vibration, f is the operating frequency and f_n is the natural frequency of the system. It is evident from the comparison that the present numerical prediction reveals a close match with the available experimental results. The difference between the outputs of the present analysis and of Swain and Ghosh (2016) (Table 5) is found to be 5.81% for vertical displacement at resonance, 3.52% for f_{mr}, 4.52% for f_n and 11.6% for damping. This difference may be attributed to various factors such as heterogeneity and non-linearity in soil, uncertainties involved in experiments, and different assumptions and simplifications considered in the numerical analysis.

6 Conclusions

In this paper, the rheology of Kanpur local soil available at IIT Kanpur, India has been explored. A generalized lumped parameter model representing linear viscoelastic stress-strain behaviour is presented. The data obtained from the creep test is converted to normalized creep compliance and relaxation modulus, and the test results are used to estimate the unknown prony series parameters g_i and t_i by performing a nonlinear least square fitting. The fitting accommodates a 5-parameter model having 2 Maxwell's elements in series arranged parallel with a spring. To check the accuracy of the lumped parameter model, 2D and 3D finite element models are developed. The 2D-axisymmetric finite element viscoelastic model developed for predicting creep captures the strain versus time response quite efficiently for $\sigma_0 = 10$ kPa and $\sigma_1 = 25$ kPa. The same prony series parameters are used in a 3D FE model to capture the displacement response of a vibrating square footing. The normalized response curve (M versus f_r) matches closely with the experimental results reported in literature.

References

Bhoumik, K.: Coupled vibration of footings on elastic half-space. M.Tech thesis, I.I.T-Kanpur, Kanpur, India (1989)

Dey, A., Basudhar, P.K.: Applicability of burger model in predicting the response of viscoelastic soil beds. Geotech. Spec. Publ. **199**, 2611–2620 (2010)

Ferry, J.D.: Viscoelastic Properties of Polymers, 3rd edn. Wiley, New York (1980)
Graham, J., Crooks, J.H.A., Bell, A.L.: Time effects on the stress-strain behaviour of natural soft clays. Geotechnique **33**(3), 327–340 (1983)
Kerr, A.D.: Viscoelastic Winkler foundation with shear interactions. J. Eng. Mech. Div. Proceed. ASCE **87**(3), 13–30 (1961)
Lyakhov, G.M.: Determination of the viscous properties of soil. Prikl. Matem. Mekh (4) (1968)
Mitchell, J.K.: Fundamentals of Soil Behavior, 2nd edn. Wiley, New York (1993)
Park, S.W., Kim, Y.R.: Interconversion between relaxation modulus and creep compliance for viscoelastic solids. J. Mater. Civ. Eng. ASCE **11**(1), 70–82 (1999)
Park, S.W., Schapery, R.A.: Methods of interconversion between linear viscoelastic material functions. Part 1—a numerical method based on Prony series. Int. J. Solids Struct. **36**(11), 1653–1675 (1999)
Schapery, R.A., Park, S.W.: Methods of interconversion between linear viscoelastic material functions. Part 2—an approximate analytical method. Int. J. Solids Struct. **36**(11), 1677–1699 (1999)
Schiffman, R.L., et.al.: The secondary consolidation of clay. In: Rheology and Soil Mechanics: Proceedings of the International Union of Theoretical and Applied Mechanics Symposium, pp. 273–303. Grenoble, Berlin (1964)
Swain, A., Ghosh, P.: Experimental study on dynamic interference effect of two closely spaced machine foundations. Can. Geotech. J. **53**(2), 196–209 (2016)
Tan, T.K.: Three-dimensional theory on the consolidation and flow of the clay-layers. Sci. Sin. **6** (1), 203–215 (1957)

A Macro-element Approach for Non-linear Response of Offshore Skirted Footings

Amin Barari[1(✉)] and Lars Bo Ibsen[2]

[1] Department of Civil and Environmental Engineering, Virginia Tech, 200 Patton Hall, Blacksburg, VA, USA
abarari@vt.edu
[2] Department of Civil Engineering, Aalborg University, Thomas Manns Vej 23, 9220 Aalborg, Denmark

Abstract. In recent times, shallow foundations have become more diverse, and now include concrete or steel bucket foundations used as anchors for floating platforms or as foundation for offshore wind turbines. They are always subjected to significant cyclic lateral loads due to wind and wave action. A work-hardening plasticity model for the moment and horizontal force resultants associated with the displacements of the offshore suction caissons is herein developed. In addition, in order to better understand the performance of offshore bucket foundations under cyclic lateral loads series of cyclic lateral load tests were carried out on small scale models in a special rig developed for the cyclic tests at Aalborg University. The influence of cyclic lateral loads on the bucket lateral secant stiffness is investigated and a function is utilized to investigate it due to permanent displacements. The bucket stiffness increased with number of cycles irrespective of the load characteristics which contradicts the traditional solutions to tackle this complex situation.

1 Introduction

Political, industrial and technological drivers are promoting the development of offshore wind farms. Offshore wind turbines are steadily increasing in terms of size. The offshore location of these viable solutions is preferred due to its intrinsic merits to alleviate the noise, size and vibrations effects on human activities (Bhattacharya and Adhikari 2011). Suction caisson foundation is an option that can decrease the overall cost and increase the diffusion of the wind turbine. The horizontal load divided by the vertical load for the bucket foundations in wind turbines is very high compared with the suction caissons for offshore oil and gas industry. For example, the non-dimensional quantity to represent the vertical load, $V/\gamma' d^3$ (i.e., γ' is the soil effective unit weight) is between 0.5 and 0.8 which is substantially lower than typical value of 3.5 for offshore oil industry.

The resistance observed from lateral load on suction caissons is of importance in the design of marine structures which may be subjected to the earthquakes and wave action. The difficulties with respect to the logistics and cost causes a few small-scale load tests

H. Khabbaz et al. (eds.), *New Prospects in Geotechnical Engineering Aspects of Civil Infrastructures*, Sustainable Civil Infrastructures,
https://doi.org/10.1007/978-3-319-95771-5_11

results are available that examine the bearing capacity and deformations developed in suction bucket foundations for wind turbines (Fig. 1) (Ibsen et al. 2014a, b; Barari and Ibsen 2012, 2014; Larsen et al. 2013). Nevertheless, the results from these tests indicate that the combined capacity of buckets can be described in terms of kinematic mechanisms accompanying pre-failure and failure states (Ibsen et al. 2015).

Fig. 1. Suction bucket foundation in Horns Rev 2 mobile met mast project

Traditionally, foundations are thought to exhibit the soil-foundation stiffness reduction (API 1993). This reduction in lateral resistance depends on various conditions such as, depth of sandy soil, ratio of excess pore pressure, loading frequency, etc.

In addition, previous studies revealed that the dynamic characteristic of a wind turbine system including its natural frequency and dynamic stiffness change with either few cycles of large amplitude or millions of cycles with intermediate amplitude (Cox et al. 2011).

Recommended by Limit State Design philosophy, the design stages for a shallow foundation system during life of the structures are as follows (Arany et al. 2017):

1. Ultimate Limit state (ULS): The first step in design of offshore foundations is to estimate the capacity under in-service VHM general loading.
2. Serviceability limit state: This requires the prediction of short and long-term deformations, and the influence of cycling loading.

To date there has been no verified long-term observation of alteration of the bucket foundations stiffness. The current study summarizes the findings from a series of small-scale tests of suction caissons. The focus of this research is on dynamic characteristics of these structures. Prior to cyclic behavior, 3-D finite element analyses are presented to outline a “macro-element” model based on work-hardening plasticity theory for ULS soil-foundation interaction relevant to offshore platforms.

2 Dynamic Considerations

These structures can be described by a mathematical model where the foundation is modeled by translational and rotational springs. The stiff nature of bucket lid simplifies bucket to a rigid foundation with six degrees of freedom. For each particular harmonic excitation, the dynamic stiffness K is defined as the ratio between force (or moment) R and the displacement (or rotation) U. The impedances are defined as: two horizontal and longitudinal impedances resulting from force-displacement ratio, two rocking impedances resulting from moment-rotation ratio and one vertical resulting from a ratio between harmonic vertical force and harmonic vertical displacement of the soil-foundation interface. Similarly the tensional impedance is defined from moment moment-rotation ratio about the vertical axis (z).

3 Static Behaviour of Suction Bucket: Numerical Modeling

A three-dimensional finite element model using the program system PLAXIS was utilized to simulate the behavior of a laterally loaded bucket foundations. Sub-soil and bucket were modeled using the full-mesh system. A schematic view of bucket under horizontal loading is shown in Fig. 2.

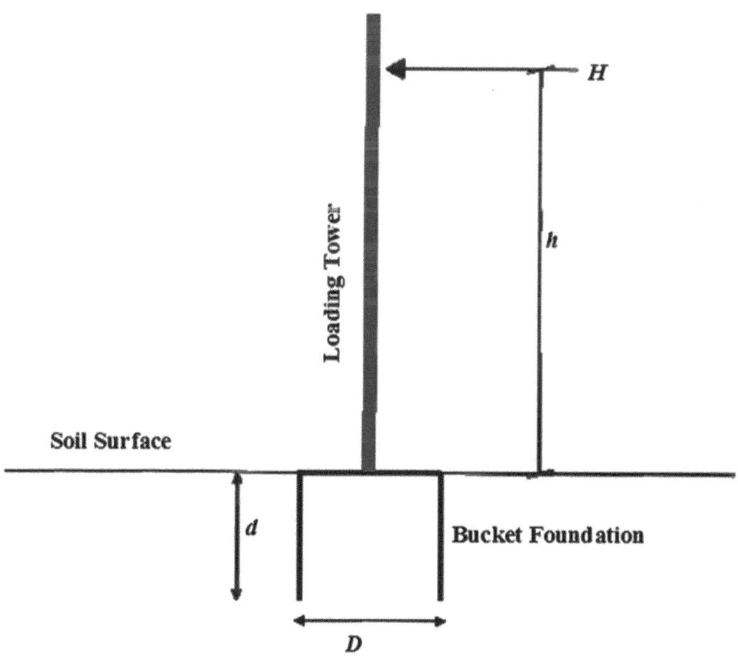

Fig. 2. A schematic view of bucket under horizontal loading

A bucket foundation supporting a 10 MN wind turbine, with $D = 12$ m and $d/D = 0.75$, withstanding a wave load with arm from the soil surface, h = $0.2D$, $1.66D$ and $8.33D$, and installed in very dense sand, is taken as real-scale model. The discredited model area had at least 6 times the bucket diameter. The bottom boundary of the model was extended twice the skirt length below the toe of the bucket. A typical mesh of the finite element model with geometrical properties is depicted in Fig. 3.

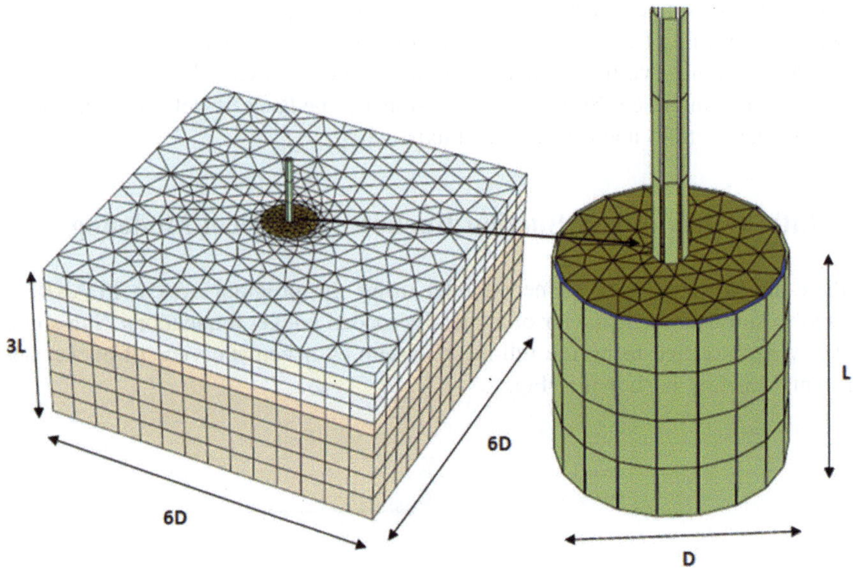

Fig. 3. Finite element model of a suction caisson system

The elasto-plastic material law with Mohr-Coulomb failure criterion was used when extended by a stress-dependency of the oedometer stiffness with the following equation (EAU 2004; Kuo et al. 2012):

$$E_s = \kappa\sigma_{at}\left(\frac{\sigma_m}{\sigma_{at}}\right)^{\lambda} \tag{1}$$

Herein $\sigma_{at} = 100$ kN/m^2 is a reference stress and σ_m is the current mean principle stress. The parameters κ and λ correspond to the soil stiffness at the reference stress state and stress dependency of the soil stiffness, respectively. The input-model parameters for the dense sand are reported in Table 1.

Accordingly, the results of FE analyses offer the opportunity to compare the force-displacement curves which are of special interest for design of wind turbines, to those already reported in the literature (Fig. 4).

Table 1. Soil parameters considered (Achmus et al. 2013)

Property	Value	Unit
Buoyant unit weight (γ′)	11	(kN/m³)
Oedometer stiffness parameter (κ)	600	(–)
Oedometer stiffness parameter (λ)	0.55	(–)
Poisson's ratio (ν)	0.25	(–)
Internal friction angle (φ′)	40	(°)
Dilation angle (ψ)	10	(°)
Cohesion (C′)	0.1	(kN/m²)

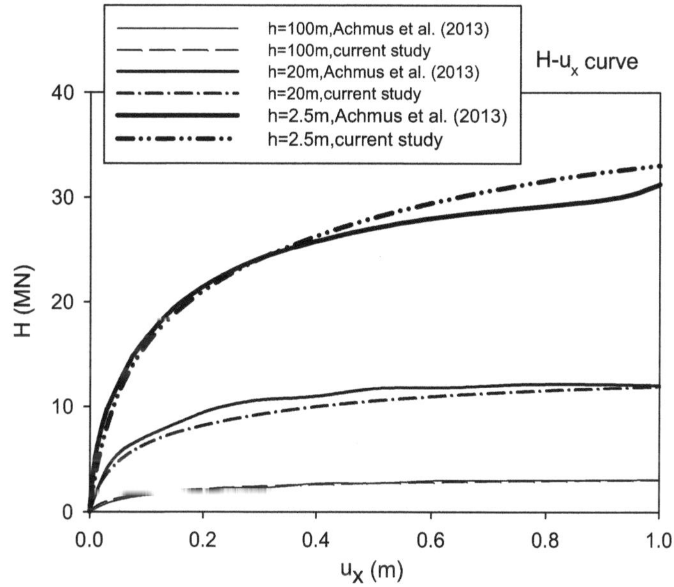

Fig. 4. Load-deformation curves for suction caisson foundations in dense sand

3.1 Failure Envelopes: Macro-Element Approach

Macro-element modeling is applicable to many kinds of geotechnical problems but its primary application is on shallow foundations. Given a shallow footing, model consists of structure, surrounding soil and displacement or load field applied to the system. For shallow foundations, the concept has perhaps its origin with Roscoe and Schofield (1957).

A series of FE calculations are presented herein to deduce the failure envelopes in H-*M* space, following a strain-hardening plasticity framework. The new model is made by giving emphasis on the relationship of applied displacement and corresponding load in yield/plastic failure state. Figure 5 depicts representative data calculated for varying embedment ratio, *L/D*. The applied horizontal load, *H* and overturning moment, *M* are normalized with the associated ultimate capacities.

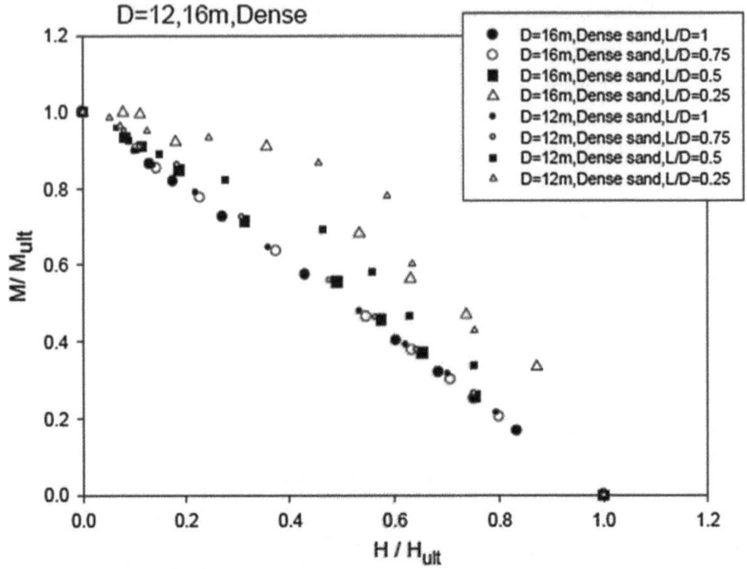

Fig. 5. H-M failure envelope normalized by corresponding pure capacities at V = 0

Given the parabolic behavior of failure envelope, the following oblique parabolic expression is adopted for offshore shallow footings to fit the numerical results:

$$f = \left(\frac{H}{H_u}\right)^2 + \left(\frac{M}{M_u}\right)^2 + n_3\left(\frac{H}{H_u}\right)\left(\frac{M}{M_u}\right) - 1 = 0 \tag{2}$$

where n_3 is a parameter controlling the shape of surface (Gerolymos et al. 2012).

A normalized presentation of the results was also sought for the effect of n^3on shape of normalized failure envelopes in Fig. 6. As a result, the shape of normalized failure envelopes remains identical for embedment ratios $L/D \geq 0.75$.

4 Performance Measure Parameters for Suction Bucket Systems

4.1 Test Equipment and Program

A loading rig is utilized to apply cyclic loading which is outlined in Fig. 7. The rig consists of a soil container (1.6 × 1.6 × 1.15 m^3), steel frame, weight hangers, lever arm with a driving motor. The rig was more recently described by Foglia et al. (2012) to carry out cyclic loading tests on bucket foundations installed in saturated sand. The tests were performed on a caisson of diameter D = 300 mm and d/D = 1. The mass of

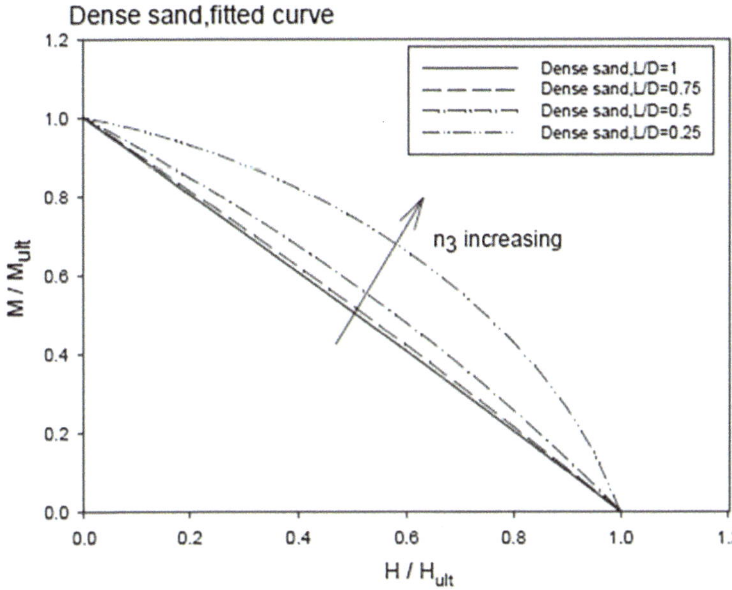

Fig. 6. Contraction of normalized yield envelopes with increasing n_3

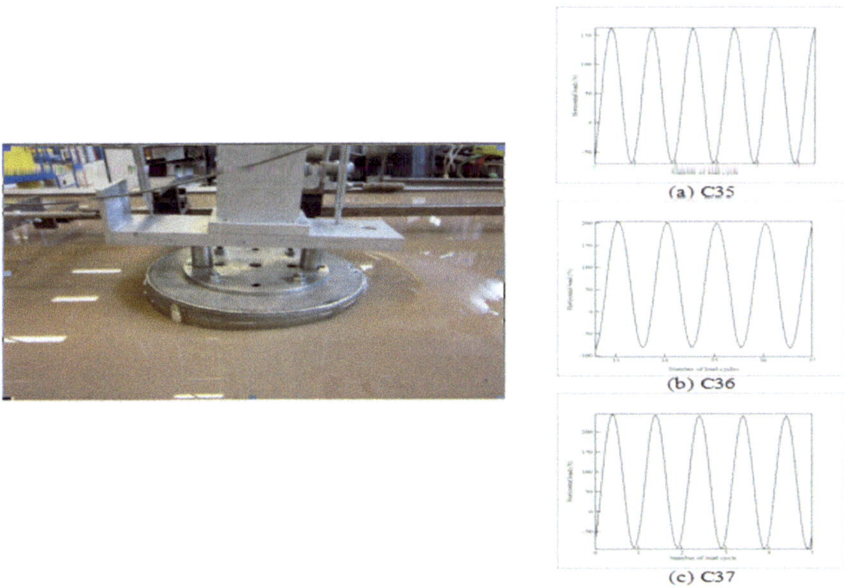

Fig. 7. Photo of experimental set-up and load-time histories of three tests **a** C35 **b** C36 **c** C37

the superstructure in real-scale suction caisson-supported wind turbine is parametrically given a typical value here corresponding to a static factor of safety $FS_V = 2$. The experimental setup and loading time histories are portrayed in Fig. 7.

4.2 Tests Results

In this section results of three force-controlled lateral loads conducted on the lever arm are presented. The loading time-histories were: (i) 9976 cycles from 162.4 N to −68.11 N (test C35) (ii) 10153 cycles from 203.3 N to −81.7 N (test C36) (iii) 10,083 cycles from 244.4 N to −90.3 N (test C37). All the information on the cyclic lateral load tests is listed in Table 2. Figure 8 displays lateral force-displacement hysteresis curves while displacements were measured at the load reference point as previously introduced by Butterfield et al. 1997 (Fig. 9).

Table 2. Information on cyclic lateral load tests

Test	Min. force	Max. force	ξ_b	ξ_c	Number of cycles
	N	N			
C35	−68.11	162.4	0.48	−0.54	9976
C36	−81.7	203.3	0.58	−0.56	10153
C37	−90.3	244.4	0.69	−0.58	10083

In a closer inspection, the final unloading in C35 test from ~160 N to ~−70 N created a recoil of ~2.6 mm prior to stopping the rig. When C36 was subsequently carried out at ~200 N to ~−80 N, it may be seen that the first reloading induced ~0.4 mm, confirming that the stop-start cycle has influence on the results. However, lateral cycles with amplitude ~200 N in C36 can be seen stiffer than C35 with in increasing permanent displacement. In later test, C37, the same phenomenon occurred but very much reduced "first cycle offset".

Three performance measure parameters are introduced herein to evaluate the response of soil-foundation system.

4.3 Performance Measure Parameters

Figure 10 depicts secant stiffness at each unloading-reloading reversal point. The non-dimensional form of secant stiffness obtained at each unloading–reloading reversal point and the sequential reversal points, respectively divided by corresponding stiffness at first cycle is outlined in Fig. 11. They are defined here as the ratio of the lateral load to the reference point lateral displacement at the soil-foundation interaction. After a tremendous increase in the first cycle, the ~$K_{s,n}$ increases slightly with the increasing number of load cycles but at a reducing rate, as shown in Fig. 11. The increase in the measured stiffness may be attributed to the soil densification during cyclic loading. It is worthy of note that the roughly nonlinear behavior during the tests causes a gradual increase in $K_{s,n}$. Given that the amplitude of cyclic loads was big enough to trigger the soil hardening, the largest value of $K_{s,n}$ in all tests is about 1694 N/mm after cyclic loading.

Plotting non-dimensional secant stiffness as a function of N^b suggests that stiffness evolves exponentially with cycle number as:

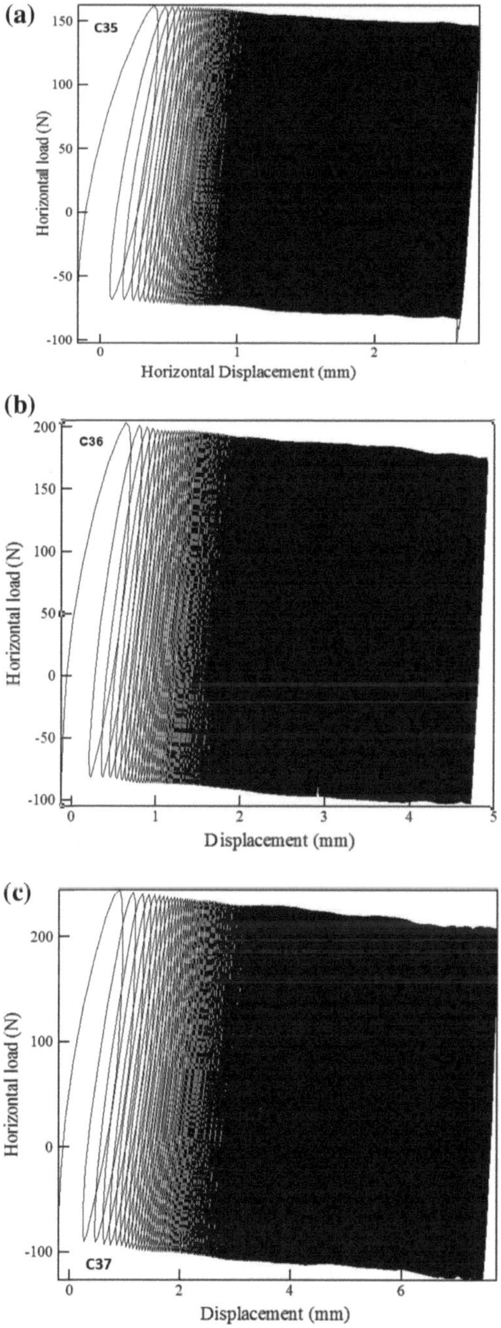

Fig. 8. The load-displacement response at load reference point during cyclic lateral load tests **a** C35 **b** C36 **c** C37

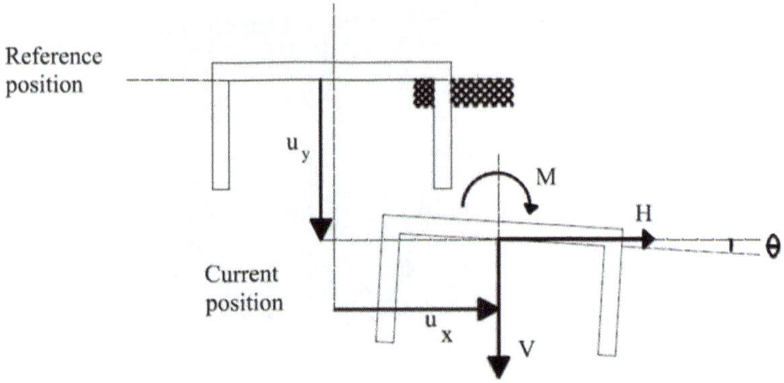

Fig. 9. Standardized sign convention for plane loading of bucket foundations (Butterfield et al. 1997)

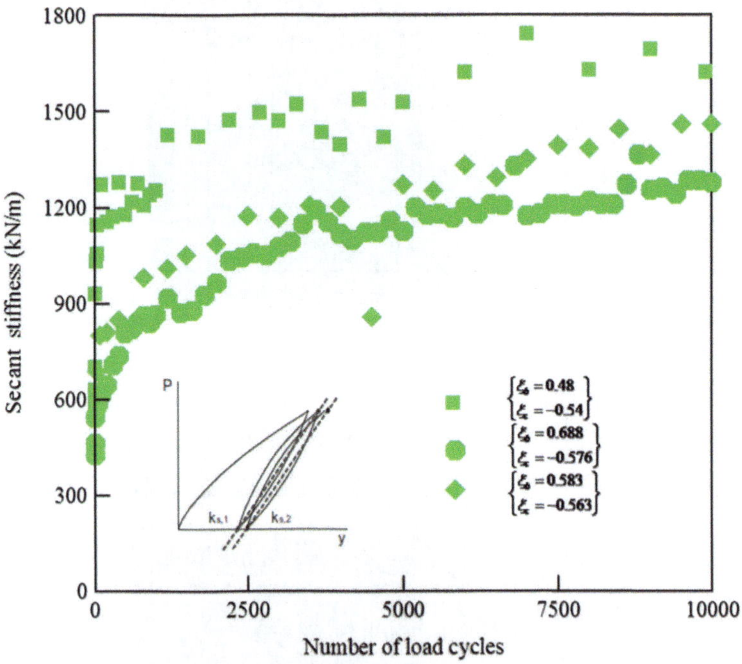

Fig. 10. Bucket secant lateral stiffness

$$k = \frac{K_{s,n}}{K_{s,1}} = aN^b \tag{3}$$

where a and b are correlation coefficients. The expression in Eq. (3) was fitted to the data in Fig. 11 and the a was determined as a function of load characteristics. It is observed from Fig. 11 that all power values are almost equal. It implies that b is

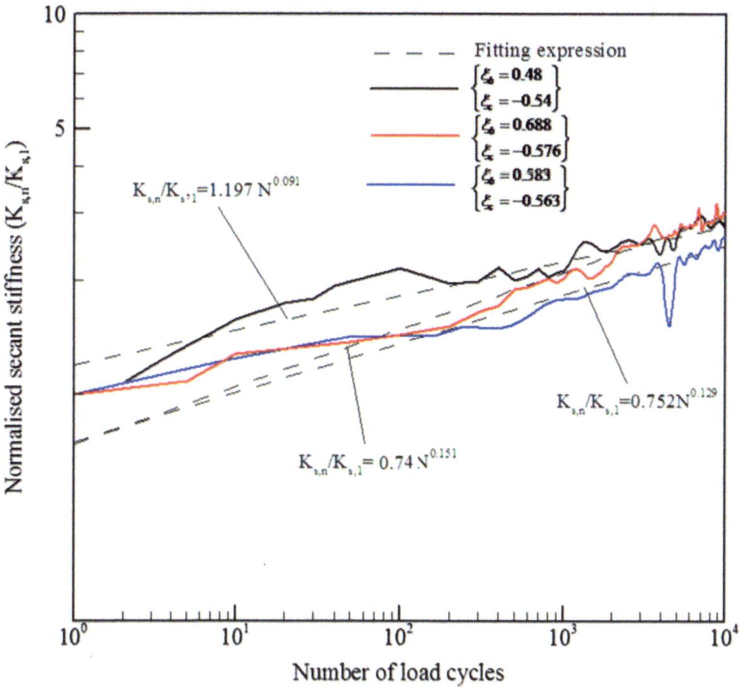

Fig. 11. Normalized bucket secant lateral stiffness

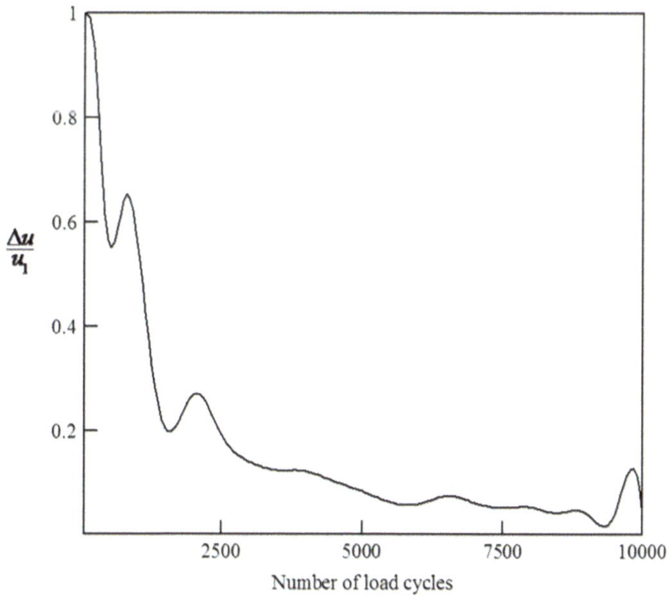

Fig. 12. Relative bucket head displacement between two consecutive re-loading unloading reversal points normalized with the one between the virgin loading and the first re-loading unloading reversal point

independent of the load characteristics within the observed range. This issue is further examined in a later publication in terms of wide range of forcing excitations.

The stability of the soil-foundation system for lateral response of buckets in sand is further investigated through analysis of accumulation rate of plastic displacement. Figure 12 presents the relative bucket head displacement between two subsequent re-loading un-loading reversal points normalized with the one between virgin loading-unloading and the first reloading -unloading reversal points. Typical results from test C36 shows that rate of accumulation of plastic displacement decreased as the number of load cycles increased, indicative of the stable response of the system.

5 Conclusions

This paper has presented selected results from finite element simulations and laboratory testing program aimed at investigating the response of suction caisson foundations to combined loads in terms of performance measure parameters. The numerical results were employed in the establishment of algebraic expressions for the failure envelopes in the M-H loading space.

The force-controlled cyclic loads induced significant accumulated displacements that may be attributed to the local densification of sand in soil-foundation interface. Considering the dramatic increase in bucket lateral secant cyclic stiffness in the first cycle, stiffness increases slightly with the increasing number of load cycles but at a reducing rate. The secant cyclic stiffness ($K_{s,n}$) values rise by around 2.8 times irrespective of the load amplitude.

Finally, the paper described in outline a theoretical framework that captures the main features of the cyclic tests—that of clear trend of hardening response, practically affected by the load path. This new soil-foundation interaction model termed performance based design model, represents a significant improvement on conventional theories, which could not capture this behavior.

References

Achmus, M., Akdag, C.T., Thieken, K.: Load-bearing behavior of suction Bucket Foundations in sand. Appl. Ocean Res. **43**, 157–165 (2013)

API: Recommended Practice for Planning, Designing, and Constructing Fixed Offshore Platforms: Working Stress Design, 20th edn. American Petroleum Institute; RP2A-WSD, Washington, D.C. (1993)

Arany, L., Bhattacharya, S., Macdonald, J., Hogan, S.J.: Design of monopiles for offshore wind turbines in 10 steps. Soil Dyn. Earthq. Eng. **92**, 126–152 (2017)

Bhattacharya, S., Adhikari, S.: Experimental validation of soil–structure interaction of offshore wind turbines. Soil Dyn. Earthq. Eng. **31**, 805–816 (2011)

Barari, A., Ibsen, L.B.: Vertical capacity of Bucket Foundations in undrained soil. J. Civil Eng. Manage. **20**(3), 360–371 (2014)

Barari, A., Ibsen, L.B.: Undrained response of Bucket foundations to moment loading. Appl. Ocean Res. **36**, 12–21 (2012)

Butterfield, R., Houlsby, G.T., Gottardi, G.: Standardized sign conventions and notation for generally loaded foundations. Geotechnique **47**(5), 1051–1054 (1997)

Cox, J., Jones, C., Bhattacharya, S.: Long term performance of suction caisson supported offshore wind turbines. Struct. Eng. **89**(19), 12–13 (2011)

Doherty, P., Gavin, K.: Cyclic and rapid axial load tests on displacement piles in soft clay. J. Geotech. Geoenviron. Eng. **138**(8), 1022–1026 (2012)

EAU: Empfehlungen des Arbeitsausschusses "Ufereinfassungen" – Häfen und Wasserstraßen. Verlag Ernst & Sohn, Berlin (2004)

Foglia A., Ibsen, L.B., Andersen, L.V., Roesen, H.R.: Physical modelling of Bucket foundation under long-term cyclic lateral loading. In: Proceedings of the International Offshore and Polar Engineering Conference, Rhodos, pp. 667–673 (2012)

Gerolymos, N., Zafeirakos, A., Souliotis, C.: Insight to failure mechanisms of caisson foundations under combined loading: a macro-element approach. In: *Second International Conference on Performance-Based Design in Earthquake Geotechnical Engineering*, Taormina, Italy (2012)

Ibsen, L.B., Barari, A., Larsen, K.: An adaptive plasticity model for Bucket Foundations. J. Eng. Mech. ASCE **140**, 361–373 (2014a)

Ibsen, L.B., Larsen, K., Barari, A.: Calibration of failure criteria for Bucket Foundations under general loading. J. Geotech. Geoenviron. Eng. ASCE (2014b). https://doi.org/10.1061/(ASCE)GT.1943-5606.0000995

Ibsen, L.B., Barari, A., Larsen, K.: Effect of embedment on the plastic behaviour of Bucket Foundations. J. Waterw. Port Coast. Ocean Eng. ASCE (2015). https://doi.org/10.1061/(ASCE)WW.1943-5460.0000284,06015005

Jardine, R.J., Standing, J.R.: Field axial cyclic loading experiments on piles driven in sand. Soils Found. **52**(4), 723–736 (2012)

Kuo, Y.S., Achmus, M., Abdel-Rahman, K.: Minimum embedded length of cyclic horizontally loaded monopiles. J. Geotech. Geoenviron. Eng. **138**, 357–363 (2012)

Larsen, K.A., Ibsen, L.B., Barari, A.: Modified expression for the failure criterion of Bucket Foundations subjected to combined loading. Can. Geotech. J. **50**(12), 1250–1259 (2013)

Roscoe, K.H., Schofield, A.N.: The stability of short pier foundations in sand. Br. Weld. J. 12–18 January (1957)

The Effect of Using Desert Sands and Cement to Stabilize the Base Course Layer of Roads in Libya

Talal S. Amhadi(✉) and Gabriel J. Assaf

Department of Civil Engineering and Construction, École de Technologie Supérieure (ÉTS), University of Québec, 1100 Rue Notre-Dame Ouest, Montréal, QC H3C, Canada
talal.amhadi.1@ens.etsmtl.ca, Gabriel.Assaf@etsmtl.ca

Abstract. This paper investigates using a mixture of natural sand, Portland cement, and manufactured aggregates to stabilize the base course layer for low-volume roads. Compaction and resistance to deformation California Bearing Ratio (CBR) results are measured and guidelines provided with consideration for hot desert areas like Libya. These tests have been conducted to provide relevant statistical conclusions and recommendations as to how much round-shaped sand may be used versus manufactured crushed sand for various traffic loads. In the last 15–20 years, Libya, along with many other developing countries in the area, has undergone a sizeable move towards modernization, including upgrading the road infrastructure that services the growing population. Such a project runs into challenges in areas with lower population density, that require that roadwork projects be kept under sharp cost constraints. Using a mixture of sand cement to stabilize the base course has been common for decades because the properties of cement improve the characteristics of natural sand; furthermore, when the road is built in a sandy desert, the use of desert sand is an obvious solution to containing costs. Nonetheless, natural sands have been rounded by years of mechanical action by environmental forces, and are therefore rounder and more prone to shifting than angular manufactured sand. Manufactured aggregates need to be imported from elsewhere in the country and so have higher costs and environmental considerations, both of which give these projects incentives to use the local desert sand. This paper only considers the use of manufactured aggregate of 0–5 mm in the base course and examines how using local desert sand in the aggregate mix can reduce costs. This solution brings with it many problems because, due to natural sand's lower inherent stability with regards to manufactured sand, the road surface will degrade faster. The solution to this problem, presented here, is to relocate some of the strength of the road from the surface layer to the base course layer. This is done by reducing the thickness of the asphalt course to a minimum, and introducing more structural strength in the base course layer by means of a cement-stabilized base layer.

Keywords: Sand · Cement · CBR · Compaction · Stabilization

H. Khabbaz et al. (eds.), *New Prospects in Geotechnical Engineering Aspects of Civil Infrastructures*, Sustainable Civil Infrastructures,
https://doi.org/10.1007/978-3-319-95771-5_12

1 Introduction

There are a number of methods for improving the geotechnical characteristics of fine-grained and soft sands that do not meet industry standards for road construction. One approach is to remove the substandard sand and instead use a higher standard, manufactured sand that has been processed chemically or mechanically. Such techniques can raise the level of the geotechnical characteristics of the whole and provide greater rigidity and strength through methods that can be applied at the construction site itself. Preloading and compacting the fine aggregate makes it denser. Other techniques such as electro-osmosis and dewatering affect how grains of sand bond together; related methods are grouting, chemical stabilization and ground freezing. Another common way to improve the stability of the fine aggregate is to use physical means to create structure, e.g. placing stone columns or geotextiles with the fine aggregate (Patel and Patel 2012). There is another approach to consider. Because of the general uniformity of the fine aggregate and the fact that there are still some gravel particles, the fine aggregate can be mixed with cement to stabilize it. This low grade, soft sand is found in great quantities in the deserts of Libya. Although there is a great financial incentive to use this sand, it can cause many problems because a base coarse layer made from desert sands tends to allow more erosion by the elements (rain, wind) and, in certain cases, will collapse when it becomes saturated by water. As the country develops, more construction is required for trade and transport, putting a strain on the supply of standard building materials such as aggregates (fine and coarse). If the desert sands could be suitably prepared for construction, it would save money and time. For this reason, this research is targeted at ways of making use of this natural desert sand while keeping to modern construction standards.

2 Background and Literature Review

2.1 A Survey of Pavement Stabilization Techniques

Stabilization, in its most basic sense, is a way of making the mechanical strength of a base layer more robust, and less affected by changes in wear resistance and swelling (from either heat or water) (Abderrahmane et al. 2013). Typically, natural sands are low grade because of years of rounding from wind; when used as a base course their loose fabric structure is prone to erosion from both rain and wind. Because of these factors, desert sands are a poor alternative for either road or dam construction; the low shear strength and the high permeability lead to higher maintenance costs and shorter life (Al-Aghbari and Dutta 2005). The process of soil stabilization is a question of altering the chemical and physical soil characteristics and thereby increasing the durability and overall strength of the base layer. Low grade soils have been brought up to acceptable standards in this way for years. The added expense of this treatment is repaid in the superior service life and improved performance of the road (Azadegan et al. 2013). A study of German road construction in recent decades found that using lime or cement to stabilize base layers had important and practical outcomes. Japan, the US and some Scandinavian countries have used the fly ash-soil stabilization method with success. On the other hand, in some countries stabilization with fly ash soil is not used simply because there are no standards for the method. When dealing with expansive or soft clayey soils, engineers must make choices to compensate for the soils that do not have sufficient stability to support the axel-loads required of the finished road. These soils must be treated to get them to a suitably stable quality of subgrade. This problem applies both to the finished road but also to the conditions during construction; if a suitable subgrade cannot support the loads during construction, then a platform must be created from which to build the pavement (Ismaiel 2006). A wide range of materials, including crushed stone and natural gravel and sand, which provide the necessary durability and strength. Alternatively, available local materials can be stabilized chemically or mechanically to achieve comparable results at a competitive price. In order to work well, the aggregates must be durable and strong and must satisfy the requirements of a particular plasticity, strength, and gradation. The roads of Libya suffer from excessive deformation (primarily cracking and rutting) in hot and droughty conditions found throughout the country, especially in the southern regions of the desert. As this overview shows, aggregate distribution and grading in road construction are already being conducted in hot and arid regions, such as in the Middle East and North Africa. This is justified because obsolete mixing designs are ineffective. There is a need for data to determine which materials can be suitable for the performance parameters defined in these regions. This is important, because lower-quality material might pass the design requirements, but may not meet other design requirements for the mixture (Almadwi and Assaf 2017).

2.2 Stabilizing Pavement Materials

The stabilizing agents that have been used for various soil types have been well studied. Lime and cement are often the best choice for stabilizing the subgrade and subbase layers. They are also used for base layer stabilization. Soils with too high a percentage of sand respond well to the stabilization created by the addition of lime. In terms of the plasticity index, and the plasticity and liquid limits, the addition of lime creates superior physical properties. Lime raise the California Bearing Ratio (CBR) from just 1% to approximately 30–50% when various percentages of lime are added to untreated soils. Furthermore, adding lime to a pavement allows for the use of pavement layers that are from 50 to 60% thinner compared with the requirements of pavement of untreated layers (Azadegan et al. 2013). This kind of stabilization has turned into standard practice nowadays and options to stabilization have been explored recently that apply different amounts of cement, lime, and bitumen to solve the problem. Aiban (1994) reported that a compacted sand mix that is then saturated with bitumen results in a more stable layer. Statistical and experimental analysis were made on the viscoelastic properties of the bitumen and lime mix (Sabbagh 1986). Sabbagh studied the effects of roads built on sand dunes with small inclines; these roads had a base layer formed from a mix of bitumen and hydrated lime; he found that this mix provided a suitable wearing course for low-volume roads and as a base layer any traffic volume. Aiban (1994) studied the effects of emulsification of fine sand aggregates using bitumen to stabilize the aggregate for low volume Saudi Arabian roads. Their conclusion was that when natural desert sands were emulsified with bitumen, the resulting asphalt was of poor quality. In contrast, adding Portland cement and manufactured fine aggregate to the mix greatly improved the road quality.

3 Methods of Stabilizing the Road Base

3.1 Mechanical Stabilization

Mechanical stabilization is the oldest technique for road building and these involve changing the physically property of the original road base soil to affect its solidity and gradation, among other factors. Of these techniques, dynamic compaction is one of the most commonly used. To do this, a heavy weight is repeatedly dropped over all points of the road base to even out any irregularities and create an evenly compacted soil. A newer variation of this is known as Vibro compaction and it works in a similar way, with vibration taking the place of kinetic force deformation (Das 2003).

3.2 Chemical Stabilization

Stabilizing soil by adding a chemical that will change its properties by physically interacting with it is another standard stabilization method. Lime, kiln dust, cement, and fly ash are common chemical compounds that are used. Depending on the nature of the existing base, the chemical reactions are pozzolanic or cementitious (Das 2003).

4 Reasons for Soil Stabilization

By stabilizing the soil at the site of the construction, projects avoid the costs of removing the soil already there and transporting new materials to the site. In areas where extreme weather conditions would slow or stop construction during certain times of the year, soil stabilization can allow work to continue by stabilizing the original soil and allowing the work to continue. Therefore, stabilization techniques are a means of cost savings because work can continue through more weather conditions (Patel and Patel 2012).

5 Problem Statement

In desert areas, quality construction materials are becoming rare. Therefore, where traffic volumes are low, roads are less cost-effective. The Sahara Desert has a lot of sand but little rock and natural gravel, which is necessary for the structure of pavement. To date, the construction of roads in the south has depended on a combination of gravel and sand, brought from the northern part of Libya, hundreds of kilometres away. Development of methods to use sand available in the south will minimize material and shipping costs. The focus of this study is to adapt the methods developed in other places, and to use natural sand in pavement road construction. Sand-cement mixture is the best option for road construction in the region.

6 Objectives

The paper investigates the possibility of improving the engineering properties of desert sands in Libya using cement. Sources of material in southern Libya are limited and are quickly being used up because of the fast pace of development and road construction. Cement stabilization of base coarse layer in roads could save costs by using the freely available materials, but normally substandard, local fine aggregate. The properties of the stabilized sand such as compaction characteristics and California Bearing Ratio (CBR), were evaluated and their variations with the content of cement of the stabilizing agent are quantified.

7 Methodology

Optimum moisture content (OMC) and maximum dry density (γd-max) compaction tests were carried out on the samples in the laboratory, following ASTM D 698 methods. Also, California Bearing Ratio (CBR) tests were carried out, following the AASHTO NO. T 193 guidelines.

7.1 Material Mix Design

The study used two kinds of sand and one kind of cement. The first sample was 100% desert sand and is referred to as Type A. It has a 2.63 specific gravity with a mean particle diameter of 1.6 mm, a passing of D60, coefficient of uniformity (Cu) of 12.35 and a 4.8 coefficient of curvature (Cc). Type A sand is thereby considered to be SP (poorly graded). The second sample, referred to as Type B, is a manufactured sand. Type B has a specific gravity of 2.6395, mean particle diameter 2.8 mm, a passing of D60, coefficient of uniformity (Cu) of 7, and a 1.75 coefficient of curvature (Cc). Type B is thereby considered to be GW (well graded). The grain size distribution for both sand samples is shown in Fig. 1. The various physical properties such as specific gravity, fineness and water absorption are shown in Table 1. The ordinary Portland cement used in the investigation was bought from the local market with properties as shown in Table 2.

7.2 Laboratory Experiments

The grain size distribution of the fine aggregate mixture is shown in Fig. 2. A laboratory investigation was carried out to determine the mechanical properties of the stabilized granular soils. 70% natural sand (SP) and 30% manufactured sand (GW), both with maximum particle size of 5 mm, have been stabilized with different percentages of cement (0, 3, 5, and 7% respectively) mixed to form bonded materials.

7.2.1 Compaction Test

The dry natural sand that passed through sieve No. 4 (4.75 mm) was mixed with the manufactured sand and then with the stated percentages of cement; this was mixed until it achieved a uniform color. At this stage, small amounts of water were added when

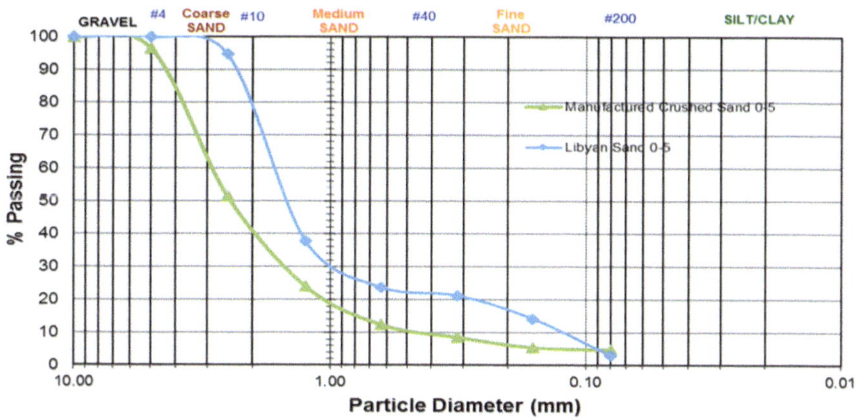

Fig. 1. The grain size distribution for both sands

Table 1. Physical properties of fine aggregates

Aggregate	Bulk specific gravity	Apparent specific gravity	Absorption (%)
Natural sand 0–5 mm	2.42	2.63	0.33
Manufactured sand 0–5 mm	2.44	2.639	0.58

Table 2. Chemical composition of ordinary Portland cement

Compound	%
Loss on Ignition (LOI)	7.91
SiO_2	18.20
CaO	59.03
MgO	1.80
Al_2O_3	5.09
Fe_2O_3	3.15
Na_2O	0.18
K_2O	0.29
SO_3	2.65
Materials not solvent	1.02

needed to help mix and compact the sample. Proctor compaction tests were run on all four mixtures: the first with only the 70/30 sand mixture; the other three with the specified cement percentages. The results of the compaction test for different percentages of cement for sand Type A and Type B are shown in Figs. 3, 4 and 5 respectively.

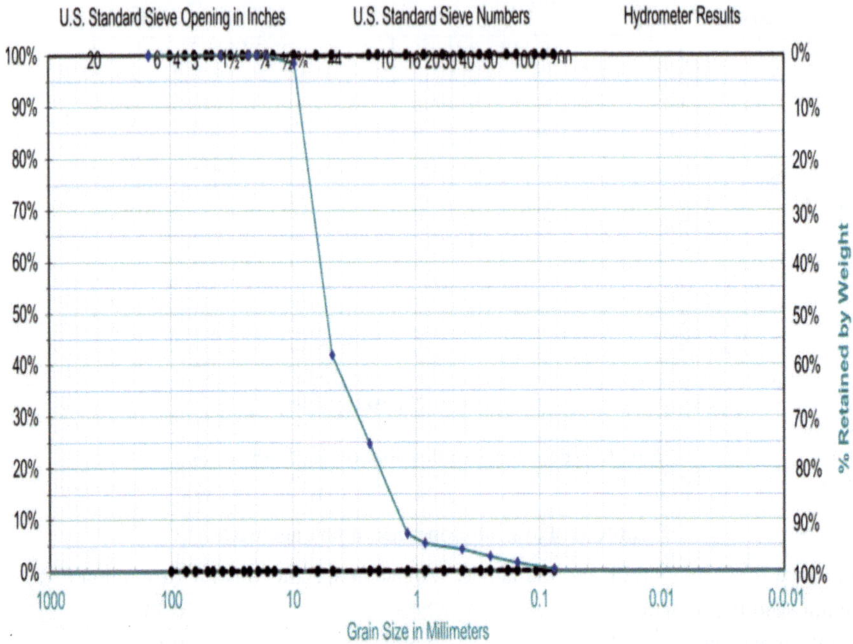

Fig. 2. The grain size distribution for fine aggregate mixture

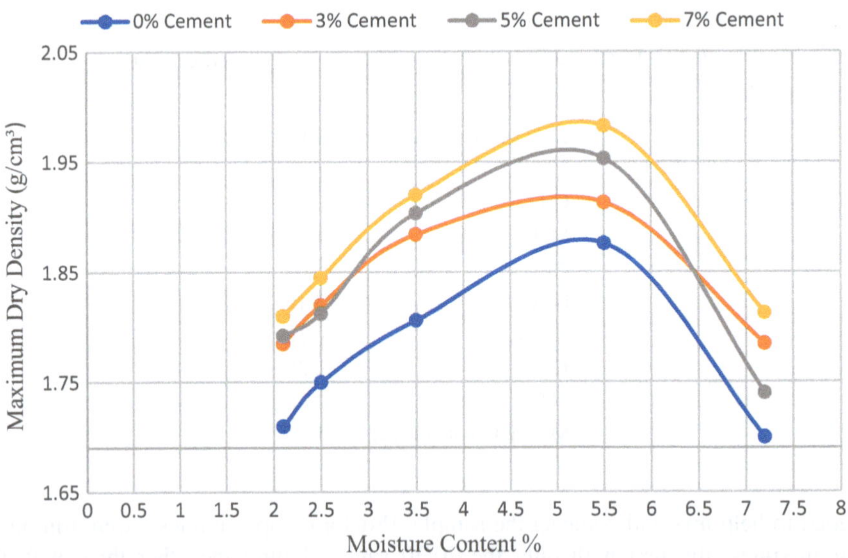

Fig. 3. Compaction curves for fine aggregate mixture with different ration of cement

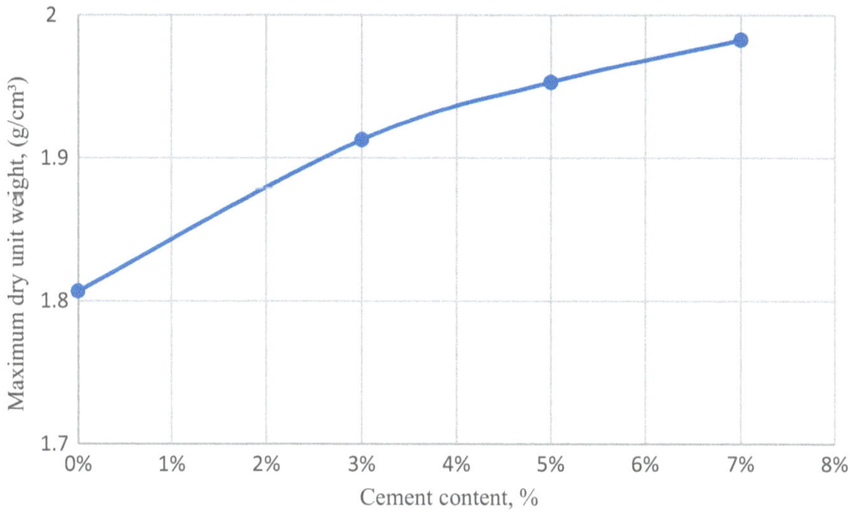

Fig. 4. Effect of cement content on the maximum dry unit weight for the mixture

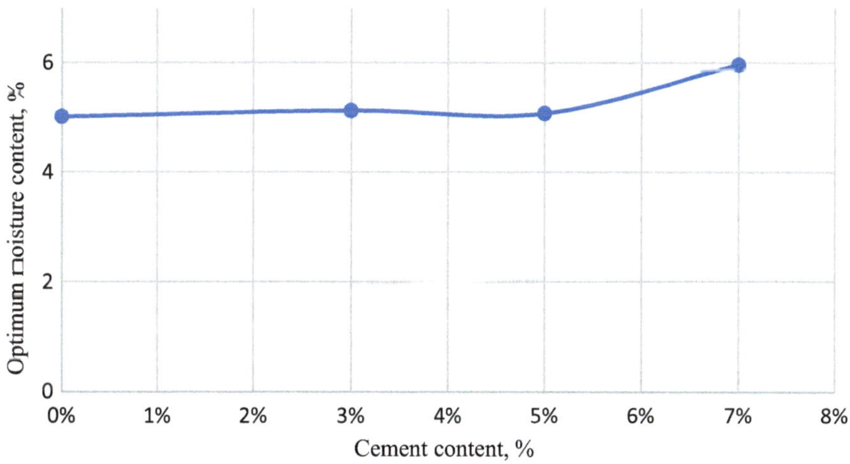

Fig. 5. Effect of cement content on the optimum moisture content for the mixture

7.2.2 California Bearing Ratio Test (CBR)

CBR test was carried out to determine the bearing capacity and the strength of the sand and sand-cement mixtures. CBR tests were run on all four mixtures: the first with only the 70/30 sand mixture; the other three with the specified cement percentages. Figure 6 shows the variation of CBR with cement contents for the mixture.

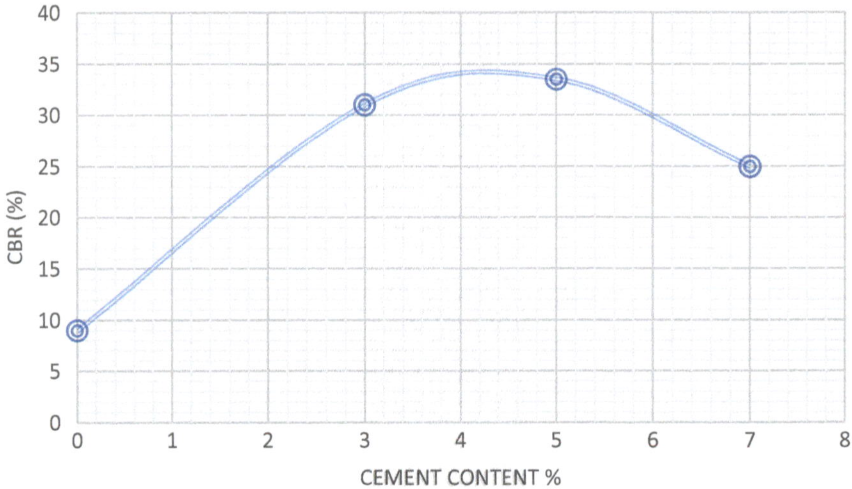

Fig. 6. Variation of CBR with cement contents

8 Results

8.1 Compaction Test Results

The compaction results of sand-cement mix are shown in Fig. 3. Figures 4 and 5 were shown results for the effect of the addition of cement to the 70/30 sand mixture in terms of maximum dry unit weight and in terms of optimum moisture content. Figure 4 shows that the 70/30 sand mixture has an increase in maximum dry unit weight. The lowest percentage of cement (3%) shows a marked increase in maximum dry unit weight over the 0% cement mix; further increases in cement percentage (5, 7%) show a further increase but not as substantial a difference as between 0 and 3%. This difference is due to the very fine nature of cement and the fact that it has greater surface area; therefore, it takes the place of the air voids between the grains of sand; this is what creates a greater dry unit weight. Figure 5 shows that the optimum moisture content for the sand mixture only increases to a marginal degree with the addition of the cement.

8.2 California Bearing Ratio Test (CBR) Results

The results of the CBR tests are summarized in Table 3. It consists of the variations of OMC, ϒd-max, CBR, and cement percentage. The CBR values increased with higher percentages of cement content, to a maximum value of 33.5 at 5% cement content, as shown in Fig. 6. As can be seen, the CBR values decreased with further increase in cement content. This suggests that cement content between 4 and 5% (on the basis of dry soil weight) can be recommended as an optimal content for improving the CBR. It can build up the CBR value up to 34% compared to the control sample.

Table 3. Compaction test and CBR test results

% Cement content	% OMC	Ɣd-max	% CBR
0	5.02	1.876	9
3	5.13	1.913	31
5	5.08	1.953	33.5
7	5.97	1.983	25

9 Summary and Conclusions

This research has the aim of improving the marginal local materials available for base course road construction with chemical stabilization. The conclusions are: material can be classified as Low quality or marginal for several reasons including unsuitable plasticity, inadequate strength and inappropriate gradation. Sources of material in southern Libya are limited and are quickly being used up because of the fast pace of development and road construction. Chemical stabilization could save costs by using the freely available, but normally substandard, local fine aggregate. Experiments show that the low-quality base materials were successfully stabilized with a mix of 70% natural sand with 30% crushed (manufactured)sand (both by weight) when mixed with various percentages of Portland cement (3, 5, 7%) to reach the necessary stability. The advantage of this research over the comparable performance of crushed sand is economic and practical because crushed sand is more expensive to transport than using the local materials with the Portland cement.

Acknowledgements. The first author would like to thank the Department of Civil Engineering at École de Technologie Supérieure (ETS), for providing all the facilities to carry out the experiments. He wishes to thank the Ministry of Higher Education in Libya for support during his Ph.D. program.

References

Abderrahmane, G., Ratiba, M.K.: Stabilized dune sand for road foundation layers-case of the dune sand of the region of Djelfa (Algeria). In: Applied Mechanics and Materials, vol. 319, pp. 263–277. Trans Tech Publications (2013)

Aiban, S.A.: A study of sand stabilization in eastern Saudi Arabia. Eng. Geol. **38**(1–2), 65–79 (1994)

Al-Aghbari, M.Y., Dutta, R.K.: Suitability of desert sand cement mixes for base courses in highway pavements. Electron. J. Geotech. Eng. **10**(4) (2005)

Almadwi, F.S., Assaf, G.J.: Performance testing of paving mixes for libya's hot and arid conditions, using marshall stability and SUPERPAVE gyratory compactor methods. In: International Congress and Exhibition Sustainable Civil Infrastructures: Innovative Infrastructure Geotechnology, pp. 313–323. Springer, Cham (2017)

Ansi, B.: ASTM D698-test methods for moisture-density relations of soils and soil-aggregate mixtures. Method A (Standard Proctor)

Azadegan, O., Yaghoubi, E., Li, J.: Effect of surface layer thickness on the performance of lime and cement treated aggregate surfaced roads. Electron. J. Geotech. Eng. **18**, 1081–1094 (2013)
Das, B.M.: Chemical and mechanical stabilization. Transportation Research Board (2003)
Ismaiel, H.A.H.: Treatment and improvement of the geotechnical properties of different soft fine-grained soils using chemical stabilization. Shaker (2006)
Patel, M.A., Patel, H.S.: A review on effects of stabilizing agents for stabilization of weak soil. Civ. Environ. Res. **2**(6), 1–7 (2012)
Sabbagh, A.O.: Design and viscoelastoplastic characterization of a lime-dune sand-asphalt mix (replacing aggregate, material laws, creep compliance, rutting) (1986)

Shaft Capacity Assessment of Recharge Impulse Technology Piles

Wafi Bouassida[1], Essaieb Hamdi[1], Mounir Bouassida[1(✉)], and Yuri Kharine[2]

[1] Ecole Nationale d'Ingénieurs de Tunis, Université de Tunis El Manar, LR14ES03-Ingénierie Géotechnique, BP 37 Le Belvédère, 1002 Tunis, Tunisia
wafy.bouassida@gmail.com,
mounir.bouassida@enit.utm.tn,
mounir.bouassida@fulbrightmail.org
[2] RITA FONDATIONS, Tunis, Tunisia

Abstract. The performance of Recharge Impulse Technology (RIT) in increasing the shaft capacity of deep foundation systems regarding classical set-processes is discussed. Based on the major contributors behind such a performance: the shape effect and the electrical discharge task, a case study is presented in view of highlighting that the coupling of these two components leads to a reasonable estimation of RIT shaft capacity with reference to in situ loading tests. A related design chart estimating the contribution of the shape effect in the increase of the shaft capacity had been drawn and a mean power law had been deduced through a parametric study on the adhesion factor.

1 Introduction

The idea of improving the shaft capacity (with reference to bored piles) by means of a vibration source (i.e. driving) dates since more than thirty years in pile technology. Recent techniques in this context had coupled the classical driving task with new procedures to further enhance this performance. As an example, inventors of full displacement piles technique as described by Pucker et al. (2012) had added the rotation task during the installation of their columns with a little enlargement of the pile section. This induced a remarkable improvement of relative densities of sandy layers crossed by such a setup process and so on the shaft capacity. A more recent technique than full displacement piles, named concrete pipe pile (PCC), had also gain its use in China, peculiarly in highways and railways embankments. This is majorly due to the use of expansive concrete after the driving task, which further ameliorates the shaft capacity compared to classically driven columns. For deeper information in this context, the framework of Hang et al. (2017) rigorously details how the expansion of concrete could increase the shaft capacity by more than 50% compared to the installation task of classical PCC piles. However, in an urban construction context, driving piles in sands or in clays could cause harmful effects to nearby structures if they exist. This is naturally due to the fact that the frequencies of vibrations developed during driving could approach one of the natural frequencies of the nearby structures and hence a resonance phenomenon starts to be generated. Serious harmful accidents had

H. Khabbaz et al. (eds.), *New Prospects in Geotechnical Engineering Aspects of Civil Infrastructures*, Sustainable Civil Infrastructures,
https://doi.org/10.1007/978-3-319-95771-5_13

occurred in the context of driving piles. For example, we state the construction project of the Damietta branch bridge (Egypt) in which the driving task had yielded to the pier crack of the delta barrage (Abdel Rahman 2012). Detailed discussions explaining the risks of pile driving to nearby structures are clearly highlighted in Lewis and Davie (1993), Marr (2001) and Abdel Rahman (2012). In this context, the geniuses of recharge impulse technology came to encounter these issues. More precisely, RIT proposes the improvement of the shaft capacity of a given deep foundation system without harming nearby constructions. This is clearly due to the high frequency range of its vibration source (shock waves by electrical discharges) sufficiently away from the natural frequencies interval of any building structure. Thus, when exerting the electrical discharge task, nearby structures "filtrate" the signal coming from RIT and remain stable. Consequently, RIT is considered as a fashionable technique, which suitably can be used in urban constructions and modern buildings development without risks of harming historical monuments and nearby constructions. This paper studies the shaft capacity performance of such a pile technology with reference to its two major contributors: the shape effect and the electrical discharge task. As a novel contribution, this work proposes a design chart for the documented case study allowing one to directly deduce the shape effect contribution in the increase of the shaft capacity of a RIT pile compared to a bored cylindrical pile. A clarified description of the installation process of RIT piles with referenced projects could be easily documented in Bouassida et al. (2016).

2 Performance of RIT Piles

2.1 Electrical Discharge

When exerting a lateral shock wave on a soil layer by means of an electrical discharge, relative density increases and, accordingly, the shear strength improves too (RITA 2005). In this context, RIT experience in saturated silt sands had shown that the horizontal compaction by the electrical discharge task yields to an increase of relative density, friction angle, and cohesion (RITA 2005). Equally, Alexandrovitch (2006) had studied the effects of electrical RIT discharges on the relative density of sand samples and clearly remarked a decrease in the void ratio (hence an increase in the relative density) after such a treatment. Thus, frictional contact between a RIT pile and a set of surrounding soil layers is clearly more important than that of a bored circular pile and the same horizons due to the electrical discharge effect. This naturally yields to an increase in shaft bearing capacity of RIT piles regarding bored piles. Compared to driven piles, Recharge Impulse Technology differs in two main points. The first one is the compaction task, done at constant volume for RIT piles, contrary to the driving task, which allows heave during the set-up process (Poulos and Davis 1980) and therefore volume change. The second point is related to the type of compaction itself. Indeed, the dynamic loading force transmitted to surrounding soil layers is different. For driven piles, the vibration transmitted to surrounding layers could be modeled as a classical wave producing pore pressure dissipation. Conversely, the electrical discharge, which can be modeled as an impulse, does not produce pore pressure dissipation as the duration of such a blasting load is extremely short (An et al. 2011). Hence, although driven piles and RIT foundations yield to a densification of surrounding soil layers and therefore an improvement of shaft bearing capacity, the two set-up processes differ and do not produce the same effect. Nowadays, as pile design codes do not propose recommendations for modeling the effect of blasting loads on surrounding soil layers and therefore on pile bearing capacities, it is obvious to find a gap between in situ load capacities of RIT piles and computed ones according to such standards. Hence, it should be stated that an accurate modeling of the electrical discharge task brings us closer to RIT piles behavior and, consequently, allows a more realistic estimation of RIT load capacities (Fig. 1).

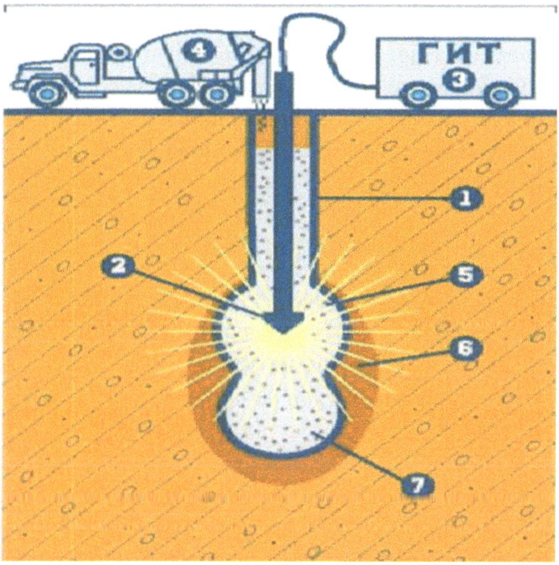

Fig. 1. Set-up process of RIT piles (RITA 2005)

2.2 Shape Effect

As the final shape of an installed RIT pile is not cylindrical but rather constituted by bulbs, the frictional mechanism developed around the loaded pile differs from that developed around a cylindrical loaded pile. Indeed, when a vertical force is applied to a cylindrical pile, the maximum mobilized frictional contact along the interface is governed, by:

$$\tau = C_a + K_s \cdot \sigma \tan(\delta) \quad (1)$$

τ is the shear stress.

σ is the vertical stress depending on short or long term analysis.

c_a: pile soil adhesion that is lower than soil cohesion (Poulos and Davis 1980).

K_s: Coefficient of lateral earth pressure at rest determined as $K_s = (1 - \sin(\varphi))\sqrt{OCR}$

δ: Friction angle between pile and soil. It varies between 1/3 φ (Pucker and Grabe 2012) and 2/3 φ (Plaxis 2016) where φ is the soil friction angle.

Clearly, the shear stress along the contact surface between the pile and the soil is characterized by an interface friction angle lower than φ and a pile soil adhesion lower than the cohesion of surrounding soil. For the case of RIT piles, and without considering the installation effect (electrical discharge compaction) improving both the friction and cohesion of the surrounding medium, the confined zones between the bulbs of a RIT pile, do not develop a shear stress along the pile-soil interface during the loading. Indeed, the frictional mechanism generated during the loading task will mainly be developed at the contact boundaries (CB) of these zones, with the surrounding soil and not with the pile interface (Fig. 2). Such a mechanism produces therefore a vertical shear plan governed at failure by the friction angle and the cohesion of the surrounding soil (taking account that the thickness of bulb summits are much lower than the length of contact boundaries). Hence, the shear stress mobilized at yield during the vertical pile loading can be written as:

$$\tau = C + K_s \cdot \sigma \tan(\varphi) \tag{2}$$

Consequently, according to expressions (1) and (2) the frictional contact between a RIT pile and a set of surrounding soils is greater than that of a cylindrical pile and the same soil layers due to the curvy shape of RIT piles. Such a frictional mechanism, developed when loading a RIT pile, is similar to the frictional contact of under reamed

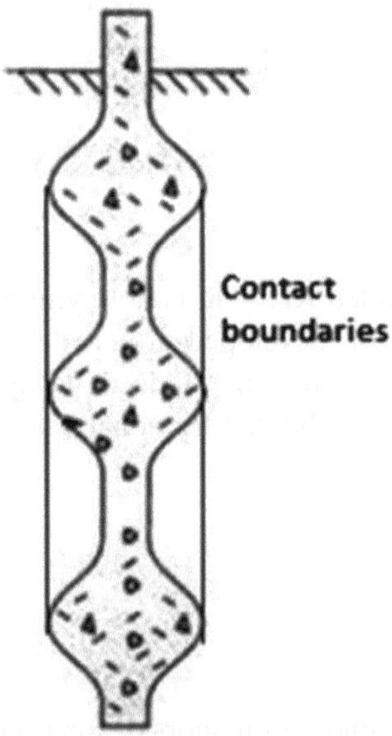

Fig. 2. Frictional mechanism developed for loaded RIT pile

piles, composed by bulbs, and widely used in India (Poulos and Davis 1980). Indeed, uplift in situ tests carried by Mohan in 1977 (Poulos and Davis 1980) had confirmed that the frictional mechanism related to a given pile composed by bulbs, is developed during the loading task, along the cylinder encountering the bulbs. Consequently, the shear stress distribution will be governed by friction angles and cohesions of surrounding soil layers (C) and not the shear parameters characterizing the frictional contact between cylindrical piles and the surrounding soil layers (c_a).

To confirm the consistency of such a mechanism for RIT piles, an uplift test had been realized on a 6 m length RIT pile with an initial drilling diameter of 0.32 m. Figure 3 presents the uplifted pile with portions of soils confined between the bulbs. Further, the measurement of pile perimeter indicated that the diameter of the uplifted pile equals 0.4 m. This confirms the 25% increase in diameter brought by the electrical discharge as stated by Bouassida et al. (2016). Hence, and similarly to the electrical discharge, it should be stated that the curvy shape of a RIT pile is a major parameter increasing its shaft bearing capacity compared to cylindrical piles. In the following, the contribution of these two parameters on the performance of RIT piles is investigated through a case study.

Fig. 3. Coring concrete samples of the uplifted RIT pile (L = 6 m, D = 0.4 m, Essalema City, Tunis)

3 Case Study

3.1 Site Conditions

In Essalema City (Capital Tunis Suburbs), the investigated site is characterized by superposed silt and clayey layers having poor pressuremeter characteristics up to 60 m depth. As exception, between 25 and 29 m depth, it is noted the presence of a relatively stiff layer clued by a significant increase in the pressuremeter modulus. Thus, the deep foundation solution was decided to be composed by a raft resting on a floating RIT piles group embedded one meter depth in this stiff layer. Table 1 and Fig. 4 present the results of the corresponding geotechnical campaign.

Table 1. Cohesions and friction angles of crossed layers

Layer	Thickness (m)	Cohesion (kPa)	Friction (°)
Brown Clayey sand	5	0	25
Silt Tunis soft clay	10	20	10
Dark Silt clay	12	30	10

3.2 Bearing Capacity

3.2.1 Bearing Capacity According to Pressuremeter Test Data (PMT)

According to PMT results plotted in Fig. 4, the load capacities of a single cylindrical pile (D = 0.4 m, L = 20.5 m) in the service state (factor of safety = 2), are presented in Table 2.

3.2.2 Bearing Capacity According to C-φ Method

Using values of undrained cohesions and friction angles (obtained from undrained simple shear tests) presented in Table 1, the bearing capacity of the same single pile studied in Sect. 2.1 according to C-φ method was evaluated. Adhesion factors had been chosen to be equal to 0.5 for all the layers in contact with the designed piles. This suitably agrees with the previous studies and recommendations of Poulos and Davis (1980) for normally consolidated and slightly over-consolidated clays. Moreover 66% of undrained friction angles had been employed to estimate the shaft and tip terms (NAFVAC DM 1982). Table 3 presents the obtained results.

3.3 Loading Test

Figures 5 illustrates the results of the load-settlement curve of an axially loaded RIT pile (D = 0.4 m, L = 20.5 m).

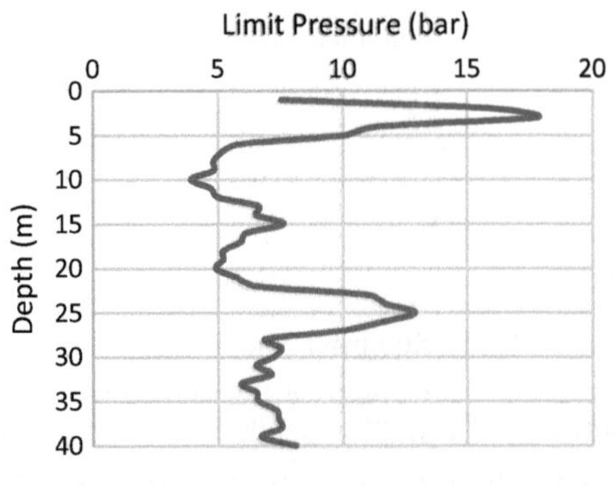

Fig. 4. Limit pressure variation in depth

Table 2. Service load capacity for a single pile (D = 0.4 m, L = 20.5 m) according to the PMT method (French code D.T.U. 13.2)

Type of loading	Bearing capacity (kN)
Shaft capacity	453
Tip capacity	57
Total load capacity	510

Table 3. Load capacity for a single pile (D = 0.4 m, L = 20.5 m) according to C-φ method

Type of loading	Bearing capacity (kN)
Shaft capacity	430
Tip capacity	42
Total load capacity	472

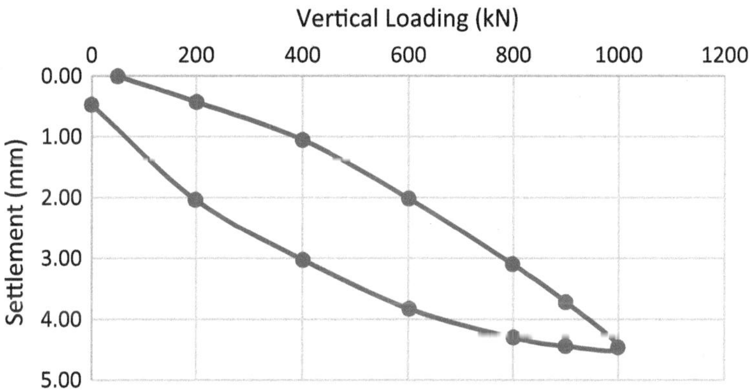

Fig. 5. Load-settlement curve from static load test executed on a RIT pile (D = 0.4 m, L = 20.5 m)

According to Fig. 5, a vertical force of 1000 kN yields to 4.4 mm of settlement. Hence, this load can fairly be adopted for service design since it corresponds to an admissible settlement. However, according to both PMT and C-φ methods, the load capacities should not exceed 510 kN. Consequently, one can clearly remark that the load capacity of a RIT pile is underestimated with reference to experimental results. Meaning that the behavior of a RIT pile, especially the frictional one, is quite different to be compared to a cylindrical pile, hence, cannot be correctly estimated using standardized codes. In order to confirm this statement, a numerical model was built up to simulate the loading test of a cylindrical pile having the same equivalent characteristics (outer diameter and length) of the described RIT pile (D = 0.4 m, L = 20.5 m) and in contact with the same surrounding layers. This will be detailed in the following.

3.4 Numerical Modeling of Static Load Test

In axisymmetric condition, a conventional cylindrical pile having the equivalent dimensions of the studied RIT pile (D = 0.4 m, L = 20.5 m) and in contact with the same layers described in Sect. 1 had been subjected to a compressive loading test. Figure 6 shows the load-displacement curves corresponding respectively to the static loading test of a RIT pile, the elastic method (Poulos and Davis 1980), and the numerical loading test. According to the plot of the bored pile (obtained numerically), the load capacity should not overpass 400 kN to guarantee an admissible settlement. In the same context, the elastic load settlement curve drawn following to the Poulos method indicates a full mobilized shear resistance along the pile interface beyond 470 kN and a limit settlement of 5.5 mm. Taking account of the pressuremeter load capacity (510 kN), the results lead to conclude that the design load is limited to 470 kN. Conversely, the installed RIT pile seems to offer a higher frictional performance as the slope of its load-settlement curve does not present a drastic decrease in the vertical loading range between 0 and 1000 kN. Equally, such a range offers a set of largely admissible settlements with a maximum value of 4.4 mm. Hence the design load of a RIT pile in this case equals at least 2 times the load capacity of a bored floating pile having an equivalent geometry. As mentioned in the two first sections, the major parameters behind such a performance are the electrical discharge and the shape effect. More precisely, these two components should be taken into consideration to estimate the load capacity of RIT piles. In the following, the contribution of these two parameters in the enhancement of RIT load capacities compared to bored cylindrical piles is quantified.

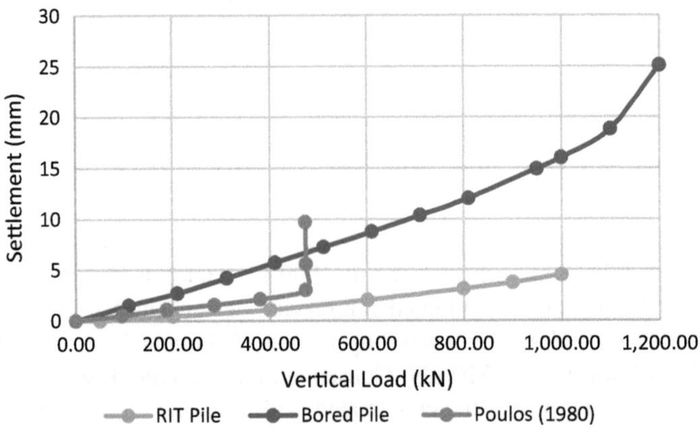

Fig. 6. Comparison between the behavior of RIT pile (RP) and a cylindrical bored pile (CP)

3.5 Shape Effect Contribution

According to Fig. 2, estimating the shaft capacity of a RIT pile is equivalent to the estimation of the shear load developed around the cylindrical surface around the studied RIT pile. Following the c-φ method the shaft load capacity can be written as:

$$Q_{su} = P\int_{0}^{L1}(C + Ks\cdot\sigma\tan(\varphi))dz + P\int_{0}^{L2}(C_a + Ks\cdot\sigma\tan(\varphi_a))dz \qquad (3)$$

where

L1: Length between the bulbs summits in which there is a soil-to-soil shear.
L2: Length along which there is a soil-pile shear (L2 is equally the sum of vertical heights of the bulbs summits)
P: Perimeter of the pile circular section.

Considering that L2 is much lower than L1 (L2 << L1), the shaft load capacity expression can be reduced to:

$$Q_{su} = P\int_{0}^{L}(C + Ks\cdot\sigma\tan(\varphi))dz \qquad (4)$$

Integrating this relationship over the length of a RIT pile (L = 20.5 m, D = 0.4 m) embedded in the studied geotechnical strata yields to a shaft load capacity of 727 KN with respect to a safety factor of 2.

On the other hand, the shaft capacity of a cylindrical pile can be obtained using the integral form of Eq. (3) between zero and L (i.e. along the length of the pile). For normally consolidated and slightly over-consolidated clays, the pile soil adhesion factor Ca/Cu varies between 0.25 and 0.7 (Poulos and Davis 1980). Since no experimental measurement of this parameter had been carried during the studied project, Fig. 7

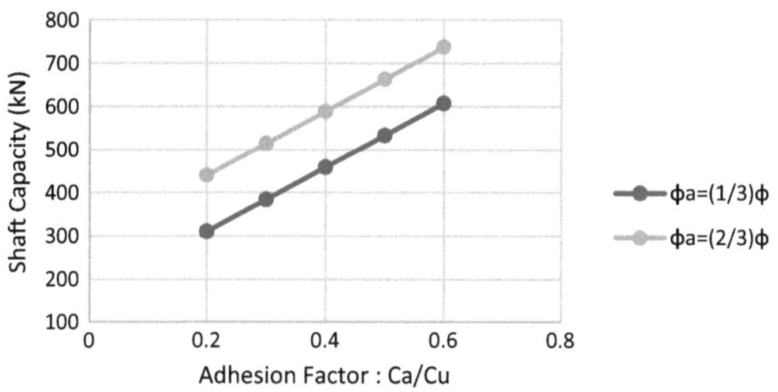

Fig. 7. Variation of shaft capacity with respect to the adhesion factor Ca/Cu for a bored pile (L = 20.5 m, D = 0.4 m)

presents the envelope evolution of the shaft capacity of a bored cylindrical pile (embedded in the strata described above, D = 0.4 m, L = 20 m) as a function of the adhesion factor Ca/Cu for interface friction angles φa varying between one third and two third of the surrounding layers friction angles.

It is clear herein that the more adhesions are close to cohesions of surrounding layers, the more the shaft capacity increases as the contact between surrounding soils and the pile employs greater shear resistance. To highlight the contribution of the shape effect in the enhancement of the shaft load capacity for a RIT pile compared to a cylindrical pile, Fig. 8 presents the variation of the shaft capacity ratio (defined by the shaft load capacity of a RIT pile divided by the shaft capacity of a bored cylindrical pile) as a function of the adhesion factor. Pile-soil friction angles are in between (1/3)φ and (2/3)φ.

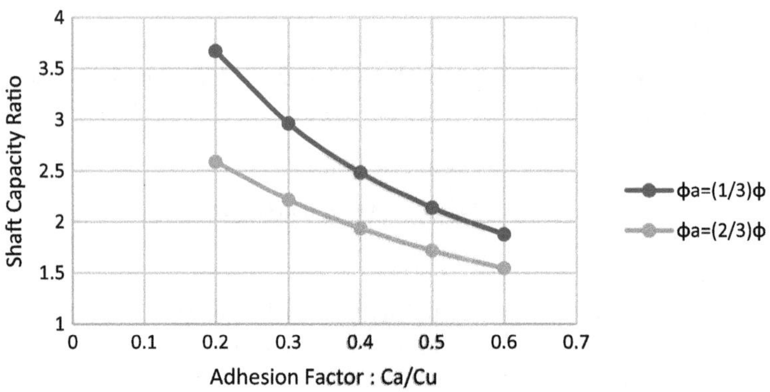

Fig. 8. Envelope of shaft capacity ratio as a function of the adhesion factor

According to Fig. 8, it is remarkable that below an adhesion factor of 0.4, the shaft capacity ratio drastically increases. Following the proposed average adhesion factor stated by Poulos and Davis (1980) for bored cylindrical piles in normally consolidated clays (equal to 0.5), we obtain a shaft capacity 2 times (an average value between the two curves) the shaft capacity of a circular pile having the same dimensions thanks to the soil-to-soil contact mechanism developed during the loading of a RIT pile. In this context, Salgado et al. (2007) had conducted extensive parametric analysis on the adhesion factor and found that such a parameter varies between 0.38 and 0.73 and also increases with the percentage of clay content. This finding is in agreement with both the envelope (0.25–0.7) and the average value (0.5) proposed by Poulos and Davis (1980). For the studied case, the average variation of the shaft capacity ratio as a function of the adhesion factor can be described by the following equation:

$$\frac{Qs(RIT\,Pile)}{Qs(Bored\,Pile)} = 1.314\left(\frac{C_a}{C_u}\right)^{-0.549} \quad (5)$$

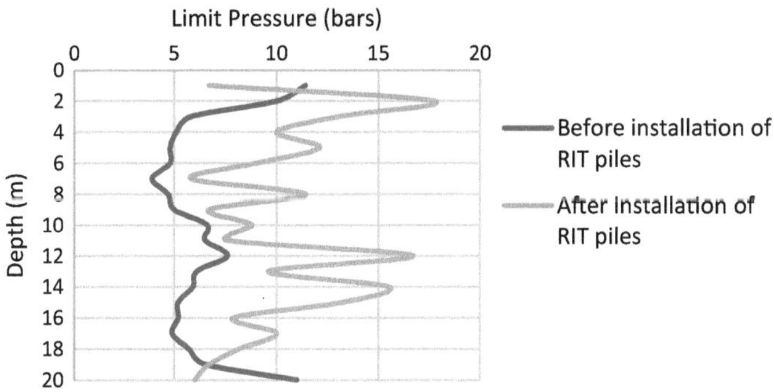

Fig. 9. Limit pressure profiles before and after the installation of RIT piles

3.6 Electrical Discharge Contribution

The contribution of the electrical discharge task to the frictional performance of RIT piles had been investigated in this project through a pressuremeter test carried nearby an executed RIT pile, located at the corner of the projected building area. Figure 9 presents the limit pressure profile before and after the electrical discharge task.

According to Fig. 9, the improvement rate brought by the electrical discharge task is clearly visible. Indeed, the limit pressure profile has shifted by a mean average value of more than 65% compared to the initial profile. Hence the shaft capacity will naturally increase. Following the pressuremeter method, the load capacities of a cylindrical pile having a length of 20.5 m and a diameter of 0.4 m in the service and ultimate states are presented in Table 4. Alternatively, it is possible to estimate the new undrained cohesions (after the electrical discharge task) of the surrounding layers to the pile, from the new limit pressure profile, based on the correlations of Cassan (Kammoun et al. 2016).

$$0.3\,\mathrm{MPa} < p_l < 1\,\mathrm{MPa}\; C = \frac{p_l}{12} + 0.03 \tag{6}$$

$$1\,\mathrm{MPa} < p_l < 2.5\mathrm{MPa}\; C = \frac{p_l}{35} + 0.085 \tag{7}$$

Correlations between limit pressure and undrained cohesion (Kammoun et al. 2016).

Taking account of the shape effect contribution presented in the previous section, Table 5 shows the new undrained cohesions and the related shaft capacity of a RIT pile installed in the studied geotechnical strata.

Table 4. Loading capacities for ultimate and service states after the electrical discharge task

State of loading	Shaft load capacity (kN)
Service loading (FS = 2)	936
Ultimate state (FS = 1.4)	1310

Table 5. Shaft Load Capacity of a Single RIT Pile

Layer	Cohesion (kPa)
Brown Clayey sand	40.7
Silt Tunis soft clay	36.6
Dark Silty clay	60.2
Shaft load capacity	925 kN

The shaft capacity (925 kN) deduced after taking account of both, the electrical discharge task and the shape effect is fairly in agreement with the results of the static loading test. Indeed, the loading test results indicate that the design load should not overpass 1000 kN. Meaning that the use of the two presented parameters (shape effect and electrical discharge) yields to a consistent estimation of the shaft capacity (925 kN) regarding in situ results (1000 kN). In terms of ratio between the shaft capacity of a RIT pile and a bored cylindrical pile installed in the studied geotechnical strata, it is stated that the electrical discharge task and the shape effect had increased the frictional performance of a RIT pile by a 2.45 factor compared to a cylindrical bored pile.

4 Conclusions

In this paper, the shape effect contribution to the increase of the shaft capacity of a RIT pile, regarding a bored cylindrical pile, has been discussed. A practical design chart allowing to estimate the ratio of such a contribution had been drawn for the studied case and a related mean power law was suggested. In addition, the electrical discharge contribution had been investigated experimentally and further combined with the shape effect to highlight that the coupling of these two components leads to a reasonable estimation of RIT shaft capacity. Future works should develop a rigorous theoretical modeling of the cavity expansion due to the electrical discharge task (taking account of the emergence of plastic shock waves in a porous medium) to be combined with the shape effect and hence elaborate a complete framework estimating the shaft capacity of RIT piles.

Acknowledgements. Authors acknowledge Ahmed Amine Kammoun (Civil Engineer) for his contribution to the shape effect work. They equally express their gratitude to AfricSoil Tunisia, which had carried the in situ tests after RIT piles installation with a fair reasonable cost.

References

Abdel-Rahman, S.M.: Vibration Associated with Pile Driving and its Effects on nearby Historical Structures. Mechanical and Electrical Research Institute, National Water Research Center, Delta Barrage, Egypt (2012)

Alexandrovitch, B.A.: A study on the deformations of Malovlazhnogoed sand around RIT piles, 220pp. PhD thesis, Moscow State University of Construction (2006)

An, J., Tuan, C.Y., Cheesman, B.A., Gazonas, G.A.: Simulation of soil behavior under blast loading. Int. J. Geomech. **11**(4), 323–334 (2011)

Bouassida, W., Essaieb, H., Bouassida, M., Kharine, Y.: On the use of RITA pile technique in Tunisia. In: Fourth International Conference on New developments in Soil Mechanics and Geotechnical Engineering, Near East University, Nicosia, North Cyprus, 2–4 June 2016

DTU 13.2: Fondations profondes pour le bâtiment, Norme Afnor P 11-212 (1992)

Hang, Z., Jian Min, H., Xuan Ming, D., Jian, C.: Theoretical model for the improved PCC pile using expansive concrete. Sci. China Technol. Sci. **60**(5), 772–791 (2017)

Kammoun, A.A, Bouassida, W., Bouassida, M.: Analyse de stabilité d'une excavation prévue pour la réalisation de l'OA 13 situé sur la ligne D du projet RFR Tunis. SIMPRO Tunisie. Rapport de validation, 22pp (2016)

Lewis, M.R., Davie, J.R.: Vibrations due to pile driving. In: Third International Conference on Case Histories in Geotechnical Engineering, St. Louis, Missouri, 1–4 June 1993. Paper No. 4.01 (1993)

Marr, W.A.: Dealing with vibration and noise from pile driving. Pile Driv. Contract. Assoc. **2**(1), 17–20 (2001)

Murthy, V.N.S.: Geotechnical Engineering, vol. I, 1056pp. Taylor & Francis Inc. NY, USA (2002)

Plaxis.: Plaxis 2D, reference manual. www.plaxis.com/2D-2-Reference.pdf (2016)

Poulos, H.G., Davis E.H.: Pile Foundations Analysis and Design, vol. I, 397pp. Wiley (1980)

Pucker, T., Grabe, J.: Numerical simulation of the installation process of full displacement piles. Comput. Geotech. **45**, 93–106 (2012)

Recharge Impulse Technologies and Apparatus (RITA): Building on the Basement of Knowledge, vol. I, 58pp. RITA Press, Moscow (2005)

Salgado, R., Prezzi, M., Seo, H.: Advanced modeling tools for the analysis of axially loaded piles. Advances in Deep Foundations—Kikuchi, Otani, Kimura & Morikawa, 438pp, Taylor & Francis Group (2007)

Mechanistic Analysis of Subgrade Soil Reinforced with Modified Jute Geotextile

Vinod Kumar Adigopula[1(✉)], Chandra Bogireddy[1], Sunny Deol Guzzarlapudi[2], Radha Gonawala[1], and Rakesh Kumar[1]

[1] Department of Civil Engineering, S.V. National Institute of Technology, Surat, Gujarat, India
{vinodkumarmtech, hari2006chandra, radhagonawala, rakesh1999}@gmail.com
[2] Department of Civil Engineering, National Institute of Technology-Raipur, Raipur, Chhattisgarh, India
sdguzzarlapudi.ce@nitrr.ac.in

Abstract. Design and construction of flexible pavement on soft subgrade soil is a daunting task to achieve a target strength level due to its expansive characteristics. Application of modified jute fibre reinforced subgrade layer is a promising technique for enhancing subgrade soil strength, keeping its biodegradability intact during the design period. In this study, locally available natural jute geotextile (NJGT) and modified jute geotextile with bitumen (BJGT) and low-density polyethene sheet (PJGT) were used for reinforcing natural clayey and silty soil. Series of laboratory strength related tests such as resilient modulus (M_r) using repeated triaxial tests, California Bearing Ratio (CBR), Unconfined Compressive Strength (UCS), were performed on parent soil, natural and modified jute geotextile reinforced soil samples. Tests results depict that CBR values were increased up to 1.4–5.3 times for three layers of natural and modified reinforced soil samples. Whereas, M_r values were increased up to 1.2–3.0 times for two layers of natural and modified reinforced soil samples respectively as compared with parent soil. The proposed soil reinforcing models were simulated by designing crust composition for the vehicular loading of 1–7 million standard axle load repetitions. The performance of the proposed crust composition was assessed by using KENLAYER program. Results show that the performance of reinforced pavement models was enhanced regarding rutting repetitions as compared with parent models. Thus, this study intended to overcome the biodegradability aspect to improve the durability characteristics of JGT reinforced subgrade soil for paving low volume road in the expansive soil region.

Keywords: Subgrade · Jute geotextile · Resilient modulus (M_r) California Bearing Ratio (CBR)

H. Khabbaz et al. (eds.), *New Prospects in Geotechnical Engineering Aspects of Civil Infrastructures*, Sustainable Civil Infrastructures,
https://doi.org/10.1007/978-3-319-95771-5_14

1 Introduction

In India, rural roads are most accessible road network and second largest in the world, having a road network length of 4.86 million kilometres next to the United States of America. The Government of India introduced Pradhan Mantri Gram Sadak Yojana (PMGSY) and National Highway Development Programs (NHDP) for providing access to rural areas and increasing mobility among metropolitan cities. The escalating cost of available material with the lack of resources motivated highway engineers to explore new alternative materials in road construction in developing countries like India (Tapash Kumar 2010). Reinforcing weaker soil with conventional materials such as crushed stone, stone dust, crushed concrete and fly ash, etc., are traditional practices for the construction of road networks in India. However, in the current scenario, availability of local material is cumbersome and is being exhaustible in nature and quality being degraded gradually (Al-Wahab and Heckel 1995). Further use of the local availability of material without compromising on quality is a challenging task for pavement design engineers. In comparison with other synthetic geotextiles, the natural jute geotextiles (NJGT) are proven more feasible for low volume roads. The advantage of using natural fibre is environmentally eco-friendly, biodegradable, non-abrasive, exhibit high initial modulus and high moisture absorption. They have non–uniformity in their physical and mechanical properties (Murugesan 2004; Bijayananda and Mahipal 2011). In recent times, the use of jute geotextile reinforcement has emerged as a promising technique to improve the performance of low volume roads and unpaved roads (Chandra and Viladkar 2008; Saride and Suraj 2014). However, the lifespan of jute products naturally decomposes faster than expected and desirable for some applications like Geotextiles, sandbags, etc. To resist the decomposition and to increase the lifespan of jute in pavement application, bitumen along with suitable chemical recipe may be viable (Ghosh et al. 2014).

2 Literature Review

Numerous efforts were made by several researchers in reinforcing the natural subgrade soils with various materials few of them were discussed in detail. Lawton and Fox (1992), performed tests on sandy soil reinforced with multi-oriented geosynthetics and results depict that highest strength regarding CBR and resilient modulus (M_r). Al-Wahab and Heckel (1995) observed that the fibre reinforcement significantly increases ductility, toughness, and energy absorption capacity of the soil. Fibres range from 0 to 5% and fibre length from 6.35 to 50.8 mm were decided to use optimum fibre content. Dhawan and Goswami (2009) studied the effect of fly ash, lime to the expansive soil used for the construction of road base, subbase, and embankment. Murugesan (2004) investigated on subgrade soil reinforced natural subgrade soil with coconut, jute, and nylon fibre at various percentages and reported an overall increase in CBR by 60% due to fibre reinforcement at an optimum percentage of fibre content lies between 0.5 and 0.6. Gosavi et al. (2004) conducted tests on weaker subgrade soil (black cotton soil) reinforced with synthetic fibre. The optimum moisture content (OMC) increases the value of maximum dry density (MDD) decreases with the increase in the quantity of woven fabrics and fibreglass, the maximum value of reinforced quantity may be limited to 2%. Soaked CBR test of black cotton soil shows good results when reinforced Gosavi et al. (2004). Ravishankar and Raghavan (2004) conducted the study on coir-stabilized laterite soils and stated that CBR increases up to 10% of coir content and the further increase in coir quantitative results in the reduction of values. In India, Central Road Research Institute (CRRI) is carrying research on Bituminized Jute Paving Fabric (BJPF) application on distinct traffic volumes with an assessment of performance-based design considerations Swapna Kumar and Bhattacharyya (2014). Jahan et al. (2012) and Mohanty and Misra (1995) stated that the jute fibres could be used as reinforcement in thermoplastics, e.g., polyethylene, polyvinyl chloride and polypropylene and thermosets like unsaturated polyester and epoxy resin, and it may also increase durability. Tang (2007) assessed the influence of discrete short polypropylene fibre (PP-fibre) on the strength characteristics of natural and cement treated clayey soil. The test results depicted that the fibre reinforcement within natural and cement treated clayey soil improved the Unconfined Compressive Strength (UCS), shear strength and failure axial strain, stiffness reduction and loss of post-peak strength, and change in the behaviour of cement treated soils from brittle to ductile. Bijayananda and Mahipal (2011) performed various laboratory tests such as CBR tests in soaked and unsoaked conditions at OMC and MDD on randomly oriented fibre reinforced and unreinforced specimens of a clayey soil. Coir fibre has been used as a reinforcing material to investigate its beneficial use in rural road subgrade soil and reported that overall increase in CBR values.

Few studies were reported on enhancement techniques of life expectancy of jute fabrics with various materials or coatings Ghosh et al. (2014). However, limited research was reported on the usage of modified jute geotextile with bitumen and polymer in pavement subgrade strength improvement and its overall structural integrity. Therefore, this study intends to propose a novel technique for reinforcing clayey and silt subgrade soil with natural jute geotextile (NJGT), modified jute geotextile with bitumen (BJGT) and Low-density polyethylene sheet (PJGT) in flexible pavements to extend the life of jute geotextile from its biodegradability and poor microbial attack. Application of Jute Geotextile (JGT) for road pavement is not a rare phenomenon. However, its practice is not vast, compared to other alternatives. In recent years, JGT has become popular in India on low volume road construction practices. Since synthetic geotextiles are expensive and non-eco-friendly, more economical and green substitutes like jute geotextiles are gaining acceptance. In the present scenario, in India, untreated jute layer was laid in between subgrade and subbase layers. The lifespan of untreated jute will be 4–6 years. In the road construction, separation is needed to segregate the subbase and subgrade for prevention of the excessive pavement deflection under axle loads of moving vehicles. This phenomenon will be applicable until the jute was biodegradable to prolong the lifespan of jute, the proposed methodology may be adopted for low volume roads. Thus the primary objective of this study is to assess the influence of NJGT, BJGT, PJGT on strength characteristics such as resilient modulus (M_r) California Bearing Ratio (CBR) value, and Unconfined Compression Strength (UCS) values of parent subgrade clayey and silty soils.

3 Materials and Methods

3.1 Parent Subgrade Soil

Two parent soils clay and silt type were selected along the proposed low volume stretch in this study for performing laboratory investigations. The selected soil types clay and silt with intermediate compressibility were designated as CI and MI groups as per Indian soil classification system. The index and volumetric properties of selected parent soils were assessed as per Indian standard specifications, IS 2720, and the corresponding test results were represented in Table 1. The grain size distribution of curves obtained for selected soil is shown in Fig. 1.

Table 1. Properties of parent subgrade soil

Sl. no.	Property	Number of samples	Test results	
			CI	MI
1	Sieve analysis	33 (11[a])		
	Gravel (%)		3.30	13.24
	Sand (%)		28.00	34.82
	Silt + clay (%)		68.70	51.94
2	Liquid limit, w_L (%)	33 (11[a])	43	32
3	Plastic limit, w_p (%)	33 (11[a])	16	24
4	Plasticity index, I_p (%)	33 (11[a])	27	8
6	Free swell index (%)	33 (11[a])	45	20
7	Degree of expansion	33 (11[a])	Low	Low
8	Maximum dry density (kN/m^3)	33 (11[a])	17.93	18.42
9	Optimum moisture content (%)	33 (11[a])	16.00	8.00

[a]Number of test locations

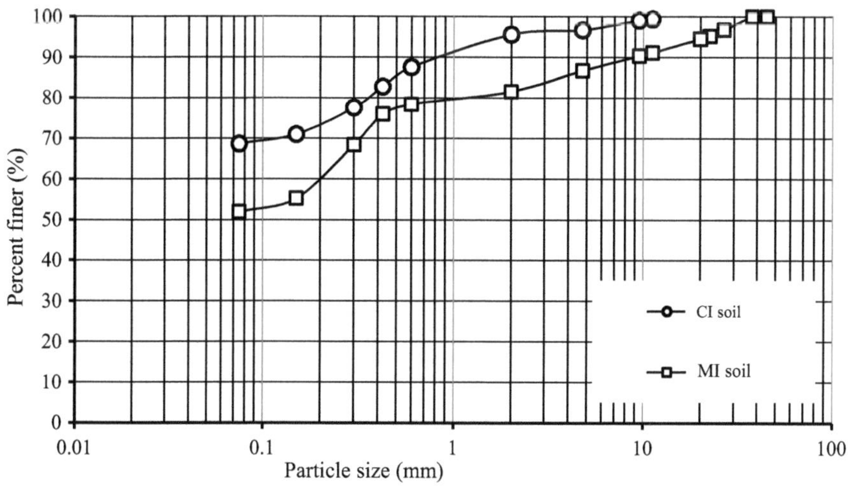

Fig. 1. Particle size distribution of soils

3.1.1 Natural Woven Jute Geotextile

Natural Jute geotextile (NJGT) was collected from local markets in Tirupati, Andhra Pradesh, India. Figure 2a shows a photograph of NJGT fabric selected for the study. The conventional engineering properties of selected NJGT were determined as per American Standards for Testing Materials (ASTM) standards, and the corresponding tests results were presented in Table 2. This NJGT is used to reinforce the parent soil to assess the NJGT influence on strength characteristics as discussed in subsequent sections.

(a) (b)

Fig. 2. **a** Photograph of Natural Woven jute geotextile fabric. **b** Woven Jute geotextile treated with polyethylene sheet

Table 2. Physical properties of Natural Woven jute geotextile

Thickness (mm)	Weight (gm/m^2)	Tensile strength (kN/m^2)	Punching strength (kN)	Failure strain (%)	Permittivity (m/s)	Transmissivity (m^2/s)	Apparent opening size (A.O.S) (mm)	Tenacity (gm/denier)
3	400	2.81	0.5	30	3.36×10^{-3}	4×10^{-6}	0.05	4.2

3.1.2 Modified Jute Geotextile with Bitumen

The bitumen of viscosity grade (VG-10) type is used in this study for coating the NJGT to protect from microbial attack and degradation for increasing lifespan and strength of the jute fibre Swapan Kumar and Bhattacharyya (2014). The property of the bitumen used in this study is shown in Table 3. Bitumen is sprayed on NJGT sheet at a temperature of 160 °C and further used as a reinforcing layer for parent subgrade in order to increase the strength characteristics such as tensile properties, bursting strength and puncture resistance.

Table 3. Physical properties of Bitumen

Grade of bitumen	Softening point (°C)	Flash point (°C)	Viscosity ate at 60 °C	Ductility at 25 °C, cm	Specific gravity @ 27/27 °C
80/100	45	285	3.5	180	1.1

3.1.3 Polymer Modified Jute Geotextile

The NJGT is also treated with low-density polyethylene (LDPE) by placing NJGT between two LDPE sheets as shown in Fig. 2b. The application of polyethylene in jute fibre is to increase its durability and subsequently to enhance the overall strength of a pavement.

4 Testing Program

4.1 Sample Preparation

Detailed laboratory experimentation program was developed on 33 prepared samples collected from 11 different locations for determining the strength characteristics such as CBR value, UCS value and M_r Value of parent subgrade soil, NJGT reinforced soil, BJGT and PJGT reinforced soil. An attempt has been made to various reinforced soil samples and prepared by placing the natural and modified jute geotextile at various heights and simultaneously by increasing fibre content in terms of some layers. CBR, UCS and M_r tests were performed on reinforced soil samples for a number of trials by increasing the soil samples were prepared by reinforcing the parent soil with NJGT, BJGT, and PJGT at different depths by increasing number of layers for each trial. Each test sample was prepared for NJGT, BJGT and PJGT at varying depths and by increasing layers at predefined OMC and MDD levels. The testing protocols were adopted as per IS (Indian standards), ASTM and AASHTO standards and the same is discussed in subsequent sections.

4.1.1 CBR and UCS Test

The conventional subgrade strength tests most commonly termed as soak CBR and UCS test were performed as per IS specifications, IS 2720 (Part-XVI) (1985) for the selected parent CI, MI soil and reinforced soil samples. The tests were conducted for different trials by increasing number of layers of at varying depths, and the corresponding CBR test results were presented in Fig. 7. UCS test was performed on a test specimen of 38 mm diameter and 76 mm height. The results of UCS test for reinforced soils are shown in Fig. 3. The stress-strain curves of soils are nonlinear since the onset of loading. Modulus of elasticity was calculated corresponding to the first tangent of the stress-strain curve. The failure stress and corresponding strain were found an increase in with a rise in jute geotextile fibre.

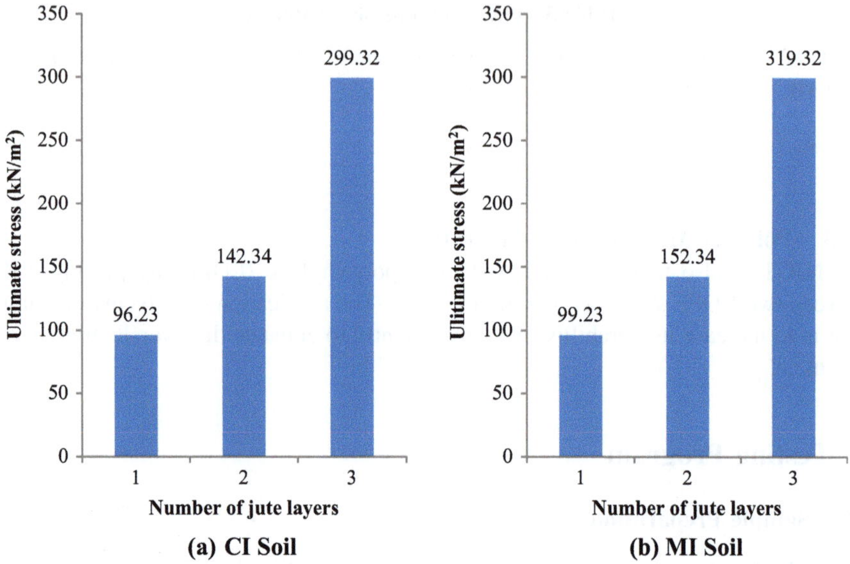

Fig. 3. UCS test results of reinforced CI and MI soil with jute geotextile

4.1.2 Repeated Triaxial Test

Laboratory estimated resilient subgrade moduli (M_r) was experimentally determined by applying repeated axial load on natural and reinforced soil samples in a triaxial cell. The resilient modulus test in this study was performed as per the test procedure suggested by AASHTO Standard (AASHTO T 307-99 2003). The deformations of soil samples were monitored using two linear variable differential transducers (LVDTs) mounted to the chamber. The application of load repetitions during conditioning stage and postconditioning stage of the prepared specimen for each load sequence and subgrade soils physical properties were shown as per the Table 1. Repeated triaxial test was performed as per AASHTO standard for subgrade soil (AASHTO T 307-99 2003). The prepared test specimens were tested under repeated triaxial test apparatus, and the corresponding meant deviator stress and meant recovered deflection/strain were recorded and subsequently M_r is calculated using the constitutive model as shown in Eq. 1.

$$M_r = \sigma_d / \varepsilon_r \tag{1}$$

where M_r is the resilient modulus, σ_d is the repeated deviator stress, ε_r is the recoverable axial strain. The measured M_r values for all the test specimens as shown in Fig. 8. The range of axial stress (13.8–68.9 kPa) and confining pressure (13.8–41.4 kPa) values shown in Fig. 8 were as per AASHTO T307-99 (2003) guidelines (Table 4).

Table 4. Modulus of elasticity, failure stress for different soils

Property	CI		MI	
	Natural	Reinforced	Natural	Reinforced
E value (MPa)	3.94	9.96	5.16	10.23
Failure stress ($\sigma 1$) (kPa)	52.36	184.29	67.24	217.49

5 Pavement Modeling and Design

Authors attempted to assess the performance of parent and reinforced subgrade soil by developing pavement models under dynamic vehicular loadings. Pavement crust thickness was estimated as per the guidelines suggested by Indian road congress IRC 37 (2012) for a typical road section. The crust thicknesses were adopted for the parent and reinforced soil based on the calculated CBR and estimated traffic loadings as shown in Fig. 4, 5, and 6.

5.1 Performance Analysis Using KENLAYER Software

Performance analysis was carried out using KENLAYER program. Complex stress, strain analysis on flexible pavements under loading condition using elastic layer theory are most familiar using KENLAYER software Ziari and Khabiri (2007). Modified Jute layer calculated strains are tabulated in Table 5. Cross-sectional details of the proposed case for distinct soils were shown in Fig. 4, 5, and 6. Lekha et al. 2015 performed KENLAYER performance analysis over lateritic soil stabilized with Arecanut coir for Indian low volume roads. Similarly, the pavement design life observed to be significantly enhanced for stabilized soil (Arecanut coir). In this study, performance analysis was carried out using Indian fatigue and rutting models (IRC 37 2012). Based on the empirically based performance analysis the proposed methodology may be used for low volume roads.

$$N_f = 2.021 \times 10^{-4}\left[\frac{1}{\epsilon_t}\right]\left[\frac{1}{E_{bs}}\right]^{0.854} \tag{2}$$

N_f = Number of cumulative standard axles to produce 20% cracked surface area; ϵ_t = Tensile strain at the bottom of bituminous surfacing (microstrain); E_{bs} = Elastic modulus of bituminous surfacing (MPa).

$$N_R = 4.1656 \times 10^{-8}\left[\frac{1}{€_z}\right]^{4.5337} \tag{3}$$

N_R = Number of cumulative standard axles to produce rut-ting of 20 mm; $€_z$ = vertical subgrade strain (microstrain).

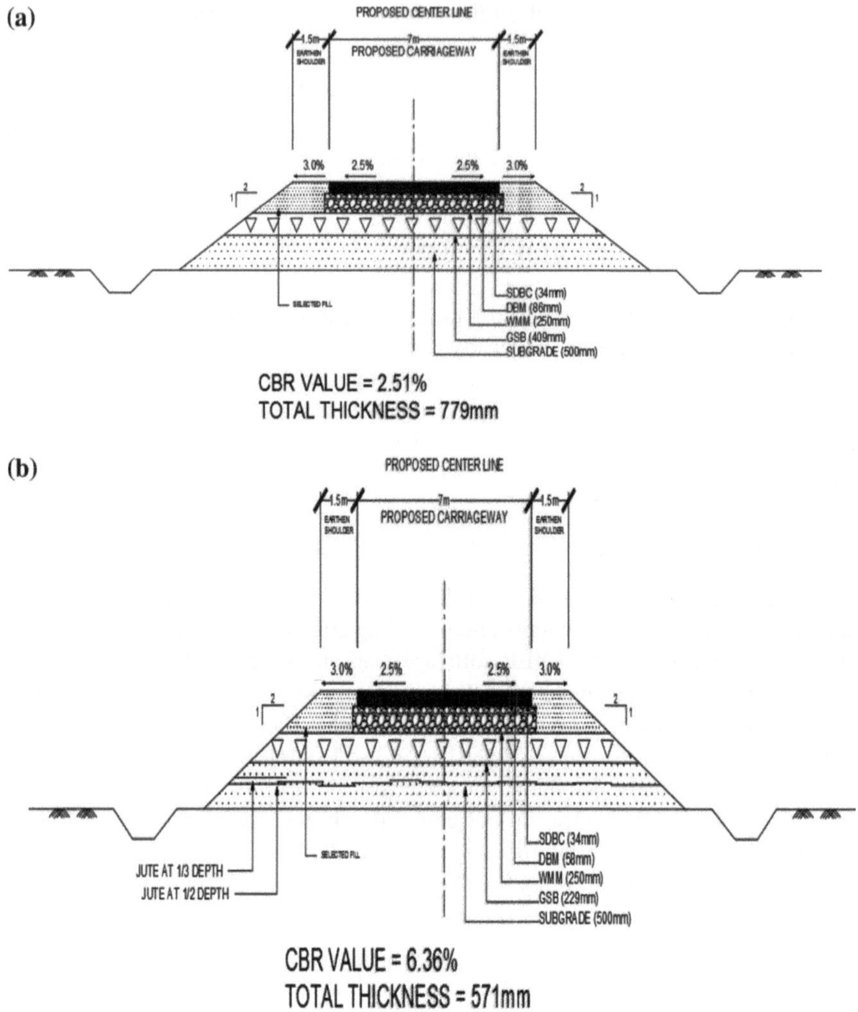

Fig. 4. Design of flexible pavements for **a** CI soils without reinforcement. **b** CI soils with Reinforcement

6 Results and Discussion

JGT membrane layer prevents granular materials from penetrating into the soft underlying subgrade as well as prevent fine-grained subgrade soil from being pumped up into permeable granular materials. It may be good in tension and weak in compression. Combination of JGT and soil ensures excellent bond and form a composite material. It is hypothesized that mechanism involved with increasing the stiffness of weak subgrade due to the use of JGT. Initially, JGT layer does not decompose through membrane action. Similarly, JGT takes some extra load via its in-plane tensile resistance. Normal untreated JGT have moisture absorption capacity of 3–4 times their dry

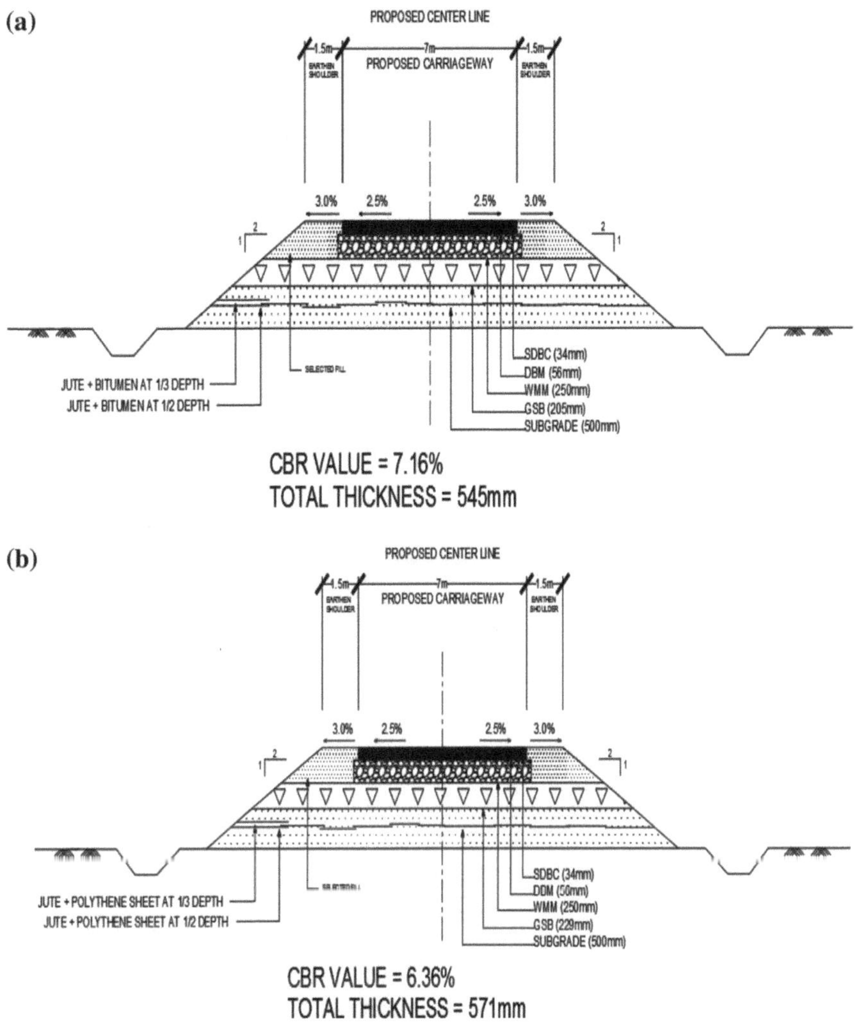

Fig. 5. Design of flexible pavements for CI soils **a** reinforced with jute + bitumen, **b** reinforced with jute + polyethylene

density, i.e. for example 1 m^2 of 626 gsm JGT is capable of absorbing about approximately 2 litres of water. However, this absorption of moisture form subgrade increases the dry density and shear strength of subgrade layer increases. This phenomenon occurs only until JGT is not fully decomposed. To increase the strength phenomenon of subgrade layer the proposed methodology may be used for weak soils. Application of JGT for road pavement is not a rare phenomenon. But its practice is not vast, compared to other alternatives. Laboratory experiments are performed to determine the analytical studies on subgrade soil regarding the strength of pavement and reduce the thickness of the subbase and base layers. From the results of experiments on

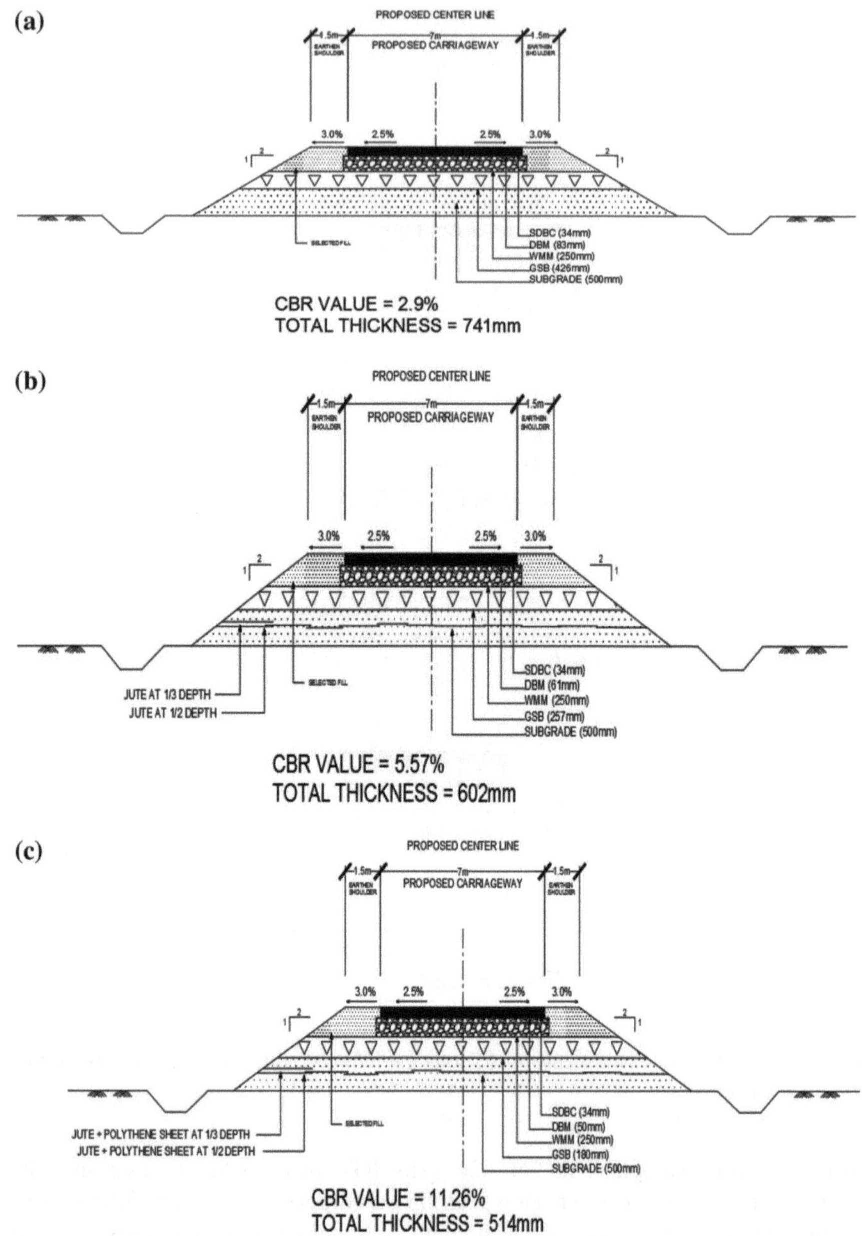

Fig. 6. Design of flexible pavements for MI soils **a** without reinforcement, **b** reinforced with jute **c**, reinforced with of jute + polyethylene

weaker soils, it is observed that UCS for reinforced soil layer using Jute for a CI shows 9.96 and MI 10.23 MPa. The ultimate stress was increased with an increase in the number of Jute layers. Natural subgrade soaks CBR value for a CI soil is 2.5%.

Table 5. Performance analysis using KENLAYER

Case	Tangential strain	Vertical compression strain
A	5.91E−04	2.40E −04
B	4.85E−04	3.81E−04
D	4.85E−04	3.81E−04
E	5.69E−04	3.28E− 04
F	5.00E−04	3.64E−04
G	3.91E−04	3.53E−04

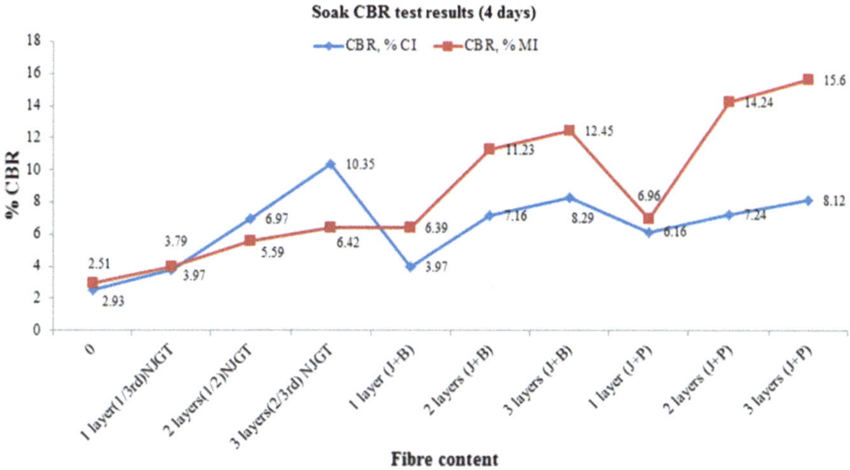

Fig 7. Soaked CBR test results. *Note* NJGT = Natural jute geotextile, J + B = Natural jute geotextile + bitumen, J + P = Natural jute geotextile + low-density polyethylene

Whereas in CI soil the ultimate stress for three layers shows 299.32 KN/m^2 and MI soil for three layers shows 319.32 kN/m^2. The untreated Jute with three layers of CI soil displays an average of 10.00% of soak CBR similarly; MI soil is 6%. Modified Jute with bitumen coating of three layers shows an average soak CBR of 8% for CI soil and 8% for jute with polyethylene sheets. No significant change was observed in soak CBR for modified Jute for a CI soil. Natural soak CBR in MI soil is 2.93%. Three layers of untreated Jute soak CBR in MI soil shows an average value of 6%, and three layers of modified jute with bitumen shows an average of 12%. Similarly, three layers of modified jute with polyethylene sheets have shown an average soak CBR of 15%. Most of the flexible pavement design requires a resilient modulus value as an input parameter. A cyclic triaxial test was conducted to determine the resilient modulus for modified Jute layers. Untreated jute with three layers shows an average value of 80 MPa for CI soil and 60 MPa for MI soil. Modified jute layer with bitumen shows maximum average value for two layers is 75 MPa. Similarly, Jute modified with polyethylene for two layers indicates an average value of 75 MPa. The proposed

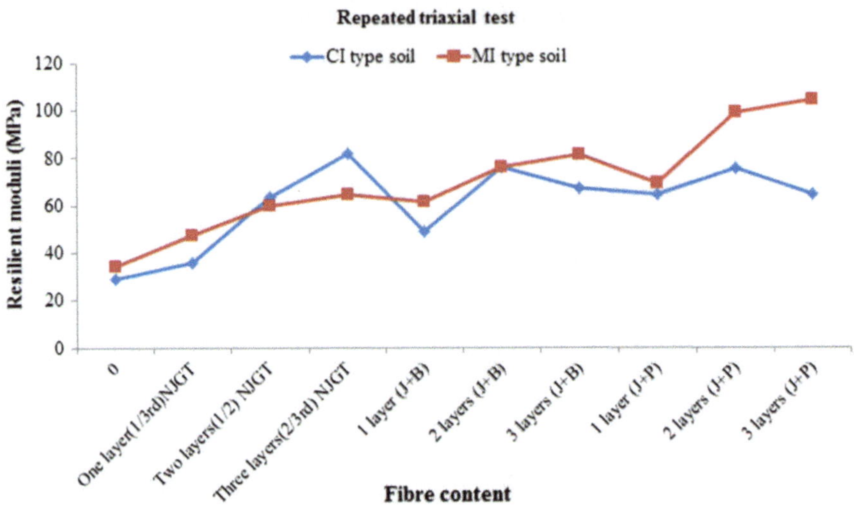

Fig. 8. Repeated triaxial test results. *Note* NJGT = Natural jute geotextile, J + B = Natural jute geotextile + bitumen, J + P = Natural jute geotextile + low-density polyethylene

cross-sectional details of pavement with and without reinforcement are shown in Figs. 4, 5, and 6.

7 Conclusion

From the experimental analysis and test results, the following conclusions are derived,

1. The naturally available material was jute bags might be used as reinforcement for weaker pavement subgrades to increase the stiffness characteristics of subgrade layer.
2. Untreated jute geotextile shows good results with jute containing three layers of CI soil with an increase in CBR and M_r Value. Treated jute geotextile shows good results with jute + Polyethylene with two layers with an increase in CBR and M_r Value.
3. Jute is biodegradable material; to increase its durability jute can be treated with bitumen spray or polyethylene sheets.
4. Even though untreated Jute Geotextile shows high CBR and resilient modulus, the modified Jute layer is preferred to increase the durability and strength of subgrade soil layer.
5. This methodology can be used for traffic intensity is less than or equal to 1–7 million standard axle loads.

References

AASHTO 307-99: Standard Test Method for Determining the Resilient Modulus of Soils and Aggregate Materials. American association of state highway and transportation officials (AASHTO), Washington, DC, USA (2003)

Al-Wahab, R.M., Heckel: Static and dynamic strength properties of a fiber-reinforced compacted cohesive soil. In: Proceeding of the 3rd International Conference on Recent Advances in Geotechnical Earthquake and Soil Dynamics, vol. II, pp. 1065–1072. St. Louis (1995)

Bijayananda, Mahipal: Permanent strain of randomly oriented fibre reinforced rural road subgrade soil under repetitive triaxial loading. In: Geo-Frontiers, ASCE, pp. 646–656 (2011)

Chandra, S., Viladkar: Mechanistic approach for fibre-reinforced flexible pavements. J. Transp. Eng ASCE, 15–23 (2008)

Dhawan, P.K., Goswami: Utilization of flyash in road construction. In: National Workshop on Utilization of Fly Ash, Roorkee, pp. 73–76 (2009)

Ghosh, S.K., Ray Gupta, K., Bhattacharya, R., Sahu, R.B., Mandol, S.: Improvement of Life Expectancy of Jute Based Needle punched Geotextiles Through Bitumen Treatment, pp. 111–121. Springer (2014)

Gosavi, M., Patil, K.A., Mittal, S., Saran, S.: Improvement of properties of black cotton soil subgrade through synthetic reinforcement. J. Inst. Eng. (India) **84**, 257–262 (2004)

IRC 37: Guidelines for the Design of Flexible Pavements. Indian Road Congress (IRC), New Delhi, India (2012)

IS 2720 Part V: The Method of Test for Soil, Laboratory Determination of Liquid and Plastic Limits. Bureau of Indian Standards, New Delhi, India (1985)

IS 2720 Part VIII: The Method of Test for Soil, Laboratory Determination of Moisture Content and Dry Density. Bureau of Indian Standards, New Delhi, India (1985)

IS 2720 Part XVI: Method of Test for Soil, Laboratory Determination of CBR. Bureau of Indian Standards, New Delhi, India (1985)

Jahan, A., Mahbubar Rahman, M., Kabir, H., Gafur, M.A.: Comparative study of physical and elastic properties of jute and glass fiber reinforced LDPE composites. Int. J. Sci. Technol. Res. **1**(10), (2012)

Kumar, Tabor: Strength characteristics of silty clay reinforced randomly oriented nylon fibre. Electron. J. Geotech. **29**(7), 55–64 (2008)

Lawton, E.C., Fox: Field experiment on soils reinforced with multi-oriented geosynthetic inclusions. Transportation Research Record, Transportation Research Board, Washington, D. C., 1369, pp. 44–53 (1992)

Lekha, Goutham, Shankar: Evaluation of lateritic soil stabilized with Arecanut coir for low volume pavements. Transp. Geotech. **2**, 20–29 (2015)

Mahipal Singh, Satyendra: Performance evaluation of silty and subgrade reinforced with fly ash and fibre. Geotext. Geomembr. **26**, 429–435 (2008)

Mohanty, Misra: M. Polym. Plas. Technol. Eng. **34**(5), 729 (1995)

Murugesan, S.: A study of fibers as a reinforcement for subgrade of flexible pavement. In: Proceedings of International Conference on Geosynthetics and Geo-environmental Engineering, Mumbai, India, pp. 163–164 (2004)

Ravishankar, A.U., Raghavan, K.S.: Coir stabilized lateritic soil for pavements. In: Proceedings of the Indian Geotechnical Conference, Ahmedabad, India, pp. 45–52 (2004)

Saride, S., Suraj: Elasto-plastic behaviour of jute-geocell reinforced sand subgrade. In Geo-Congress 2014 Technical Papers, GSP 234@ ASCE, pp. 2911–2920 (2014)

Special Report 21: State of the Art: Use of Jute Geotextiles in Road Construction and Prevention of Soil Erosion/Landslides. IRC Highway Research Board, New Delhi (2012)

Swapan Kumar, Bhattacharyya, R.: Design and engineering of an innovative bituminized jute paving fabric for potential application in road construction. Int. J. Eng. Sci. Res.Technol. 608–621 (2014)

Tang: Study on effects of sand content on strength of polypropylene fibre reinforced clay soil. Chin. J. Rock Mech. Eng. 26. Suppl. **1**(2007), 2968–2973 (2007)

Tapash Kumar: Application of jute geo-textile in the morrum-sand mixed layer as granular subbase of flexible pavement. In: Traffic and Transportation Studies, ASCE, pp. 1269–1275 (2010)

Ziari, H., Khabiri, M.M.: Interface condition influence on prediction of flexible pavement life. J. Civ. Eng. Manag. **XIII**(1), 71–76 (2007)

A Parametric Study of Deep Mixing Columns and Fibre Reinforced Load Transfer Platform Supported Embankments

Liet Chi Dang[1(✉)], Cong Chi Dang[2], and Hadi Khabbaz[1]

[1] School of Civil and Environmental Engineering, University of Technology Sydney, 2007 Ultimo, NSW, Australia
lietchi.dang@student.uts.edu.au, hadi.khabbaz@uts.edu.au

[2] Hau Giang Department of Environment and Natural Resources, 910000 Vi Thanh, Hau Giang Province, Viet Nam
chicongct1912@gmail.com

Abstract. The ground improvement technique using deep cement mixing (DCM) columns combined with geosynthetic reinforced traditional angular layer as a load transfer platform supported embankments has recently been adopted widely in construction of roads, railways, and highways over soft soils. This modification technique provides a practical and efficient measure for such infrastructure construction projects as it brings significant savings in construction costs and construction time. In this numerical study, a novel ground modification technique utilising fibre reinforced load transfer platform (FRLTP) and DCM columns supported embankment constructed on top of multilayers of soft soils is proposed and investigated based on the finite element method incorporated in PLAXIS. A series of numerical analyses was firstly carried out on the full geometry of a DCM columns supported (CS) embankment reinforced without or with an FRLTP in a two-dimensional plane strain condition. The main objective of this analysis is to examine the effectiveness of the FRLTP inclusion into the CS embankment system in terms of maximum settlement and lateral displacement. Subsequently, an extensive parametric study was conducted to further investigate the influence of the FRLTP thickness on the performance of the CS embankment by comparing the maximum and differential settlements, and the lateral displacement. The numerical results reveal that the embankment with FRLTP inclusion can effectively enhance the total settlement and the lateral displacement of the embankment. The findings of the extensive parametric study reveal that the platform thickness has significant influence on the embankment behavior, especially in improving the total and differential settlements, the rigidity and stability of the embankment, and the more load transfer from the embankment to DCM columns. Meanwhile it significantly minimises the lateral displacement and the embankment loads transferred to soft foundation soils.

H. Khabbaz et al. (eds.), *New Prospects in Geotechnical Engineering Aspects of Civil Infrastructures*, Sustainable Civil Infrastructures,
https://doi.org/10.1007/978-3-319-95771-5_15

1 Introduction

The fast development of essential infrastructure including roads and rail networks being constructed in many countries worldwide is to meet the demand of the dramatic increase in population and economic growth. As a result, many countries are suffering from lack of readily available stiff grounds in support to such transport infrastructure projects. Thus, many road and highway embankments have to be founded on soft grounds. This practice is highly risky because soft ground has low bearing capacity, insufficient shear strength and high compressibility. Therefore, in order to ensure the stability of embankment during the construction process and long term services, appropriate ground improvement techniques are needed to be adopted in enhancing the engineering properties of soft soil or even in order to transfer embankment and traffic loads to a deeper and stiffer soil stratum.

A growing number of ground modification approaches have been applied to improve soft soil properties to support embankment construction such as preloading with the vertical drain application (Liu and Rowe 2015; Parsa-Pajouh et al. 2015), excavating the existing soft ground and substituting it with high shear strength and bearing capacity backfill soil, reducing embankment load using lightweight fill materials, constructing in stages and leaving time for consolidation, improving soft ground underneath embankment by chemical treatment (Dang et al. 2016b; Fatahi and Khabbaz 2013, 2015), stone columns (Fatahi et al. 2012) and geosynthetic-reinforced and pile supported earth platform (Dang et al. 2016a; Han and Gabr 2002; Liu et al. 2007). On the top of those methods, the ground improvement technique using deep cement mixing (DCM) columns has been widely used in construction practice because it provides an economical and fast ground improvement solution for the construction of road and highway embankments over soft soil (Chai et al. 2015; Venda Oliveira et al. 2011; Yapage et al. 2014; Yapage et al. 2015). In order to improve the soft soil characteristics such as bearing capacity, shear strength and compressibility, in wet mixing method, slurry cement is mixed with soft ground at a designated water/cement ratio during the DCM columns construction process, while dry cement is mixed with in situ soils to establish artificial DCM columns in dry mixing method. However, most of road and railway embankments built on end bearing piles/columns are most likely to be subjected to large differential settlements between the embankment and the adjacent roads due to the different stiffness level (Chai et al. 2015; Liu and Rowe 2015). It is known that floating soil columns provide less stiff ground foundation, are more cost-effective and technically feasible than the end bearing columns when soft foundation soils reach greater depth. Furthermore, one of the most advantages of the floating columns solution is that when a sand layer acting like an aquifer layer is directly located below soft ground layers, the soft ground improvement solution using floating DCM columns could be the best choice for not contaminating the underground water since the floating DCM columns would leave an intact clay sub-layer below the columns tip and immediately above the aquifer. As expected, such unimproved soft clay sublayer would serve as a barrier in hindering hazardous chemicals from spreading out the deep cement mixing soil improvement area into the underground water. Nevertheless, the analysis and design of floating soil columns are complex, involve considering complicated soil-structure interaction, and there are no clear standards and guidelines readily available for the design of embankments supported on floating cement-soil columns.

Numerical modeling based on the finite element method (FEM) has been used as an effective tool in predicting the performance of DCM columns supported embankments founded on soft ground in terms of the total and differential settlements, the horizontal movement, the rate of settlement, the slope stability and the degree of consolidation over a long period of time. The finite element numerical modeling can reasonably simulate the load transfer mechanism between DCM columns and foundation soil, and the generation and dissipation of excess pore water pressure, consequently assist in predicting the total settlement, the lateral displacement, the bending moment of DCM columns with depth under the embankment using three-dimensional FEM model or even with an equivalent two-dimensional (2D) plane strain FEM model (Chai et al. 2015; Dang et al. 2016a; Tan et al. 2008; Yapage et al. 2014). Nevertheless, most of numerical investigations have been recently performed to investigate the behaviour of embankments over DCM columns (Chai et al. 2015; Yapage et al. 2014), the performance of geosynthetic reinforced traditional angular layer as a load transfer platform and piles/columns supported embankments over soft soils (Han and Gabr 2002; Liu et al. 2007; Yapage et al. 2015). However, not much attention has been paid to the behaviour of load transfer platform (LTP) reinforced with lime-fibre-soil as a replacement of geosynthetics and DCM columns supported highway embankments constructed on soft ground. According to Zhang et al. (2013), the popular applications of the geosynthetic reinforced traditional angular platform over columns supported embankments built on top of soft soils has come up with many geotechnical difficulties, namely intolerable total and differential settlements, larger lateral earth pressure and displacement, local or global instability, and potential failures due to over bearing or bending capacity of DCM columns. Therefore, a novel ground modification technique utilising eco-friendly, low cost and recycled materials as lime-soil-fibre reinforced load transfer platform (FRLTP) to be used a replacement of geosynthetic reinforced traditional angular platform layer combined with DCM columns supported embankments constructed on top of multilayers of soft soils is required to overcome those aforementioned geotechnical difficulties. Recently, a number of researchers (Dang et al. 2016a; Okyay and Dias 2010) conducted numerical investigations on the mechanical behaviour of lime-soil or lime-fibre reinforced soil layer to be adopted as a load transfer platform (LTP) positioned on top of piles supported embankments. They concluded that the application of the FRLTP reduced the embankment load transfer to soft foundation soil between adjacent piles and consequently curtailed the generation of excess pore water pressure in the soft soil stratum. In addition to that, the use of the FRLTP under embankments alleviated effectively the total and differential settlement of the embankment. Eventually, the authors indicated that the shear strength properties of the FRLTP had a remarkable influence on the load transfer mechanism between piles and surrounding soft soils as well as the embankment settlement. Nonetheless, more investigations on the mechanical performance of the columns/pile supported embankments with FRLTP inclusion are crucial to provide an insight and promote the application of eco-friendly and recycled fibre-lime reinforced soil to be used in engineering practice as a replacement of geosynthetic reinforced traditional angular LTP layer supported embankments for purpose of sustainable construction development.

This study presents a 2D numerical simulation to investigate the behavior of floating DCM columns supported embankment without or with FRLTP built on top of soft soils, using an equivalent 2D FEM model with proper modified parameters of structure and soil models. An extensive parametric study on the influential factors such as the FRLTP thickness of the FRLTP was conducted to comprehend the performance of the floating DCM columns combined with FRLTP supported embankment. Based on the numerical findings of the detailed parametric study, comparisons and comprehensive discussions on the total and differential settlements, the stress transfer mechanism between DCM columns and foundation soils with the embankment height, and the lateral displacement with depth are undertaken to have a deeper understanding of the embankment behavior with FRLTP inclusion constructed over multilayers of soft soils. This investigation aims in enhancing the design code for the floating DCM columns and FRLTP supported embankment found on multilayered soft soils.

2 Case Study

A hypothetical construction of natural fibre-reinforced load transfer platform (FRLTP) and deep cement mixing (DCM) columns supported highway embankment over soft clay layers is considered in this numerical investigation. The embankment geometry is shown in Fig. 1 representing the only right half of the domain of the embankment since the embankment is symmetrical along its centreline. As can be seen from Fig. 1, the embankment is 20.8 m wide and 6 m high with a 1 V:1.8H side slope. The embankment is made of good quality soil with a cohesion of 20 kPa, a friction angle of 35° and an average unit weight of 19 kN/m^3. It is constructed on a 1 m thick fill material as a surface layer overlaying an 11 m thick deposit of soft clay. This deposit of soft clay overlies a 3 m thick stiff clay stratum followed by a 15 m thick sand layer. The ground-water table is located at a depth of 1.0 m below the ground surface. Details of these soil layers are summarised in Table 1. A fibre-lime-soil layer as adopted in the previous study (Dang et al. 2016a) having an effective cohesion and internal friction angle of 75 kPa and 42°, respectively, an average unit weight of 12.5 kN/m^3, a Poisson's ratio of 0.32, a Young's modulus of 125.8 MPa and tensile strength of 240 kPa is used in this numerical study and serves as a fibre reinforced load transfer platform of 0.5 m that is placed on the top of DCM columns.

To improve the engineering properties such as bearing capacity, shear strength, and compressibility of the thick soft soil strata of 12 m, deep cement mixing columns with 1.2 m diameter and 10 m length are used, which yield an improvement depth ratio (β) of roughly 83%. It can be noted that the improvement depth ratio (β) is defined as the depth of the DCM columns to the depth of soft soils from the embankment base to the top of a stiffer soil layer which is the stiff clay in this numerical analysis. $\beta = 0$ implies that soft soils is not improves with DCM columns, whereas $\beta = 1$ means that soft soils is fully improved with DCM columns. With the improvement depth ratio of $\beta = 0.83$ chosen in this case, the DCM columns are considered as floating columns supported

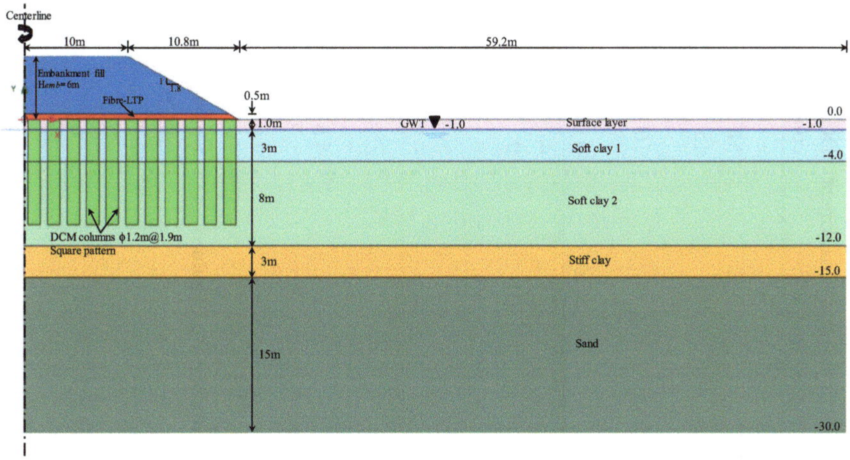

Fig. 1. Cross section of the fibre reinforced load transfer platform and DCM columns supported embankment

embankment. In addition, the DCM columns are arranged in a square grid pattern with centre-to-centre spacing of 1.9 m, which results in an area replacement ratio (α) of approximately 31% corresponding to those aforementioned geometric properties. It is noted that for a hypothetical embankment modelling, the DCM column parameters were selected to represent the typical soil-cement columns properties from published data available in the literature. Therefore, the unconfined compressive strength (q_{u}) was assumed to be 1000 kPa (Chai et al. 2015) and the 100 MPa young modulus (E) of DCM columns was determined using the correlation E = 100 q_u (Yapage et al. 2014). The 150 kPa tensile strength of the DCM columns was considered to 0.15 q_u (CDIT 2002; Jamsawang et al. 2016) and the undrained shear strength of 500 kPa of DCM columns was assumed to be s_u = 1/2 q_u (Filz and Navin 2006). The average unit weight of 15 kN/m^3 was considered to be in range of 3–15% higher than that of soft soils (CDIT 2002) and a Poisson's ratio of 0.15 (Chai et al. 2015) was selected as of the typical propeties of DCM columns.

The construction sequence of the embankment is assumed to be in 0.5–1.0 m lifts at average filling rate of 0.06 m/day to a total height of 6 m including the FRLTP with a height of 0.5 m. Following the completion of embankment construction, the consolidation period is left for 2 years. Table 2, exhibiting the simulated construction sequence of the highway embankment, has been adopted in this numerical analysis.

3 Numerical Modeling

3.1 Finite Element Models and Parameters

A two-dimensional plane strain model was built using commercial finite element software PLAXIS 2D version 2015 adopting the equivalent 2D numerical analysis method proposed by previous researchers (Chai et al. 2015; Dang et al. 2016a; Okyay

Table 1. Material properties of the embankment, FRLTP, DCM columns and subgrade soil layers

Parameters	Surface layer	Soft clay 1	Soft clay 2	Stiff clay	Sandy clay	Fibre-lime-soil	Embankment fill	DCM columns
Depth (m)	0–1	1–4	4–12	12–15	15–30	–	–	–
Material model	MCC	MCC	MCC	MCC	MC	MC	MC	MC
Unit weight γ (kN/m^3)	16	13.4	14.3	18	19	12.5	19	15
Young's modulus, E (MPa)	–	–	–	–	20	125.8	1	100
Poisson's ratio, ν	0.15	0.15	0.15	0.15	0.10	0.32	0.40	0.15
Effective cohesion, c′ (kPa)	–	–	–	–	20	75	20	$c_u = 500$
Effective friction angle, $\varphi'\varphi'$ (°)	–	–	–	–	35	42	35	0
Compression index, λ	0.25	0.87	0.43	0.12	–	–	–	–
Swelling index, κ	0.025	0.087	0.043	0.012	–	–	–	–
Over consolidation ratio, OCR	1.5	2.5	1.2	1.0	–	–	–	–
Slope of the critical state line (M)	1.2	1.2	1.2	1.4	–	–	–	–
Initial void ratio (e_0)	1.5	3.1	2.49	0.8	0.7	–	–	–
Vertical Permeability coefficient, k_v (m/day)	6×10^{-4}	4.4×10^{-4}	4.6×10^{-4}	2.5×10^{-3}	2.5×10^{-2}	–	–	4.6×10^{-4}
Horizontal Permeability coefficient, k_h (m/day)	9.1×10^{-4}	6.6×10^{-4}	6.9×10^{-4}	2.5×10^{-3}	2.5×10^{-2}	–	–	4.6×10^{-4}
Material behaviour	Undrained	Undrained	Undrained	Undrained	Drained	Undrained	Drained	Undrained type B

MC mohr-coulomb; *MCC* modified cam clay

Table 2. Construction stages in the FEM simulation of embankment construction procedure

Stage	Description	Thickness (m)	Duration (days)
1	Generation on the initial stresses (Ko–condition)	–	–
2	Installation of the DCM columns	–	–
3	Construction of a 0.5 m high embankment	0.5	8
4	Construction of a 1.0 m high embankment	0.5	8
5	Construction of a 2.0 m high embankment	1.0	16
6	Construction of a 3.0 m high embankment	1.0	16
7	Construction of a 4.0 m high embankment	1.0	16
8	Construction of a 5.0 m high embankment	1.0	16
9	Construction of a 6.0 m high embankment	1.0	20
10	Consolidation period of 2 years	–	730

and Dias 2010; Tan et al. 2008) to simulate the performance of the natural fibre reinforced LTP and DCM columns supported highway embankment. The equivalent 2D model was selected because of less analysis time consuming, while generating results with reasonable accuracy. The DCM columns were simulated by continuous plane strain walls of 0.60 m thickness for the entire columns length of 10 m to maintain the same area of replacement ratio of approximately 31%, taking into account of the equivalent normal stiffness (EA) as implemented and recommended for numerical simulations of columns supported embankments by many researchers (Chai et al. 2015; Dang et al. 2018; Yapage et al. 2015), while the centre to centre spacing between two adjacent walls in this numerical simulation was remained the same as the 1.9 m centre to centre spacing between two adjacent DCM columns.

With regard to the constitutive modeling, the DCM columns, as mentioned previously, were modeled as a linear elastic-perfectly plastic material using Mohr-Coulomb (MC) model (Huang and Han 2010; Liu and Rowe 2015). Similarly, the fibre-lime reinforced load transfer platform, embankment and fill material were simulated using a linear elastic-perfectly plastic model with Mohr-Coulomb failure criterion. The Mohr-Coulomb material model requires the Young's modulus (E), Poisson's ratio (v), the effective cohesion (c'), the angle of internal friction (φ'), the dilation angle (ψ) and the tensile strength. Subsequently, the soft soil layers were represented by Modified Cam Clay (MCC) model. The required parameters for the MCC model are slope of the virgin consolidation line (λ), the slope of swelling line (κ), the initial void ratio (e_0), the slope of the critical state line (M), and Poisson's ratio (ν). It is assumed that the values of horizontal permeability (k_h) are about 1.5 times the corresponding values of vertical permeability (k_v) of the subgrade soils, whereas the horizontal and vertical permeability of sand and DCM columns are equal. A summary of the constitutive model parameters is presented in Table 1. During consolidation process due to an increase in embankment load, the hydraulic permeability was changed attributable to the relationship between the void ratio change and the corresponding embankment load; therefore, the permeable change index $C_k = 0.5e_0$ was adopted in this investigation.

Referring to Fig. 2, only right half of the embankment is represented in this numerical simulation since the embankment is symmetrical along its centreline. The foundation soil was taken to 30 m depth from the ground surface overlying a stiff clay stratum. The horizontal length of the FEM model was taken to be 80 m, which was almost three times the half width of the embankment base, in order to minimize the boundary effect. Both the left and right boundaries were considered to be impermeable, meanwhile pore fluid flow was permitted from both the ground surface and the bottom boundaries.

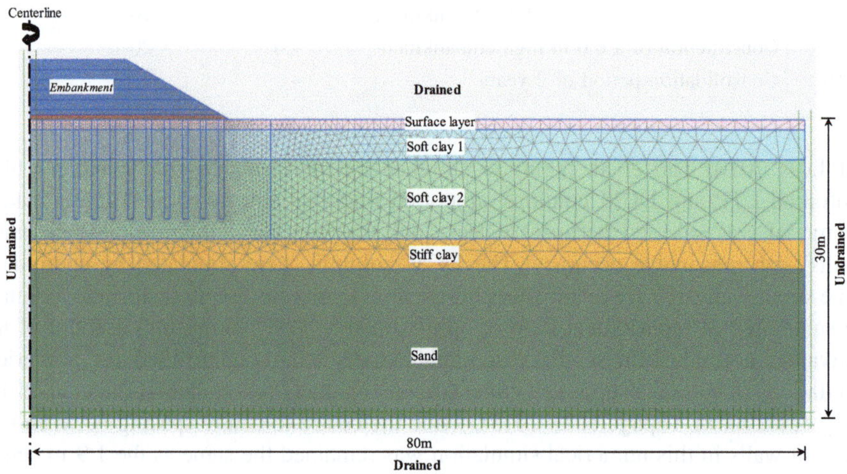

Fig. 2. Mesh and boundary conditions for a 2D FEM analysis of embankment

In this numerical analysis, for the 2D plane strain FEM model, the horizontal displacement at the left and right boundaries was not permitted but vertical movement was allowed, whereas both the vertical and horizontal displacements were prevented at the bottom boundary. On one hand, due to the relatively high permeability, the embankment fill and sand stratum were assumed to behave in a drained condition. On the other hand, natural fibre reinforced load transfer platform, DCM columns, and other four foundation soils were assumed to act as undrained materials. In this numerical analysis, fifteen-node triangular elements with excess pore water pressure degrees of freedom at all nodes were adopted to simulate the foundation soils, DCM columns and FRLTP, while fifteen nodes triangular elements without excess pore water pressure degrees of freedom at all nodes were applied to model the embankment fill and sand material.

4 Analysis of Results and Discussion

4.1 Influence of FRLTP Inclusion on the Maximum Settlement and the Lateral Displacement

The influence of the FRLTP inclusion on the behavior of a DCM columns supported embankment was assessed by comparing the maximum total settlement at the base of the embankment centre and the maximum lateral displacement of a column head under the embankment toe with an increase in the embankment height from 0.5 to 6.0 m as illustrated in Figs. 3 and 4, respectively. In overall, the inclusion of FRLTP combined with DCM columns considerably reduced the maximum settlement and the lateral deformation of the embankment. As observed in Fig. 3, the maximum settlement of the ground surface under the embankment without or with FRLTP increased linearly with increasing the embankment height. However, the maximum settlement of the embankment without FRLTP was obviously larger than that of the embankment reinforced with an FRLTP. Moreover, it can be seen from Fig. 3, the difference in the magnitude of the maximum settlement between the embankment without FRLTP and with FRLTP appeared to increase gradually with an increase in the height of the embankment fill. For example, during the embankment construction, the maximum settlement increased from about less than 0.01–0.5 m for the embankment without FRLTP as the embankment height increased from 0.5 to 6 m, meanwhile the corresponding increase in the maximum settlement of the embankment reinforced with an FRLTP was from less than 0.01–0.4 m. In comparison with the embankment without FRLTP, the maximum settlement was curtailed by 20% for the embankment reinforced with an FRLTP, having its thickness of 0.5 m, at the completion of the embankment

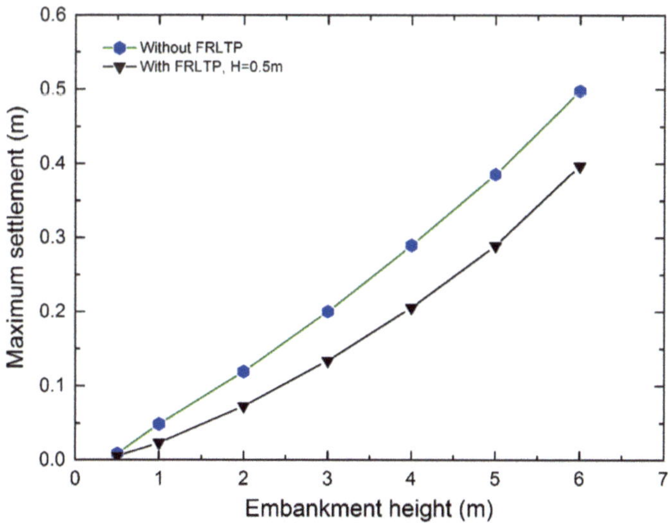

Fig. 3. Variation of the maximum settlement at the ground surface under the embankment centre

construction. The improvement in the total settlement demonstrates the effectiveness of the FRLTP inclusion into the DCM columns-embankment system. Referring to Dang et al. (2018), the improvement in the vertical settlement could be attributable to enhancing the stiffness and the arching effect of the entire embankment system contributed from the FRLTP inclusion.

Similar to the trend of the maximum settlement, as presented in Fig. 4, the lateral displacement at the columns head under the embankment toe increased with an increase in the embankment height. It is of interest to note that a nonlinear relationship between the lateral displacement and the embankment height was clearly depicted as the predominant mechanism for the DCM columns supported embankment either without or with FRLTP. However, as observed in Fig. 4, the inclusion of the FRLTP into the embankment resulted in the significant reduction in the lateral displacement. For examples, at the completion of the embankment construction, the maximum lateral displacement dramatically reduced from 117 mm for the embankment without FRLTP to 63 mm for the embankment reinforced with an FRLTP. As a result, the reduction in the lateral displacement was approximately 46% for the embankment with FRLTP. In addition, as evident in Fig. 4, the lateral displacement of the embankment with FRLTP initially remained almost zero (or marginally increased) when the embankment height increased from 0.5 to 3 m and then started to increase with further increase in the embankment height beyond 3 m. Whereas, the gradual increase in the lateral displacement of the embankment without FRLTP was visually observed from a small increase in the embankment height of 0.5 m to the medium embankment height of 3 m and afterward the dramatic increase in the lateral deformation was seen in Fig. 4 with further increase in the embankment height exceeded 3 m. Therefore, it can be concluded that the introduction of the FRLTP combined with DCM columns supported embankment can effectively alleviate the lateral displacement and consequently enhance the stability of the entire embankment system built over multilayers of soft soils.

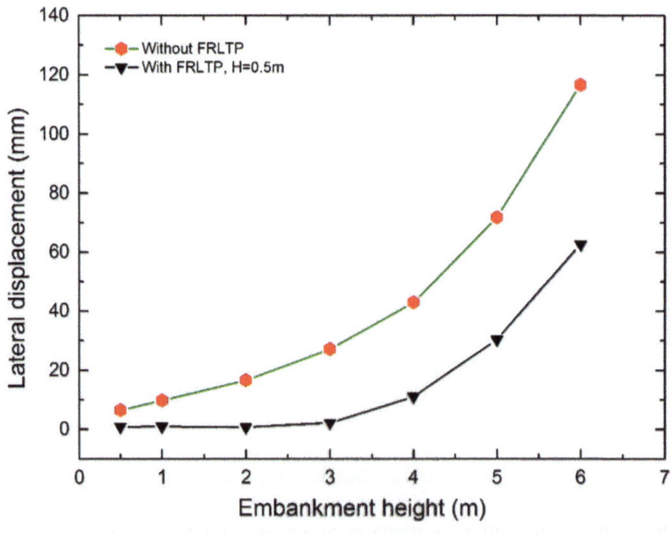

Fig. 4. Variation of lateral displacement at the embankment toe

4.2 Variation of the Total and Differential Settlements for Various FRLTP Thickness (H)

The thickness of the FRLTP plays an important role in facilitating the embankment load transferred efficiently to the DCM columns, meanwhile reducing the embankment load transferred to soft foundation soil between DCM columns as well as alleviating the possibility of the FRLTP punching failure induced by column heads. To investigate the effect of various FRLTP thickness on the total and differential settlement of the floating DCM columns supported embankment, the FRLTP thickness (H) was changed from 0.0 m for the embankment without FRLTP to 3.0 m for the embankment reinforced with an FRLPT. Figure 5 shows the development of the ground surface settlement with elapsed time between two adjacent DCM columns, under the embankment centre for various FRLTP thickness (H). As visually illustrated in Fig. 5, the ground surface settlement with time under the embankment increased significantly during the first period of 100 days due to the increase in embankment load followed by a gradual increase in the surface settlement up to two years after the completion of the embankment construction due to the evolution of consolidation with elapsed time. However, it is observed in Fig. 5 that there was significant reduction in the ground surface settlement under the embankment when the FRLTP thickness increased from H = 0.0 to H = 3.0 m. For example, the surface settlements of approximately 0.52 m were numerically predicted at the end of the 2 years post-construction period for the FRLTP thickness of H = 3.0 m, which resulted in the considerable reduction of about 32% in comparison with the total settlement of the embankment without FRLTP (or the FRLTP thickness of H = 0.0 m). This phenomenon indicates the beneficial effect of the FRLTP inclusion in diminishing the final surface settlement of the entire DCM

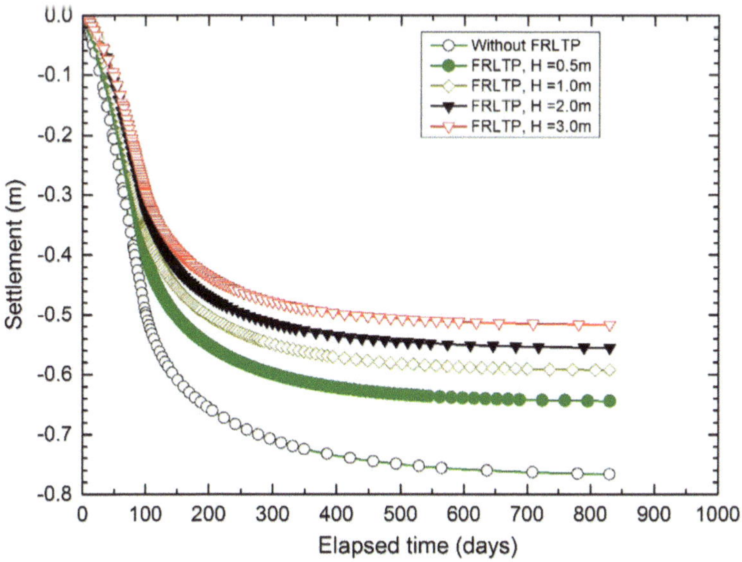

Fig. 5. Variation of total settlement with time for different FRLTP thickness

columns-embankment system. In addition, the predicted settlement results exhibit that the embankment with FRLTP not only reduced the final surface settlement, but also accelerated the consolidation process of soft soil layers. As evident in Fig. 5, the maximum total settlement under the base of the embankment with an FRLTP thickness of 0.5 m reached 95% of its total settlement after an investigation period of 335 days, meanwhile the corresponding total settlement of the embankment without FRLTP was numerically predicted to achieve after approximately 385 days of elapsed time. This reconfirms the FRLTP inclusion in the embankment was certainly effective in accelerating the consolidation process of underlying soft soils subjected to the embankment loading. However, it is found the improvement in the consolidation process of soft soils with further increasing in the FRLTP thickness up to H = 3 m appeared to be insignificant.

Figure 6 illustrates fluctuation of the differential settlement at the ground surface under the centre of the embankment base with respect to the embankment height for different FRLTP thickness (H). It should be noted that the differential settlement is defined as the settlement difference between the top of DCM columns and the ground surface at the embankment base. As shown in Fig. 6, during the embankment construction, the differential settlement at the ground surface increased with increasing the height of the embankment fill but decreased with an increase in the thickness of FRLTP from H = 0 m (the embankment without FRLTP) to H = 3.0 m. For example, As observed in Fig. 6, during the embankment construction, the differential settlement of the embankment without FRLTP (FRLTP thickness of H = 0.0 m) linearly increased from 1.5 to 125 mm as the embankment height increased from 0.5 to 6 m. However, when the FRLTP thickness slightly increased to H = 0.5 m, the corresponding increase in the differential settlement was ranging from about 0.1 to 33 mm, resulting in the

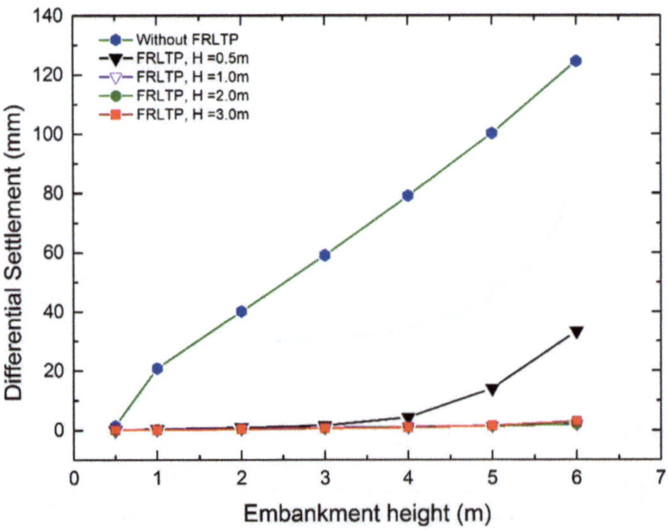

Fig. 6. Variation of differential settlement with embankment height for various FRLTP thickness

differential settlement was reduced by approximately 73% at the completion of the embankment construction. The reduction in the differential settlement confirms the highly positive effect of the FRLTP inclusion on the embankment behavior. Another 0.5 m increment of the FRLTP thickness to H = 1.0 m resulted in the considerable decrease in the maximum differential settlement during the embankment construction down to almost zero, meanwhile further increase in the FRLTP thickness up to 3.0 m revealed the zero differential settlement at the ground surface as illustrated in Fig. 6. Hence, it can be concluded that the influence of the FRLTP thickness on the differential settlement was significant. The differential settlement under the embankment base significantly enhanced with the increase in the FRLTP thickness could be attributed to effective contributions from the high shear strength, the high tensile strength and the high stiffness of the FRLTP inclusion. Therefore, the higher thickness of FRLTP, the higher strength and the greater stiffness of the FRLTP will be, which efficiently promote the load transfer mechanism between DCM columns and soft foundation soil, as a consequence, alleviate the differential settlement of the embankment or even totally prevent the development of the differential settlement during the embankment construction.

4.3 Variation of Lateral Displacement with Depth for Various FRLTP Thickness

As shown in Fig. 7, the lateral displacement of the columns head under the embankment toe at the two years post-construction significantly decreased with an initial increase in the FRLTP thickness from 0 to 1.0 m followed by a slight increase in the lateral displacement of the columns head but a gradual decrease in the lateral displacement at the columns tip. As expected, the lateral displacement of the embankment without FRLTP (FRLTP thickness of H = 0 m) was larger than those of the embankment reinforced with an FRLPT (H = 0.5–3.0 m), which confirms the FRLTP inclusion was effective in controlling the lateral deformation and consequently increasing the stability of the embankment. As observed in Fig. 7, it should be noted that at the shallow depth, the effect of the FRLTP was significant in improving the lateral resistance to the embankment deformation. For example, the difference in the lateral displacement was visually observed to a depth of around 4–5 m from the ground surface. The lateral displacement reduced by almost 6 times as the FRLTP thickness from H = 0 to H = 1.0 m (equivalent to about 83% reduction in the lateral displacement). However, it can be seen from Fig. 7 that the lateral displacement linearly increased with greater depth (i.e., 5–10 m) but decreased with the increase in the FRLTP thickness from 0 to 3 m. The difference in the lateral displacement was insignificant for the greater depth between 5 and 10 m compared to those of the shallow depth of less than 5 m. The decrease in the lateral movement observed at the greater depth as the FRLTP thickness increased up to 3 m could be mainly due the replacement of the embankment fill by the lighter weight material of lime-fibre-soil, which resulted in reduction of the lateral earth pressure applied on the DCM columns with the greater depth where the lateral resistance effect of the FRLTP inclusion was insignificant. As evident in Fig. 7, the FRLTP thickness has notable influence on the final (post-construction) lateral displacement of the embankment.

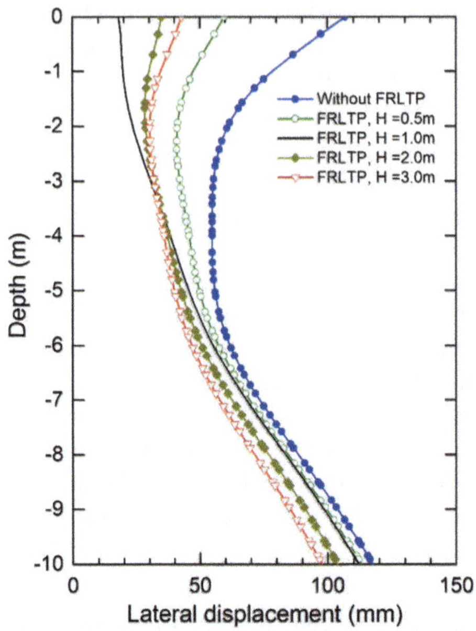

Fig. 7. Variation of lateral displacement with depth for various FRLTP thickness at 2 years post-construction

5 Conclusions

A fibre reinforced load transfer platform (FRLTP) and deep cement mixing (DCM) columns supported embankment over soft soils were simulated in this numerical study. An equivalent two-dimensional (2D) plane strain finite element method (FEM) model was used to evaluate the behaviour of the embankment without or with FRLTP. An extensive parametric study on the full geometry of the DCM columns supported embankment was also numerically carried out to further investigate the influence of the FRLTP parameters such as the platform thickness on the performance of the embankment during construction and post-construction periods. The predicted results were analysed and discussed in detail. The main findings of this study can be summarized as follows:

- Comparing with the DCM columns supported embankment without FRLTP, the numerical results indicated that the application of the FRLTP combined with DCM columns supported embankment can minimize the total settlement significantly, meanwhile effectively alleviate the lateral displacement, and consequently increase the stability of the entire embankment system built over multilayers of soft soils.
- The findings of the parametric study confirm that the thickness of the FRLTP had significant effects on the total and differential settlements, and the lateral displacement of the embankment via the effective enhancement the load transfer mechanism between DCM columns and foundation soils.

- More importantly, this numerical investigation has explored an interesting potential for making use of agricultural waste by-products such as coconut coir fibre, bagasse fibre and jute fibre as construction fill materials for sustainable development in the area of civil infrastructure foundations.
- Further investigation is required to assess the efficiency and the interaction between DCM columns, FRLTP and soft foundation soils varied with the other properties of FRLTP (e.g. the stiffness, the tensile strength and the shear strength parameters), the columns stiffness, the columns distance ratio, and the improvement depth ratio of soft soils during embankment construction and serviceability.

References

CDIT: The Deep Mixing Method: Principle Design and Construction. CRC Press/Balkema, The Netherlands (2002)

Chai, J.-C., Shrestha, S., Hino, T., Ding, W.-Q., Kamo, Y., Carter, J.: 2D and 3D analyses of an embankment on clay improved by soil–cement columns. Comput. Geotech. **68**, 28–37 (2015)

Dang, L.C., Dang, C., Fatahi, B., Khabbaz, H.: Numerical assessment of fibre inclusion in a load transfer platform for pile-supported embankments over soft soil. In: Chen D., Lee J., Steyn W. J. (eds.) Geo-China 2016, vol. GSP 266, ASCE, pp. 148–55 (2016a)

Dang, L.C., Dang, C.C., Khabbaz, H.: Numerical analysis on the performance of fibre reinforced load transfer platform and deep mixing columns supported embankments. In: Bouassida, M., Meguid, M.A. (eds.) Ground Improvement and Earth Structures, pp. 157–169. Springer, Cham (2018)

Dang, L.C., Hasan, H., Fatahi, B., Jones, R., Khabbaz, H.: Enhancing the engineering properties of expansive soil using bagasse ash and hydrated lime. Int. J. Geomate **11**(25), 2447–2454 (2016b)

Fatahi, B., Basack, S., Premananda, S., Khabbaz, H.: Settlement prediction and back analysis of Young's modulus and dilation angle of stone columns. Australian J. Civil Eng. **10**(1), 67 (2012)

Fatahi, B., Khabbaz, H.: Influence of fly ash and quicklime addition on behaviour of municipal solid wastes. J. Soils Sediments **13**(7), 1201–1212 (2013)

Fatahi, B., Khabbaz, H.: Influence of chemical stabilisation on permeability of municipal solid wastes. Geotech. Geol. Eng. **33**(3), 455–466 (2015)

Filz, G.M., Navin, M.P.: Stability of column-supported embankments. Virginia Transportation Research Council, Charlottesville, VA., Report No. VTRC 06-CR13 (2006)

Han, J., Gabr, M.: Numerical analysis of geosynthetic-reinforced and pile-supported earth platforms over soft soil. J. Geotech. Geoenviron. Eng. **128**(1), 44–53 (2002)

Huang, J., Han, J.: Two-dimensional parametric study of geosynthetic-reinforced column-supported embankments by coupled hydraulic and mechanical modeling. Comput. Geotech. **37**(5), 638–648 (2010)

Jamsawang, P., Yoobanpot, N., Thanasisathit, N., Voottipruex, P., Jongpradist, P.: Three-dimensional numerical analysis of a DCM column-supported highway embankment. Comput. Geotech. **72**, 42–56 (2016)

Liu, H., Ng, C., Fei, K.: Performance of a geogrid-reinforced and pile-supported highway embankment over soft clay: Case study. J. Geotech. Geoenviron. Eng. **133**(12), 1483–1493 (2007)

Liu, K.W., Rowe, R.K.: Numerical modelling of prefabricated vertical drains and surcharge on reinforced floating column-supported embankment behaviour. Geotext. Geomembr. **43**(6), 493–505 (2015)

Okyay, U.S., Dias, D.: Use of lime and cement treated soils as pile supported load transfer platform. Eng. Geol. **114**(1–2), 34–44 (2010)

Parsa-Pajouh, A., Fatahi, B., Khabbaz, H.: Experimental and numerical investigations to evaluate two-dimensional modeling of vertical drain-assisted preloading. Int. J. Geomech. **16**(1), B4015003 (2015)

Tan, S., Tjahyono, S., Oo, K.: Simplified plane-strain modeling of stone-column reinforced ground. J. Geotech. Geoenviron. Eng. **134**(2), 185–194 (2008)

Venda Oliveira, P.J., Pinheiro, J.L.P., Correia, A.A.S.: Numerical analysis of an embankment built on soft soil reinforced with deep mixing columns: parametric study. Comput. Geotech. **38**(4), 566–576 (2011)

Yapage, N., Liyanapathirana, D., Kelly, R., Poulos, H., Leo, C.: Numerical modeling of an embankment over soft ground improved with deep cement mixed columns: case history. J. Geotech. Geoenviron. Eng. **140**(11), 04014062 (2014)

Yapage, N., Liyanapathirana, D., Poulos, H., Kelly, R., Leo, C.: Numerical modeling of geotextile-reinforced embankments over deep cement mixed columns incorporating strain-softening behavior of columns. Int. J. Geomech. **15**(2), 04014047 (2015)

Zhang, J., Zheng, J.-J., Chen, B.-G., Yin, J.-H.: Coupled mechanical and hydraulic modeling of a geosynthetic-reinforced and pile-supported embankment. Comput. Geotech. **52**, 28–37 (2013)

Flexural Behavior of RC Two Way Slabs Made With Crushed Melted Bricks as Coarse Aggregate

Ihab S. Saleh(✉), Saddam Kh Faleh, and Aqeel H. Chkheiwer

Civil Engineering Department, Engineering College, University of Basrah, Basra, Iraq
{dr.ihabsabri, saddam_alzuwair, aqeelcivil}@yahoo.com

Abstract. This paper provides the results of an experimental investigation into the use of crushed melted clay brick waste as coarse aggregate in producing concrete. Fast development in the structural concrete constructions has increased; therefore the high consumption of gravel as a coarse aggregate in concrete works has increased in alarming rate. Furthermore, the growth of the brick industry has also produced lots of product mainly the waste materials affected on the environment, so that the production of concrete with crushed melted brick as a suitable aggregate will face the countries in the coming few years and affect its sustainability and economy. In order to fulfil the objective of this investigation, an array of tests were conducted to determine concrete performance, including the compressive strength tests, the flexural tests and the tests to determine the modulus of elasticity. The data were collected and analyzed to satisfy the study on melted brick waste as an alternative in coarse aggregate as a partial or full replacement in concrete production. To study the effect of using the melted brick on structural behavior of concrete, five reinforced concrete slabs were tested, the key variable was, the replacement ratio of crushed melted brick, ranging from 25 to 100% of the total coarse aggregate, the mix proportions and W/C ratio were constant. Behavior of slabs under loading was observed, cracking and ultimate strength and deflections were recorded. The results reveal that the mixes with melted crushed brick aggregate show decrease in concrete compressive strength about 3–13%, flexural strength about 11–33% and modulus of elasticity about 10–15% than gravel concrete, at age of 28 days depending on the ratio of crushed melted brick replacement. Slab with replacement ratio 25% exhibit ultimate load 95% the ultimate load of reference slab made with gravel. The results also indicate that the ultimate load of full melted brick waste concrete has achieved 80% of ultimate load of reference slab.

1 Introduction

The economic and environmental gain is very important in this study by recycling the waste (melting brick) resulting from manufacturing process of clay brick and then maintain the resources of natural gravel and get cheapest and moderate density concrete used in slabs because concrete is basically a mixture of two components: aggregates

H. Khabbaz et al. (eds.), *New Prospects in Geotechnical Engineering Aspects of Civil Infrastructures*, Sustainable Civil Infrastructures,
https://doi.org/10.1007/978-3-319-95771-5_16

and paste. The paste comprised of Portland cement and water, binds the aggregates (usually sand and gravel or crushed stone) and then the aggregate occupy about 75% of the absolute volume of concrete (PCA-Chap09 2002).

Kallak (2009) was carried out to study the feasibility of using crushed bricks to substitute the coarse aggregate (gravel) in concrete. Two types of concrete mixing are prepared. The first one was a mixture of 1:2:4 without crushed bricks and is used as a reference mixture in the equivalent strength of grades 30 MPa. The second one was made of different weight of crushed bricks (as a percentage from the weight of the coarse aggregate 0, 25, 50, 75 and 100%). compression and split tension as per relevant to British standard specifications were conducted. Test results indicated that using crushed bricks reduces the strength of concrete about 11–87% at age of 28 day for replacement ratio 25 and 100% respectively. Also, the percentage of water to cement ratio increases for constant slump when the percentage of crushed bricks increased due to high absorption characteristics of crushed brick.

Reddy (2012) investigated the influence of recycled aggregate on the structural behaviour of two way slabs under punching shear and flexural loading. The designed strength of concrete for 28 days is 25 MPa. Six concrete mix proportions were used with 0, 20, 40, 60, 80 and 100% replacement of natural coarse aggregate with recycled coarse aggregate. Total 18 number of 150 mm standard cubes, 36 numbers of 150 mm diameter and 300 mm height standard cylinders and 72 number of 1100 × 1100 50 mm square slabs had been cast. The results showed that the compressive and splitting strengths of recycled coarse aggregate concrete (replacing of natural coarse aggregate with recycled coarse aggregate from 20 to 100%) is decreased about 2–23% and 3–18% than natural coarse aggregate concrete respectively. The first crack load of simply supported recycled from 20 to 100% coarse aggregate concrete slab specimen under flexure decreases by 0.00–19.23%, when compared with natural coarse aggregate concrete slab specimen. The flexural ultimate load of simply supported natural coarse aggregate concrete slab specimen under flexure is 186 kN. The ultimate load of simply supported recycled coarse aggregate concrete slab specimens is in the range of 184–166 kN. The central deflection in flexure of simply supported recycled coarse aggregate concrete slab specimens was increased when compared with natural coarse aggregate concrete specimens. From the results of experimentation, it was concluded that replacement of natural coarse aggregate with recycled coarse aggregate up to 40% is quite suitable.

Khudair (1987) conducted an investigation on the properties and structural behavior of concrete made with Abu-Ghar crushed lime stone as coarse aggregate. The research dealt with aggregate properties, testing fresh and hardened concrete and testing R.C beams. Two types of mixes were made, one by using the crushed limestone as coarse aggregate, other by using gravel. Slump and vebe time tests were carried out on fresh concrete, while compressive, flexural, and splitting tensile strengths, modulus of elasticity, and linear change tests were carried out on hardened concrete at different ages. It was concluded that the concrete with crushed limestone generally show lower workability and modulus of elasticity than gravel concrete, while for same mix proportions, limestone concrete generally, show higher compressive, flexural and splitting tensile strengths than gravel concrete at all ages. In addition, the analysis of test results of R.C beams were exhibit that beams with lime stone give almost the same ultimate moment strength as that of beams with gravel concrete, also limestone beams show more deflection than gravel concrete beams.

The purpose of this study is to give information on the results of a series of tests aimed at investigating the effect of use crushed melted brick as a coarse aggregate by replacement ratios of 25, 50, 75, 100% of gravel aggregate in concrete mix on the properties of harden concrete. The structural behavior, failure mode, crack pattern and load deflection response of slabs failed by flexure were also studied to provide a thorough assessment of failure mode and overall response. Also the paper is aimed to achieve a low cost with responsible density concrete about 2000 kg/m^3, and then decrease the loads transmitted to the soil beneath the foundations.

2 Experimental Program

The experimental program consists of preparing and testing 5 reinforced concrete slabs, first slab was a reference slab with natural gravel concrete and four slabs were made with crushed melted bricks as coarse aggregate. These slabs were made with 5 different replacement crushed melted bricks 0, 25, 50, 75, 100% of natural aggregate, the symbols of tested slabs were S0, S25, S50, S75 and S100 respectively.

2.1 Materials

The choice of materials like cement and aggregate was based on locality, convenience, availability and consistence of supply. Optimum proportions must be selected according to the mix design method, considering the characteristics of all used materials.

2.1.1 Cement

The cement used in this study is Iraqi ordinary Portland cement Type (I). It is stored in airtight plastic containers to avoid exposure to different atmospheric conditions. This cement is tested and checked according to the Iraqi Standard Specification (Iraqi Specification No.5:1984 2004). Tables 1 and 2 are illustrate the chemical and physical properties of this cement. It conforms to the Iraqi specifications.

2.1.2 Fine Aggregate

The grading and amount of fine aggregate are important factors in the production of concrete. Natural sand from Al-Zubair region in Basrah city was used. Table 3 shows the grading of fine aggregate and the limits of the (Iraqi Specification No.45:1984 2004). Table 4 shows the physical properties of the used fine aggregate. The fine aggregate was within zone 2 of (Iraqi specification No.45:1984 2004). Fineness modulus was 2.6.

2.1.3 Coarse Aggregate

Crushed gravel from Sanam mountain region in Basrah city with size of 5–20 mm was used. The grading of the coarse aggregate is shown in Table 5. The results indicate that, the coarse aggregate grading is within the requirements of (Iraqi specification

Table 1. Chemical composition of the cement

Compound composition	Chemical composition	Percentage by weight	Limits of IOS No. 5:1984 (2004)
Lime	CaO	62.00	–
Silica	SiO_2	21.00	–
Alumina	Al_2O_3	5.26	–
Iron oxide	Fe_2O_3	3.00	–
Magnesia	MgO	2.70	<5
Sulfate	SO_3	2.10	<2.8
Loss on ignition	L.O.I	1.10	<4
Insoluble residue	I.R	0.49	<1.5
Lime saturation factor	L.S.F	0.92	0.66–1.02
Main Compounds (Bogue's equation) percentage by weight of cement			
Tricalcium silicate (C3S)		47.11	
Dicalcium silicate (C2S)		30.81	
Tricalcium aluminate (C3A)		8.87	
Tetracalcium aluminoferrite (C4AF)		9.12	

Table 2. Physical properties of the cement

Physical Properties	Test result	Limit of IOS No. 5:1984 (2004)
Fineness using Blaine air permeability apparatus (m^2/kg)	312	>230
Setting time using Vicat's instruments Initial (hrs:min.) Final (hrs:min)	2:10 4:00	>45 min <10 h
Compressive strength 3 days (MPa) 7 days (MPa)	20.5 29.0	>15 >23

Table 3. Grading of the fine aggregate

Sieve size (mm)	Passing (%) by weight	
	Fine aggregate	Limits of IOS No. 45/1984-Zone 2 (2004)
10	100	100
4.75	99	90–100
2.36	90	75–100
1.18	75	55–90
0.60	53	35–59
0.30	17	8–30
0.15	2	0–10

Table 4. Properties of the fine aggregate

Physical properties	Test results	Limits of IOSNo.45/1984 (2004)
Specific gravity	2.65	–
Sulfate content (SO_3) (%)	0.33	≤ 0.5
Absorption (%)	1.5	–
Loose bulk density (kg/m^3)	1645	–

Table 5. Grading of the coarse aggregate

Sieve size (mm)	Passing (%) by weight	
	Coarse aggregate	Limits of IOS No.45/1984 (2004)
20	100	95–100
14	80	–
10	37	30–60
5	2	0–10

No. 45:1984 2004). Table 6 presents the specific gravity, Sulphate and chloride content and absorption of the used coarse aggregate.

2.1.4 Crushed Melted Bricks

Melted brick from brick factories region in Amara city was used in this study after crushing and sieving on sieves 20, 10 and 5 mm and remixing of particles to give grading comforted with the requirements of (Iraqi specification No. 45:1984 2004) size of 5–20 mm, Fig. 1 shows the raw and crushed melted brick. The grading of the used melted brick is presented in Table 7. Table 8 illustrates the specific gravity, Sulphate and chloride content and absorption of the used melted brick.

2.1.5 Steel Reinforcement

Deformed bars of 10 mm diameter were used for the two directions reinforcement of slabs. Three tensile specimens of each size of bars were tested. Test results indicated that the adopted deformed bars conformed to the (ASTM A615/A615 M-04b 2004). The properties of reinforcing bar are presented in Table 9.

2.1.6 Water

Ordinary potable water was used for casting and curing of concrete.

2.1.7 Superplasticizer

For the production of high strength concrete, superplasticizer (high water reducing agent HWRA) based on polycarboxylic ether is used. One of a new generation of copolymer-based superplasticizer is the Glenium 51. It complies with (ASTM C494 2004), type A and type F.

Table 6. Properties of the coarse aggregate

Physical properties	Test results	Limits of IOS No.45/1984 (2004)
Specific gravity	2.60	–
Sulfate content (SO_3) (%)	0.073	≤ 0.1
Chloride content (Cl) (%)	0.092	≤ 0.1
Absorption (%)	0.65	–
Loose bulk density (kg/m^3)	1548	–

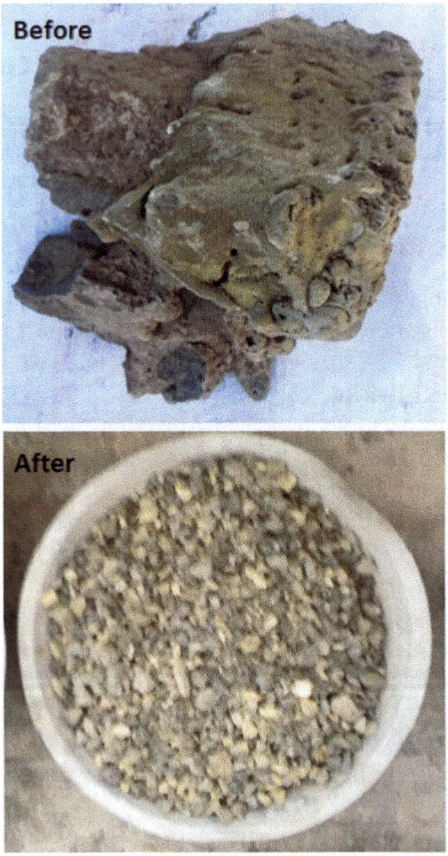

Fig. 1. The raw and graded rushed melted brick

2.2 Mix Proportions and Properties of Concrete

Table 10 illustrates mix proportions and properties of all concrete mixes used in this study depending on of crushed melted bricks fraction 0, 25, 50, 75 and 100% of coarse aggregate. The mix proportion of normal concrete mix used in this study 1: 1.8: 2.7 (cement: fine aggregate: coarse aggregate) by weight and w/c ratio 0.46. In mixes with

Table 7. Grading of the crushed melted bricks

Sieve size (mm)	Passing (%) by weight	
	Coarse aggregate	Limits of IOS No.45/1984 (2004)
20	100	95–100
14	82	–
10	44	30–60
5	0	0–10

Table 8. Properties of the crushed melted bricks

Physical properties	Test results	Limits of IOS No.45/1984 (2004)
Specific gravity	2.35	–
Sulfate content(SO_3) (%)	0.09	≤ 0.1
Chloride content(Cl) (%)	0.082	≤ 0.1
Absorption	4.3%	–
Loose bulk density (kg/m^3)	1150	–

Table 9. Properties of reinforcing bars

Bar size (mm)	Test results			ASTM A615/A615 M-06b (2004) limits		
	Yield stress (N/mm^2)	Ultimate strength (N/mm^2)	Elongation (%)	Yield stress Min. (N/mm^2)	Ultimate strength Min. (N/mm^2)	Elongation Min. (%)
10	500	650	14	420	620	9

crushed melted bricks, the weight of crushed melted bricks were calculated depending on densities of it and natural coarse aggregate to give same volume of concrete mix. All coarse Aggregate was in saturated surface dry condition at time of mixing. One variable was investigated in this study to show their effects on the flexural strength of the slabs. This variable was: crushed melted bricks fraction 0, 25, 50, 75 and 100% of coarse aggregate. Table 10 illustrates the method of mixing was according to (B.S. 1881: Part 125 1986), the density of hardened concrete was complying with (B.S. 1881: Part 114 1983) and the other properties of all mixes of the tested slabs were depending on the following tests:

2.2.1 Compressive Strength Test

Standard cubes 100 × 100 × 100 mm were used according to (BS1881: Part 116 1983). The specimens were cast with two layers and compacted by vibrating table. The specimens were covered with nylon sheets to prevent moisture loss and after 24 h from casting, the specimen was demolded and placed in a container filled with tap water until the time of testing. Testing of hardened concrete cubes was carried out at age 28 days. The machine which was used in the tests is one of the hydraulic types of 2000 kN

Table 10. Mix proportions of all the used mixes

Replacement ratio		Ref	25%	50%	75%	100%
Specimen symbol		S0	S25	S50	S75	S100
Mix proportion	Unit					
Cement	Kg/m^{-3}	400	400	400	400	400
Sand	Kg/m^{-3}	725	725	725	725	725
Gravel	Kg/m^{-3}	1080	810	540	270	0
Crushed melted bricks	Kg/m^{-3}	0	193	386	579	771
Water	Kg/m^{-3}	184	184	184	184	184
Superplasticizer	Kg/m^{-3}	2	2.2	2.5	3.0	3.8
Properties of fresh concrete						
Slump	mm	150	150	150	145	140
Properties of hardened concrete						
Density	Kg/m^{-3}	2310	2240	2160	2090	2000
Compressive strength	MPa	40.35	39.2	37.0	35.9	35.2
Flexural strength	MPa	4.5	4.0	3.6	3.4	3.0
Modulus of elasticity	MPa	26,730	24,057	23,960	23,011	22,850

capacity. Compressive strength (MPa) is obtained by averaging of the results of three specimens.

2.2.2 Flexural Strength Test

Flexural strength is calculated using the results obtained from a prism with third-point loading test according to the (ASTM C78 2002). Test prisms measuring 100 × 100 500 mm were cast and cured at the same conditions for the specimens of the compressive strength tests. A testing machine of 2000 kN capacity was used for this test.

2.2.3 Static Modulus of Elasticity Test

Modulus of elasticity is the ratio of stress to elastic strain of concrete. It is a measure of resistance of material to deformation. Modulus of elasticity of concrete tests was carried out in accordance with (ASTM C469 2002). Test specimens used are concrete cylinders of 150 mm diameter and 300 mm height. Average results obtained from three individual specimens for each concrete mixes were used.

2.2.4 Casting, Curing and Test Set-up of Slabs Failed in Flexure

In order to study the influence of the test variable on the flexural strength of the square slabs, five slabs were cast and tested. All the slabs were square of dimensions 800 × 800 × 95 mm and simply supported along the four edges with a clear span of 700 mm in each direction. A sand bag was used to apply the distribution load on the slab, and the corners of the slab were carefully clamped to prevent any possibility of slab uplift. These requirements were found important which ensured a flexural failure type occurring in the slab when the load was applied. The reinforcement used were 10 mm diameter deformed bars placed at a spacing of 175 mm c/c in two directions with 760 mm length, placed at the bottom face of the slab with a concrete cover of 20 mm, as shown in the Fig. 2. All molds were cleaned and their internal surfaces were oiled to prevent adhesion to concrete after hardening. Before placing concrete in the mold, reinforcement was placed near the bottom face of the slab's mold. Fresh concrete was placed in the mold in two equal layers. Each layer was compacted by using electrical vibrating table for two minutes to ensure removing the entrapped air as much as possible. The specimen was then removed from the vibrating table. Top surface of all slabs was well finished using a steel trowel so that the upper surface of the wooden block was kept level with the concrete surface. The slabs were covered with a damp canvas and Nylon sheets to prevent moisture loss and after 7 days from casting, the specimen was demolded and placed in the lab without any curing until the time of testing. Before the testing day, the slab was cleaned and painted with white paint on both surfaces, to achieve clear visibility of cracks during testing. The slab was labeled and carefully placed along the edges on simple supports. The dial gauge was positioned under the center of the bottom surface of the slab, so that a precise set-up of the testing equipment was achieved as shown in Fig. 3. A special supporting frame was manufactured and used inside the testing machine, as shown in Fig. 4 to provide the required span of the slab. This supporting frame was made using four steel beams welded and arranged to form a square shape. Each of these four steel beams had a 25 mm dia. steel bars are welded on its top face to provide a simple support for the slab edges. All slabs were tested using a hydraulic universal testing machine (Torsee's Universal Testing machine with a capacity of 2000 kN) as shown in Fig. 5. The load was applied gradually in increments of 10 kN.This amount of incremental loading allowed sufficient number of loads and corresponding deflections to be taken during the test which gave a good picture for the structural behavior of the slab. The load at first crack as well as the ultimate load with their corresponding deflections at the slab center were observed and recorded.

Fig. 2. Reinforcement and mold

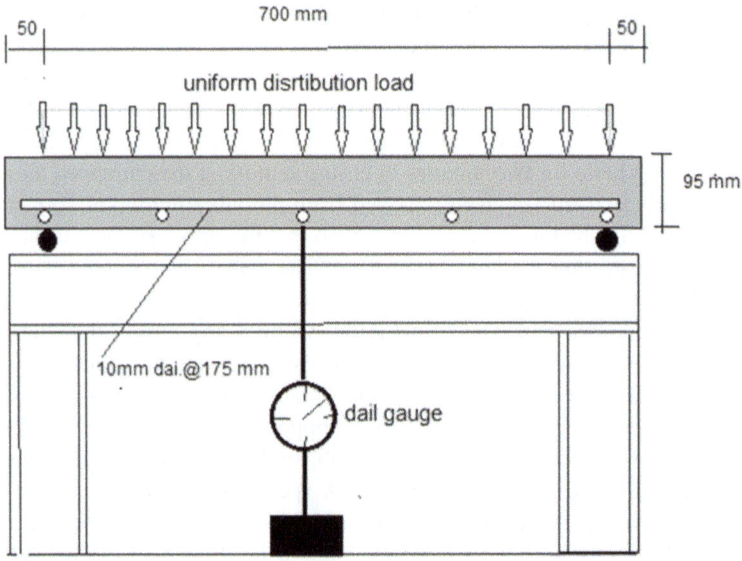

Fig. 3. Details of the slab specimen

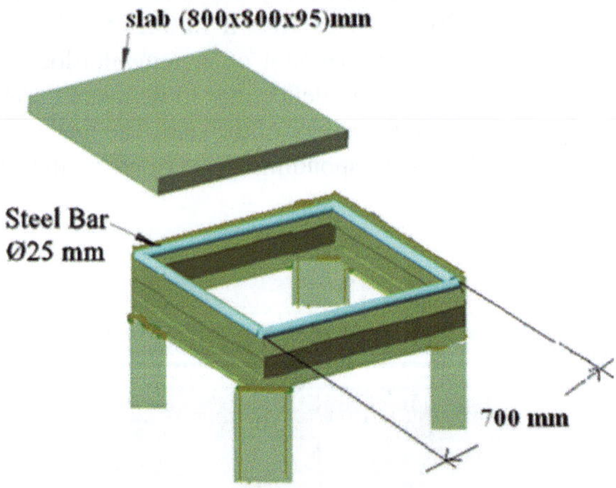

Fig. 4. Supporting slab frame

3 Test Results and Discussion

The use of crushed melted brick as coarse aggregate in concrete leads to little decrease in strength with the increase of replacement ratio of natural gravel, this may be attributed to the absorption capacity of the aggregate affects strength by influencing the effective W/C ratio (Mindess and Young 1981), and the increase of specific surface of aggregate for constant mix proportions will cause decrease in strength because increased specific surface leads to decrease in the amount of cement paste relative to

Fig. 5. Test setup

surface of aggregate, leaving more voids around surface of aggregate particles. Tables 10 and 11 are list the main test results from the observations during the tests, which included the properties of hardened concrete, first cracking load and ultimate load. The results recorded during the experiments are presented and compared.

Table 11. The cracking and ultimate loads

Slab notation	First crack load, Fcr (kN)	Ultimate load, Fu (kN)	Fcr/Fu ratio (%)	Fu of slab/Fu of ref slab ratio (%)
S0	42.2	150	28.1	100.0
S25	37.8	143	26.4	95.3
S50	34.2	131	26.1	87.3
S75	32.5	125	26.0	83.3
S100	30.0	120	25.0	80.0

3.1 Properties of Hardened Concrete

The results are exhibit that the mixes with melted crushed brick aggregate show reduction in concrete compressive strength in value 3–13%, flexural strength by 11–33% and modulus of elasticity about 10–15% than gravel concrete, at age of 28 days with respect to the replacement ratio of melted crushed aggregate from 25 to 100% respectively. This behavior may beyond to the in science of solids; there exists a fundamental inverse relationship between porosity (volume fraction of voids) and strength. Consequently, in multiphase materials such as concrete, the porosity of each component of the microstructure can become strength—limiting. Natural aggregates are generally dense and strong; therefore, it is the porosity of the cement paste matrix as well as the interfacial transition zone between the matrix and coarse aggregate, which usually determines the strength characteristic of normal-weight concrete. Although the water-cement ratio is important in determine the porosity of both the matrix and the interfacial transition zone and hence the strength of concrete (Mehta and Paulo 2006).

3.2 General Behavior, Mode of Failure and Cracks Pattern

In general, all slabs failed under pure bending. Flexural cracks started at the center of the slab, first cracks began at load of 42.2 kN for Ref. slab as illustrated in Table 11, while decrease with increasing replacement ratio of crushed melted brick for other slabs, it is obviously that the replacement by crushed melted brick effect and weakens the slab from the first stage of loading. Flexural cracks started at the center of the slab developing perpendicular to the adjacent line of support. Under increasing load, these cracks developed diagonally towards the four support corners, symmetrically located across the entire tension face as illustrated in Fig. 6. As failure load was reached, the deflection was increased and cracks were propagated quickly.

3.3 First Crack and Ultimate Loads

The cracking and ultimate loads are presented in Table 11. From this table it can be observed that, the first cracking load decreased for slabs with crushed melted brick compared with Ref slab (S0), where the first crack load of ref. slab was 42.2 kN and it reduced to 37.8, 34.2, 32.5 and 20.8 kN for slabs with crushed melted brick as a coarse aggregate by 25, 50, 75, 100% of coarse aggregate respectively. This is attributed to the flexural strength of concrete reduced with increase the replacement ratio of melted brick in concrete mix as released in Table 10. For the ultimate load, from Table 11 it can be found that, the increase in the replacement ratio for concrete in the slabs made with crushed melted brick leads to decrease in the flexural capacity of these compared with that of Ref. slab. It is found that the ultimate load of slabs made with crushed melted brick with replacement ratios 25, 50, 75, 100% of total coarse aggregate decreased to 95.3, 87.3, 83.3 and 80.0% of ultimate load of slab made with natural coarse aggregate. Therefore it can be concluded that the use of crushed melted brick in concrete has small effect on decreasing the ultimate load strength of the slabs as noticed in Fig. 7.

3.4 Deflections

The magnitude of the deflection is very important in the discussion of the load carrying capacity of the slabs. Figure 8 show comparatives in load–deflection relationships for slab. From Fig. 8 it can be observed that, the replacement process showed soft behavior than that in slab made with natural coarse aggregate. Change in slope of the curves occurred after the formation of flexural cracking because, the formation of the first major crack significantly reduced slab stiffness, therefore the deflection was increased. Also, it is found that at the same load, the deflection increased as crushed melted brick replacement ratio increases. This is attributed to that the modulus of elasticity of concrete reduced with increase the replacement ratio of melted brick in concrete mix as shown in Table 10.

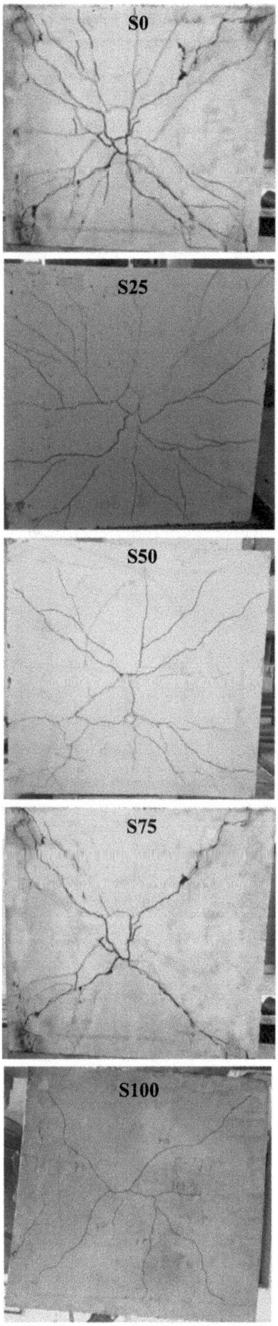

Fig. 6. Mode of failure and crack pattern of slabs

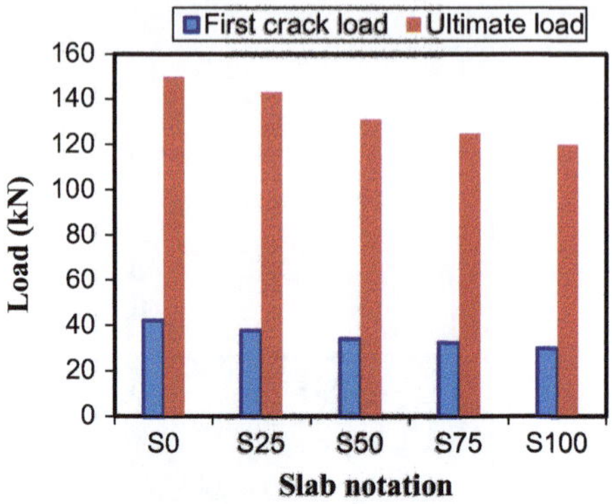

Fig. 7. Comparison first and ultimate load for slabs

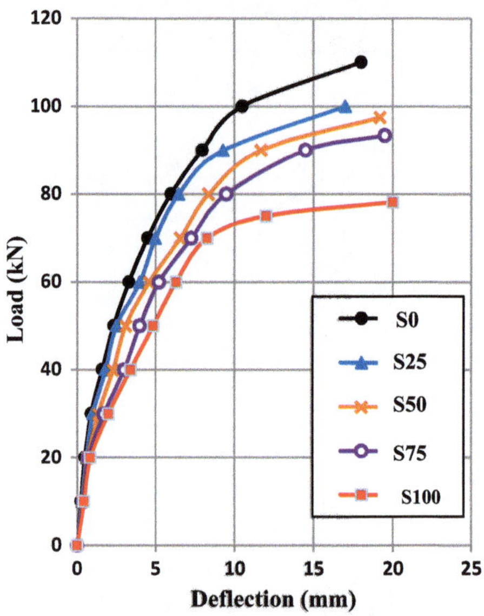

Fig. 8. Comparison load deflection curve for slabs

4 Conclusions

Based on the experimental results presented in this study, the following conclusions can be made. The crushed melted brick suitable to use in concrete as a coarse aggregate with low density up to 2000 kg/m^3, and then assists in decreasing the load transmitted to footings. The cube compressive strength of melted brick aggregate concrete (replacing of gravel coarse aggregate with melted brick aggregate from 25 to 100%) is in the range of 39.2–35.2 MPa, whereas gravel coarse aggregate concrete is having a cube compressive strength of 40.35 MPa. The flexural strength of melted brick aggregate concrete decreases about 12, 20, 25 and 33% as a ratio of replacement of crushed melted brick increases from 25 to 100%. Mixes with melted brick aggregate were exhibit lower modulus of elasticity in the range of 10–15% than gravel concrete at age 28 days. The first crack load of ref. slab was 42.2 kN and it reduced to 37.8, 34.2, 32.5 and 20.8 kN for slabs with crushed melted brick as a coarse aggregate by 25, 50, 75, 100% of coarse aggregate respectively. So that, first crack load of melted brick concrete slab specimen was decrease by 10–50% comparing with reference slab of gravel concrete. The ultimate load of slabs made with crushed melted brick with replacement ratios 25, 50, 75, 100% of total coarse aggregate decreased to 95.3, 87.3, 83.3 and 80.0% of ultimate load of slab made with natural coarse aggregate. The central deflection of slabs increased as crushed melted brick replacement ratio increases.

References

ASTM A615/A615 M-04b.: Standard specification for deformed and plain carbon steel bars for concrete reinforcement. pp. 1–6 (2004)

ASTM C469.: Standard test method for static modulus of elasticity and Poisson's ratio of concrete in compression, vol. 4.2, pp. 1–5 (2002)

ASTM C78.: Standard test method for flexural strength of concrete (using simple beam with third-point loading). vol. 4.2, pp. 1–3 (2002)

B.S. 1881: Part 125.: Methods for mixing and sampling fresh concrete in the laboratory, pp. 1– 9 (1986)

B.S. 1881: Part 114.: Methods for determinations of density of hardened concrete, pp. 1–10 (1983)

B.S. 1881: Part 116.: Methods for determination of compressive strength of concrete cubes, pp. 1–8 (1983)

Iraqi Standards No.45/1984.: Aggregate from natural sources for concrete and construction. Ministry of Housing and Construction, Baghdad (2004)

Iraqi Standards No.5/1984.: Ordinary Portland cement. Ministry of Housing and Construction, Baghdad (2004)

Kallak, F.S.: Used of crushed bricks as coarse aggregate in concrete. Tikrit J. Eng. Sci. **16**, 64–69 (2009)

Khudair, J. A.S.: Properties and structural behaviour of concrete made with abu-ghar crushed lime stone as coarse aggregate. M.Sc. Thesis, College of Engineering, Department of Civil Engineering, University of Basrah, Iraq (1987)

Mehta, P.K., Paulo, J.M.: Concrete microstructure, properties, and materials, 3rd edn. McGraw Hill, New York (2006)

Mindess, S., Young, J.F.: Concrete. Prentice Hall, New Jersey (1981)

PCA, Chap09.: Design and Proportioning Normal Concrete Mixtures. Portland Cement Association. Available at: www.ce.memphis.edu (2002)

Reddy, V., Kumar, S.: Behavior of recycled aggregate concrete two-way slabs in flexure and punching shear–an experimental investigation. Ph.D. Thesis, Faculty of Civil Engineering, Jawaharlal Nehru Technological University, India (2012)

A Novel Finite Element Model for Modeling Pile Dynamics

Reza Yaghmaie(✉)

Department of Civil Engineering, Johns Hopkins University, 3400 N. Charles Street, Baltimore, MD 21218, USA
ryaghma2@jhu.edu

Abstract. An efficient semi analytical-numerical methodology is presented for modeling pile dynamics. The medium is assumed to be a transversely isotropic infinite domain. In this paper, the mechanical state of stress and displacement fields arising from the coupling of wave propagation and the various pile effects for vertical cross sections of piles is solved analytically by utilizing the Boussinesq solution. The effective pile stiffness is obtained by finite element assembly of the displacement and force vectors due to wave propagation solution obtained at vertical pile cross sections and the resulting dynamic system is solved using the finite element method (FEM) in the frequency domain. A few numerical examples demonstrating various factors affecting the response of the pile system are presentenced. The present model predicts the pile dynamics problems accurately and efficiently for linear elastic mediums.

1 Introduction

Dynamics of single piles and pile groups have wide spread applications in structures exposed to dynamic loading such as offshore platforms, bridges, and machinery foundations, rock mechanics, oil extractions and embankments. Modeling the pile dynamics requires wave propagation analysis in anisotropic media. The medium is often assumed as homogenous elastic and isotropic (see for example Apsel and Luco (1983) and Yaghmaie and Noorzad (2010), Yaghmaie et al. (2013, 2014, 2016a, b) and Yaghmaie and Ghosh (2015, 2018a)). The most common anisotropy form observed in pile dynamics problems is the transverse isotropy with a vertical axis of symmetry. There are a few investigations, who have derived the Green's functions for either isotropic or anisotropic functionally graded materials (FGMs) (see for example Wang et al. (2006), Kashtalyan and Rushchitsky (2009) and Yaghmaie and Asgari (2016a, b, c). Computational modeling of the pile dynamics problems is a complicated phenomenon due to the anisotropy of the soil, the interaction of the group piles with the medium and vice versa, challenges in defining the size of the no-reflecting boundary and also the computational expenses in discretizing the soil and the group piles in 3D utilizing finite element method (FEM).

There are both analytical and numerical models for pile dynamics analysis. Synge (1957) and Buchwald (1961) have developed analytical wave propagation models for elastic anisotropic and transversely isotropic media respectively. Waas et al. (1985) has proposed the analytical displacement solutions for dynamic loads in transversely

H. Khabbaz et al. (eds.), *New Prospects in Geotechnical Engineering Aspects of Civil Infrastructures*, Sustainable Civil Infrastructures,
https://doi.org/10.1007/978-3-319-95771-5_17

isotropic stratified media. A mixed analytical-numerical solution to the pile-soil interaction under lateral loading using infinite and finite elements is developed by Chen and Polous (1993). Liu and Novak (1994) have developed an analytical solution for dynamic response of single piles embedded in transversely isotropic layered media. Inelastic analysis of pile-soil interaction is investigated by Kukukarsalan and Banerjie (2004). The mixed boundary element method (BEM) and finite element method (FEM) model for the dynamic analysis of piles and pile groups is developed by Padron et al. (2007). Khodar and Abdel-Mohti (2014) have developed a semi-analytical-numerical analysis of pile-soil interaction under axial and lateral loads.

In this paper, an efficient computational model for solving 3D group pile dynamics embedded in transversely isotropic half-space is developed based on a semi-numerical-analytical method. Herein, the rigid massless elements, referred to as radiation discs are modeled with exact analytical solutions to the wave propagation problems in soils while the piles are modeled through the finite element analysis. The analytical solution for the rigid disks can be obtained by using existing potential functions for general response of pile dynamics in transversely-isotropic half-space domains. For modeling the piles, the finite element discretization is only required along the length of piles. This leads to a huge computational time saving since it avoids discretizing the soil, top free surface boundary, and cross sections of piles. A few numerical results are presented to compare the accuracy of the present model with existing analytical solutions. Furthermore, the influence of excitation frequency, anisotropy and load cases are studied.

2 Formulation of the Functionally Graded Transversely Isotropic Domain

The soil-pile group system considered in this study is shown in Fig. 1, in which Ω_p and Ω_S are the regions of the three dimensional elastic space occupied by cylindrical piles and the surrounding medium, respectively with the surrounding boundary Γ. The piles are assumed to be fully embedded in, and continuously bonded to the medium. The Navier equation of motion in the Cartesian coordinate system can be written as

$$\sigma^{\alpha}_{ij,j} + b^{\alpha}_{i} = \rho^{\alpha} \ddot{u}^{\alpha}_{i} \; (\alpha = P, S) \tag{1}$$

where $\sigma^{\alpha}_{ij} = \frac{C^{\alpha}_{ijkl}}{2}\left(u^{\alpha}_{k,l} + u^{\alpha}_{l,k}\right)$ is the Cauchy stress tensor, u is the displacement field, b is the body force vector and ρ is the density. The compatibility conditions for the stress and displacement field are written as

$$u^{P}_{i} = u^{S}_{i}, \quad \left(\sigma^{P}_{ij} - \sigma^{S}_{ij}\right) \cdot n_j = \mathbf{0} \text{ on } \Gamma \tag{2}$$

where n is the unite outward normal on the boundary Γ. The equations governing the response of soil-pile system can be decomposed into a soil medium and a pile group system as shown in Figs. 2 and 3. These equations are given as

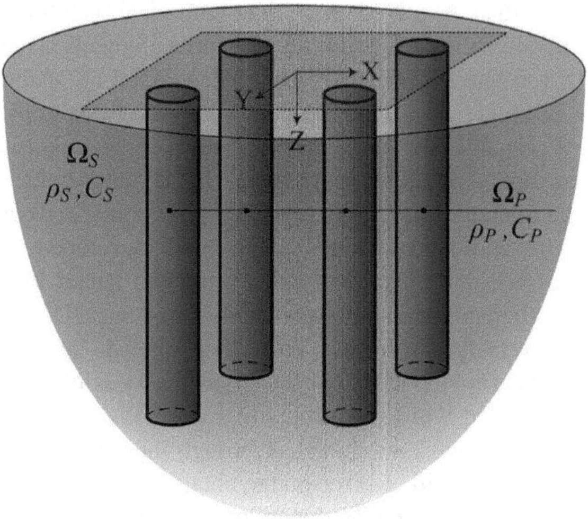

Fig. 1. The 3D model for pile-soil-pile interaction problem

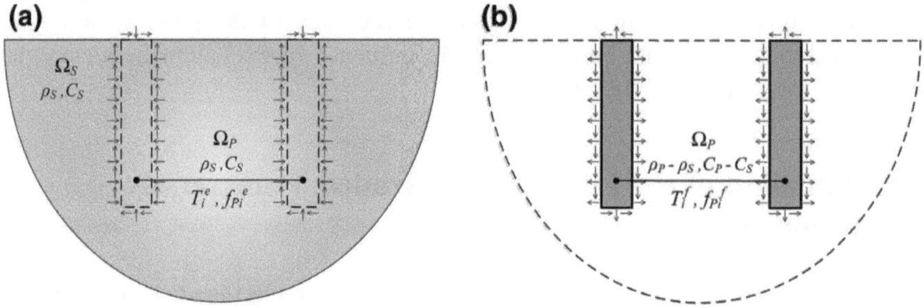

Fig. 2. The pile-soil-pile system decomposed into the **a** soil medium, and **b** pile group system

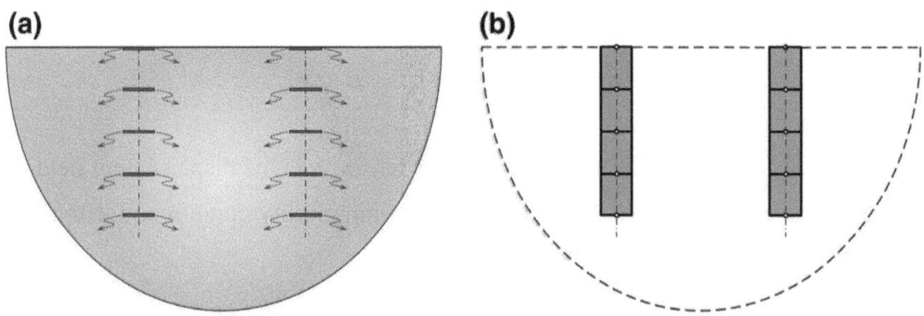

Fig. 3. The discretized pile-soil-pile system using the finite elements for the **a** soil medium, and **b** pile group system

For the soil medium:

$$\begin{cases} \sigma^S_{ij,j} + b^S_i = \rho^S \ddot{u}^S_i \text{ on } \Omega_S \\ \sigma^P_{ij,j} + b^P_i = \rho^P \ddot{u}^P_i \text{ on } \Omega_P \end{cases} \tag{3}$$

$$\begin{cases} \sigma^S_{ij} = \frac{C^S_{ijkl}}{2}\left(u^S_{k,l} + u^S_{l,k}\right) \text{ on } \Omega_S \\ \sigma^P_{ij} = \frac{C^P_{ijkl}}{2}\left(u^P_{k,l} + u^P_{l,k}\right) \text{ on } \Omega_P \end{cases} \tag{4}$$

$$u^P_i = u^S_i, \quad \left(\sigma^{P(e)}_{ij} - \sigma^S_{ij}\right) \cdot n_j = T^e_i \text{ on } \Gamma \tag{5}$$

For the pile group system:

$$\begin{cases} \sigma^{S(f)}_{ij} = \mathbf{0}, \quad b^{S(f)}_i = \mathbf{0} \text{ on } \Omega_S \\ \sigma^{P(f)}_{ij,j} + b^{P(f)}_i = (\rho^P - \rho^S)\ddot{u}^P_i \text{ on } \Omega_P \; where \; b^{P(e)}_i + b^{P(f)}_i = b^P_i \end{cases} \tag{6}$$

$$\sigma^{P(f)}_{ij} = \frac{\left(C^P_{ijkl} - C^S_{ijkl}\right)}{2}\left(u^P_{k,l} + u^P_{l,k}\right) \text{ on } \Omega_P \; where \; \sigma^{P(e)}_{ij} + \sigma^{P(f)}_{ij} = \sigma^P_{ij} \tag{7}$$

$$\sigma^{P(f)}_{ij} \cdot n_j = T^f_i \text{ on } \Gamma \tag{8}$$

Furthermore, $T^e_i + T^f_i = \mathbf{0}$ and the superscripts e and f represent the corresponding quantities of the extended soil medium and the pile group system, respectively. The body forces $\left(b^{P(e)}_i, b^{P(f)}_i\right)$ and the tractions acting on Ω_p and the tractions $\left(T^e_i, T^f_i\right)$ on Γ are defined such that they result in the same displacement field u^p_i in both systems. In this model a few assumptions are considered: small radius-length ratio of piles, i.e. $a/l \ll 1$, the large pile-soil stiffness ratio $C^P_{ijkl}/C^S_{ijkl} \gg 1$ and linear distribution of the body force $f^{P(e)}_i$ along the cross sections of the pile region in the soil medium system. These assumptions indicate that the pile cross section approximately remain plane during deformation. A Boussinesq-type force distribution is considered to act on the rigid discs to hold the rigid mode of deformation along the cross-sections of Ω_p expressed as

$$f^{P(e)}_i(r,\theta,z,t) = \begin{cases} \frac{F_i(\theta,z,t)}{2\pi a^2\sqrt{1-\frac{r^2}{a^2}}}, & r<a \\ 0, & r \geq a \end{cases} \tag{9}$$

in the Cylindrical coordinate system. $F_i(\theta, z, t)$ depends on direction and magnitude of external forces. The compatibility condition between the pile-soil systems in z-direction is satisfied by considering the same displacement field. This leads to the dynamic stiffness matrix of the pile-soil-pile system $[K_T] = [K_S] + [K_P]$ where $[K_S]$ and $[K_P]$ correspond to dynamic stiffnesses of the rigid disks and the piles respectively. The dynamic stiffness matrix of the rigid discs is obtained by inverting the flexibility matrix which is derived through applying unit dynamic load at each rigid disc, and determining the response of all the discs in the group. In this model, no simplifying assumptions are made in regard to the number of piles or pile-soil-pile interaction.

2.1 Dynamic Response of a Rigid Disk Buried in a Transversely Isotropic Half-Space

The Navier's equations for transversely isotropic half-space medium with the axis of material symmetry being in the z-axis and perpendicular to the horizontal free surface are expressed as

$$\begin{gathered}
A_{11}\left(\frac{\partial^2}{\partial r^2}+\frac{1}{r}\frac{\partial}{\partial r}-\frac{1}{r^2}\right)u+\frac{A_{66}}{r^2}\frac{\partial^2 u}{\partial\theta^2}+A_{44}\frac{\partial^2 u}{\partial z^2}+\frac{A_{11}+A_{12}}{2}\left(\frac{1}{r}\frac{\partial^2}{\partial r\partial\theta}+\frac{1}{r^2}\frac{\partial}{\partial\theta}\right)v \\
-\frac{2A_{11}}{r^2}\frac{\partial v}{\partial\theta}+(A_{13}+A_{44})\frac{\partial^2 w}{\partial r\partial z}=-\rho^S\omega^2 u \\
A_{66}\left(\frac{\partial^2}{\partial r^2}+\frac{1}{r}\frac{\partial}{\partial r}-\frac{1}{r^2}\right)v+\frac{A_{11}}{r^2}\frac{\partial^2 v}{\partial\theta^2}+A_{44}\frac{\partial^2 v}{\partial z^2}+\frac{A_{11}+A_{12}}{2}\left(\frac{1}{r}\frac{\partial^2}{\partial r\partial\theta}-\frac{1}{r^2}\frac{\partial}{\partial\theta}\right)u \\
+\frac{2A_{11}}{r^2}\frac{\partial u}{\partial\theta}+\frac{(A_{13}+A_{44})}{r}\frac{\partial^2 w}{\partial\theta\partial z}=-\rho^S\omega^2 v \\
A_{44}\left(\frac{\partial^2}{\partial r^2}+\frac{1}{r}\frac{\partial}{\partial r}+\frac{1}{r^2}\frac{\partial^2}{\partial\theta^2}\right)w+A_{33}\frac{\partial^2 w}{\partial z^2} \\
+(A_{13}+A_{44})\left(\frac{\partial^2 u}{\partial r\partial z}+\frac{1}{r}\frac{\partial u}{\partial z}+\frac{1}{r}\frac{\partial^2 v}{\partial\theta\partial z}\right)=-\rho^S\omega^2 w
\end{gathered} \tag{10}$$

where u, v and w are displacement components in r, θ, and z directions respectively and ω is the frequency for the harmonic excitation $e^{i\omega t}$. A_{ij} are the domain material constants written as

$$\begin{gathered}
A_{11}=\frac{E^S\left(1-\frac{E^S}{\tilde{E}^S}\tilde{\upsilon}^{(S)2}\right)}{(1+\upsilon^S)\Delta},\quad A_{12}=A_{11}-2A_{66},\quad A_{13}=\frac{E^S\tilde{\upsilon}^{(S)}}{\Delta} \\
A_{33}=\frac{\tilde{E}^S(1-\upsilon^S)}{\Delta},\quad A_{44}=\tilde{G}^S,\quad A_{66}=\frac{E^S}{2(1+\upsilon^S)}=G^S
\end{gathered} \tag{11}$$

where $\Delta=1-\upsilon^S-2E^S\tilde{\upsilon}^{(S)2}/\tilde{E}^S$, E^S and $\tilde{E}^S$ are the Young's modulus in the plane of isotropy and in the direction normal to it, υ^S and $\tilde{\upsilon}^S$ are in-plane and normal Poisson's ratios, and G^S and $\tilde{G}^S$ are in-plane and normal shear modulus, respectively. The positive definiteness of the strain energy (see e.g. Yaghmaie et al. (2016b), Yaghmaie and Ghosh (2017, 2018a)) requires that

$$A_{11} > |A_{12}|, \quad (A_{11} + A_{12})A_{33} > 2A_{13}^2, \quad A_{44} > 0 \tag{12}$$

A set of complete scalar potential functions introduced by Wang et al. (2006) are used to decouple the equations of motion (10). The displacement components in terms of the potential functions Ψ and $\aleph$ are expressed as

$$\begin{aligned} u &= -\alpha_3 \frac{\partial^2 \Psi}{\partial r \partial z} - \frac{1}{r}\frac{\partial \aleph}{\partial \theta} \\ v &= -\frac{\alpha_3}{r}\frac{\partial^2 \Psi}{\partial \theta \partial z} + \frac{\partial \aleph}{\partial r} \\ w &= (1+\alpha_1)\left(\nabla_{r\theta}^2 + \beta\frac{\partial^2}{\partial z^2} + \frac{\rho_0\omega^2}{1+\alpha_1}\right)\Psi \end{aligned} \tag{13}$$

where $\alpha_1 = (A_{11} + A_{12})/(A_{11} - A_{12})$, $\alpha_2 = A_{44}/A_{66}$, $\alpha_3 = (A_{13} + A_{44})/A_{66}$, $\beta = \alpha_2/(1+\alpha_1)$, $\rho_0 = \rho^S/A_{66}$ and $\nabla_{r\theta}^2 = \frac{\partial^2}{\partial r^2} + \frac{1}{r}\frac{\partial}{\partial r} + \frac{1}{r^2}\frac{\partial^2}{\partial \theta^2}$. Substituting (13) into (10) yields to

$$\begin{aligned} &\nabla_0^2 \aleph = 0 \\ &\nabla_1^2 \nabla_2^2 \Psi + B\rho^S\omega^2 \frac{\partial^2 \Psi}{\partial z^2} = 0 \end{aligned} \tag{14}$$

where

$$\begin{aligned} \nabla_i^2 &= \nabla_{r\theta}^2 + \frac{1}{s_i^2}\frac{\partial^2}{\partial z^2} + \frac{1}{\mu_i}\rho_0\omega^2, \quad (i = 0, 1, 2) \\ B &= \frac{1}{A_{11}}\left(1 + \frac{A_{33}}{A_{44}}\right) - \frac{1}{A_{66}}\left(\frac{1}{\mu_1 s_2^2} + \frac{1}{\mu_2 s_1^2}\right) \end{aligned} \tag{15}$$

$\mu_0 = 1$, $\mu_1 = \alpha_2$, $\mu_2 = 1 + \alpha_1$, $s_0^2 = 1/\alpha_2$ and s_1 and s_2 are the roots of the following equation

$$A_{33}A_{44}s^4 + \left(A_{13}^2 + 2A_{13}A_{44} - A_{11}A_{33}\right)s^2 + A_{11}A_{44} = 0 \tag{16}$$

The potential functions are expanded in a Fourier series as

$$[\aleph, \Psi](r, \theta, z) = \sum_{m=-\infty}^{+\infty} [\aleph_m, \Psi_m](r, z)e^{im\theta} \tag{17}$$

where $\aleph_m$ and Ψ_m are the m-th Fourier coefficients of the corresponding potential functions. Additionally, the Fourier coefficients of the potential functions are transformed into a fictitious coordinate ξ using the following Hankel transformation

$$\left[\aleph_m^m, \Psi_m^m\right](\xi, z) = \int_0^\infty [\aleph_m(r,z), \Psi_m(r,z)] r J_m(\xi r) dr \tag{18}$$

where the superscript m represents the m-th order Hankel transform of the Fourier coefficients $\aleph_m$ and Ψ_m, and J_m is the Bessel function of the first kind of order m. Substituting (17) and (18) into (14) yields to

$$\begin{aligned} &\nabla_{0m}^2 \aleph_m^m(\xi, z) = 0 \\ &\left(\nabla_{1m}^2 \nabla_{2m}^2 \Psi + B\rho^S \omega^2 \frac{d}{dz^2}\right) \Psi_m^m(\xi, z) = 0 \end{aligned} \tag{19}$$

where

$$\nabla_{im}^2 = \frac{\rho_0 \omega^2}{\mu_i} - \xi^2 + \frac{1}{s_i^2} \frac{d^2}{dz^2}, \quad (i = 0, 1, 2) \tag{20}$$

The general solutions of (19) are obtained as

$$\aleph_m^m(\xi, z) = A_m(\xi) e^{-\bar{\alpha}_0 z} + B_m(\xi) e^{\bar{\alpha}_0 z} \tag{21}$$

$$\Psi_m^m(\xi, z) = C_m(\xi) e^{-\bar{\alpha}_1 z} + D_m(\xi) e^{\bar{\alpha}_1 z} + E_m(\xi) e^{-\bar{\alpha}_2 z} + F_m(\xi) e^{\bar{\alpha}_2 z} \tag{22}$$

where

$$\begin{aligned} &\bar{\alpha}_0 = \sqrt{s_0^2 q_0(\xi)}, \quad \bar{\alpha}_1 = \sqrt{\frac{-M - \sqrt{M^2 - 4LN}}{2N}}, \; \bar{\alpha}_2 = \sqrt{\frac{-M + \sqrt{M^2 - 4LN}}{2N}}, \\ &M = \frac{\rho_0 \omega^2}{\mu_2}\left(1 + \frac{A_{33}}{A_{44}}\right) - \xi^2 \left(\frac{1}{s_1^2} + \frac{1}{s_2^2}\right) \end{aligned} \tag{23}$$

where $L = q_1(\xi) q_2(\xi)$, $N = 1/s_1^2 s_2^2$ and $q_i(\xi) = \xi^2 - \rho_0 \omega^2 / \mu_i$. The coefficients $A_m(\xi), B_m(\xi), \ldots, F_m(\xi)$ are solved after applying the boundary conditions. Consider a rigid disc of radius a embedded at a depth h in the transversely isotropic half-space as shown in Fig. 4. The half-space can be divided into two regions: the region between the free surface and the embedded disc, i.e. Region I ($-h \leq z < 0$), and the unbounded region II ($z \geq 0$). In Region II, the corresponding coefficients to the term $e^{\bar{\alpha}_i z}$ ($i = 0, 1, 2$) are zero as $B_m^{II}(\xi) = D_m^{II}(\xi) = F_m^{II}(\xi) = 0$ where the superscripts refer to the region of interest. The general set of Dirichlet and Neumann boundary conditions are expressed as

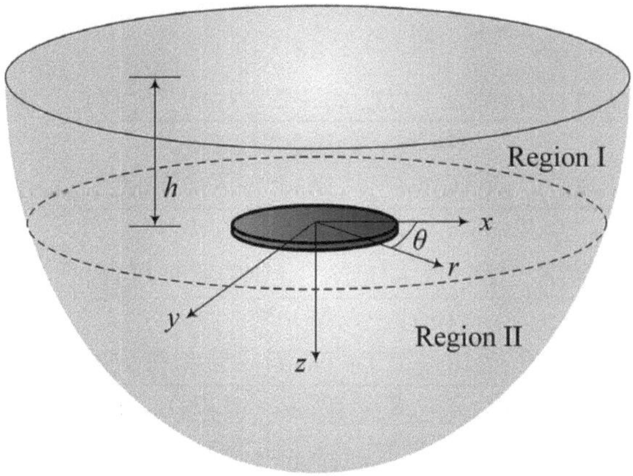

Fig. 4. The rigid disk embedded in a transversely isotropic half-space medium

$$u^{I}(r,\theta,z=0^{-})=u^{II}(r,\theta,z=0^{+}) \tag{24}$$

$$v^{I}(r,\theta,z=0^{-})=v^{II}(r,\theta,z=0^{+}) \tag{25}$$

$$w^{I}(r,\theta,z=0^{-})=w^{II}(r,\theta,z=0^{+}) \tag{26}$$

$$\sigma_{zz}^{S(I)}(r,\theta,z=-h)=0,\quad \sigma_{rz}^{S(I)}(r,\theta,z=-h)=0,\quad \sigma_{\theta z}^{S(I)}(r,\theta,z=-h)=0 \tag{27}$$

$$\sigma_{zz}^{S(I)}(r,\theta,z=0^{-})-\sigma_{zz}^{S(II)}(r,\theta,z=0^{+})=\begin{cases} f(r,\theta), & r<a \\ 0, & r\geq a \end{cases} \tag{28}$$

$$\sigma_{rz}^{S(I)}(r,\theta,z=0^{-})-\sigma_{rz}^{S(II)}(r,\theta,z=0^{+})=\begin{cases} g_r(r,\theta), & r<a \\ 0, & r\geq a \end{cases} \tag{29}$$

$$\sigma_{\theta z}^{S(I)}(r,\theta,z=0^{-})-\sigma_{\theta z}^{S(II)}(r,\theta,z=0^{+})=\begin{cases} g_\theta(r,\theta), & r<a \\ 0, & r\geq a \end{cases} \tag{30}$$

where $f(r,\theta)$, $g_r(r,\theta)$ and $g_\theta(r,\theta)$ are the external load distribution components. The implementation of the boundary conditions (24–30) and unknown coefficients are (see e.g. Yaghmaie and Asgari (2016a, b, c) for details).

$$
\begin{aligned}
A_m^I(\xi) &= \frac{1}{2}\bar{g}_1(\xi)e^{-2\bar{\alpha}_0 h}, \quad B_m^I(\xi) = \frac{1}{2}\bar{g}_1(\xi),\\
C_m^I(\xi) &= \frac{(L_2N_1 + L_1N_2)e^{-\bar{\alpha}_1 h}D_m^I(\xi) + 2L_2N_2e^{-\bar{\alpha}_2 h}F_m^I(\xi)}{(L_1N_2 - L_2N_1)e^{\bar{\alpha}_1 h}},\\
D_m^I(\xi) &= \frac{(L_1\bar{\alpha}_2 - L_2\bar{\alpha}_1)K_2\bar{g}_2(\xi) + \left(K_2\bar{\alpha}_1^2 - K_1\bar{\alpha}_2^2\right)\bar{\alpha}_2 f_m^m(\xi)}{2\left(K_2\bar{\alpha}_1^2 - K_1\bar{\alpha}_2^2\right)(L_1\bar{\alpha}_2 - L_2\bar{\alpha}_1)},\\
E_m^I(\xi) &= \frac{2L_1N_1e^{-\bar{\alpha}_1 h}D_m^I(\xi) + (L_2N_1 + L_1N_2)e^{-\bar{\alpha}_2 h}F_m^I(\xi)}{(L_2N_1 - L_1N_2)e^{\bar{\alpha}_2 h}},\\
F_m^I(\xi) &= \frac{(L_2\bar{\alpha}_1 - L_1\bar{\alpha}_2)K_1\bar{g}_2(\xi) + \left(K_1\bar{\alpha}_2^2 - K_2\bar{\alpha}_1^2\right)\bar{\alpha}_1 f_m^m(\xi)}{2\left(K_1\bar{\alpha}_2^2 - K_2\bar{\alpha}_1^2\right)(L_2\bar{\alpha}_1 - L_1\bar{\alpha}_2)},\\
A_m^{II}(\xi) &= \frac{1}{2}\bar{g}_1(\xi)\left(1 + e^{-2\bar{\alpha}_0 h}\right),\\
C_m^{II}(\xi) &= C_m^I(\xi) - \frac{(L_1\bar{\alpha}_2 - L_2\bar{\alpha}_1)K_2\bar{g}_2(\xi) + \left(K_2\bar{\alpha}_1^2 - K_1\bar{\alpha}_2^2\right)\bar{\alpha}_2 f_m^m(\xi)}{2\left(K_2\bar{\alpha}_1^2 - K_1\bar{\alpha}_2^2\right)(L_1\bar{\alpha}_2 - L_2\bar{\alpha}_1)},\\
E_m^{II}(\xi) &= E_m^I(\xi) - \frac{(L_2\bar{\alpha}_1 - L_1\bar{\alpha}_2)K_1\bar{g}_2(\xi) - \left(K_1\bar{\alpha}_2^2 - K_2\bar{\alpha}_1^2\right)\bar{\alpha}_1 f_m^m(\xi)}{2\left(K_1\bar{\alpha}_2^2 - K_2\bar{\alpha}_1^2\right)(L_2\bar{\alpha}_1 - L_1\bar{\alpha}_2)}
\end{aligned}
\tag{31}
$$

where $K_i = \alpha_2\bar{\alpha}_i^2 + \rho_0\omega^2 - \mu_2\xi^2$, $(i = 1, 2)$, $N_i = K_i - \alpha_3\bar{\alpha}_i^2$, $L_i = \bar{\alpha}_i(A_{33}K_i + A_{13}\alpha_3\xi^2)$ and

$$
\begin{aligned}
\bar{g}_1(\xi) &= \frac{i}{2\xi A_{44}\bar{\alpha}_0}\left[\left(g_{rm}^{m+1} + ig_{\theta m}^{m+1}\right) + \left(g_{rm}^{m-1} - ig_{\theta m}^{m-1}\right)\right],\\
\bar{g}_2(\xi) &= \frac{i}{2\xi A_{44}\alpha_3}\left[\left(g_{rm}^{m+1} + ig_{\theta m}^{m+1}\right) + \left(g_{rm}^{m-1} - ig_{\theta m}^{m-1}\right)\right]
\end{aligned}
\tag{32}
$$

The Fourier coefficients of the displacement fields are obtained by numerically integrating the below inverse Hankel transforms

$$
u_m = \frac{\alpha_3}{2}\int_0^\infty (J_{m+1}(\xi r) - J_{m-1}(\xi r))\xi^2\frac{d\Psi_m^m}{dz}d\xi - \frac{i}{2}\int_0^\infty (J_{m+1}(\xi r) + J_{m-1}(\xi r))\xi^2\aleph_m^m d\xi
\tag{33}
$$

$$
\begin{aligned}
v_m = &-\frac{i\alpha_3}{2}\int_0^\infty (J_{m+1}(\xi r) + J_{m-1}(\xi r))\xi^2\frac{d\Psi_m^m}{dz}d\xi\\
&-\frac{1}{2}\int_0^\infty (J_{m+1}(\xi r) - J_{m-1}(\xi r))\xi^2\aleph_m^m d\xi
\end{aligned}
\tag{34}
$$

$$w_m = (1+\alpha_1)\int_0^\infty \xi\left(-\xi^2+\beta\frac{d^2}{dz^2}+\frac{\rho_0\omega^2}{\mu_2}\right)\Psi_m^m J_m(\xi r)d\xi \tag{35}$$

The dynamic flexibility matrices of the corresponding rigid discs can be obtained for different modes of vibration from (33) to (35). Note that the above integrals are singular at $\xi_{c_1}=\omega\sqrt{\rho_0/\mu_2}$, $\xi_{c_2}=\omega\sqrt{\rho_0/\mu_1}$ and $\xi_{c_3}=\omega\sqrt{\rho_0/\mu_0}$ and also for $L_1N_2=L_2N_1$. The Gauss-quadrature integral scheme is adopted to integrate the kernels for a fictitious ζ_{max} based on a relative tolerance of 10^{-5} on the value of the kernel at ζ_{max}. The convergence criteria in the kernel integrations are developed in Yaghmaie et al. (2016b), Yaghmaie and Ghosh (2017, 2018a). The numerical results are presented in the next section.

3 Results and Discussion

Figure 5 illustrate the normalized displacement of the rigid disks undergoing vertical harmonic excitations with the magnitude P_Z and the dimensionless frequency $a_0=\omega d\sqrt{\rho^S/A_{44}}=0.1, 0.5, 1.0, 2.0$ where d is the disk diameter. In these figures, the results are compared with the results of uniform external loading distribution, i.e. $f(r,\theta)-1/\pi a^2$, $(r{<}a)$. The normalized displacement contours for the rigid disc undergoing horizontal external loads with the magnitude of P_X along the x-direction is shown in Fig. 6. The dimensionless frequencies of $a_0=0.2$ and 1.0 are utilized in this example. As can be seen, the displacement of the disc is nearly following the assumptions mentioned earlier for both the vertical and horizontal loadings with the Boussinesq-type distribution. The assumption of a uniform distribution leads to displacement incompatibility in all the cases considered. Also, the external Boussinesq-type distribution is associated with the elastostatic response of the rigid discs, this assumption may not be valid for the problems with high dimensionless frequencies $a_0>3.0$ (see e.g. Yaghmaie and Ghosh (2018b, c, d)). The dynamic stiffness matrix corresponding to the piles are obtained from the exact solutions of the wave propagation problem, therefore, the number of pile elements does not affect the accuracy of the numerical solution. Nevertheless, the number of rigid discs used in the model, affects the accuracy of the numerical integrations. Figure 7 shows the normalized compliance of a single pile undergoing vertical and horizontal harmonic excitations with different dimensionless frequencies $a_0=0.0, 0.2, 0.5, 1.0$. The pile has the parameters $E^P/\bar{E}^S=1000$, $\rho^P/\rho^S=1.5$, $l/d=15$ and is embedded in a transversely isotropic half-space with the material properties of material-II shown in Table 1. In these figures, the normalized compliance is plotted versus the number of rigid discs. As the number of rigid disks increase, the semi-analytical-numerical scheme converges rapidly. The rate of convergence depends inversely on the frequency of excitation, i.e. the lower the frequency, the higher the convergence rate. Figure 8 shows the comparison of the results of the present model for the rigid discs embedded in an isotropic elastic half-space with the results in Apsel and Luco (1983). The figures correspond to the real and imaginary parts of the normalized displacement of the half space along the z-direction for the case of uniform lateral excitations with $a_0 = 1.0$ with $h = 0, 20a$. As it can be seen, the results are in excellent agreement with the reference solutions confirming the accuracy of the present solution scheme. Figure 9 shows the vertical impedance of a 2 × 2 pile group embedded in the viscoelastic half space with $\upsilon^S = 0.4$ and the material damping coefficient $\beta^S = 0.05$ and the parameters $E^P/\bar{E}^S = 1000,\ \rho^P/\rho^S = 1.43,\ l/d = 15$. The results are compared with the reference solution in Apsel and Luco (1983). The figures illustrate the variation of the real and imaginary parts of the impedance of the pile group with dimensionless frequencies a_0 for different pile spacing ratios $s/d = 2, 5, 10$. Again, a good agreement between the results of the present model with the reference solution is achieved. Figure 10 shows the influence of pile spacing on dynamic response of a 2 × 2 pile system in transversely isotropic half-space medium. The figures show the normalized horizontal and vertical compliances of the groups with different spacing ratios $s/d = 2, 5, 10$ embedded in the half-space with the material-IV shown in Table 1. The compliances are plotted versus dimensionless frequency a_0. The response of the pile group is significantly affected by the interaction among the piles and also depends on the excitation frequency and the pile spacing ratios. The analysis shows that the frequency dependent response of the group cannot be deduced a priori from the behavior of single piles, and therefore a complete treatment of the group interaction is crucial for the correct solution. However, as the pile spacing ratio increases, the dynamic response of the group pile approaches the single-pile response.

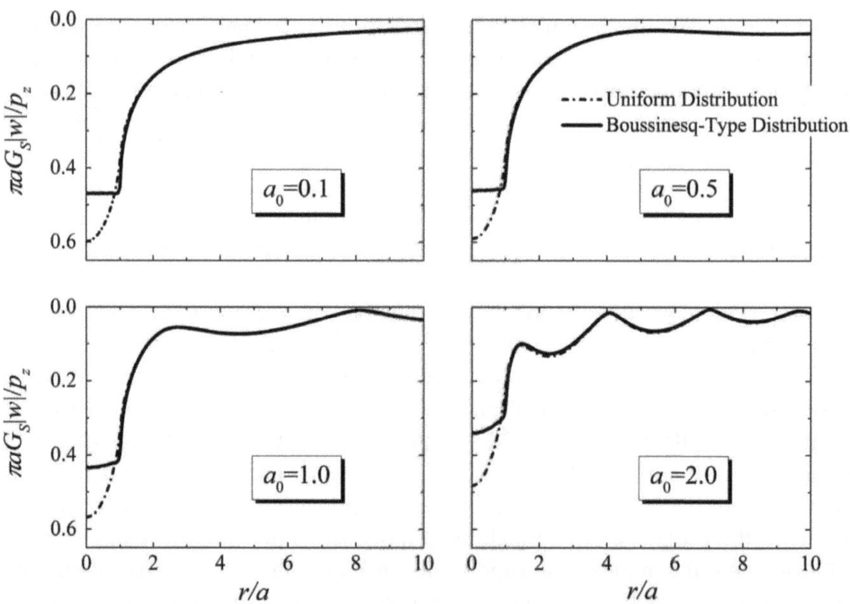

Fig. 5. The vertical displacement field distribution along normalized radial-direction for the Boussinesq-type and uniform harmonic excitations

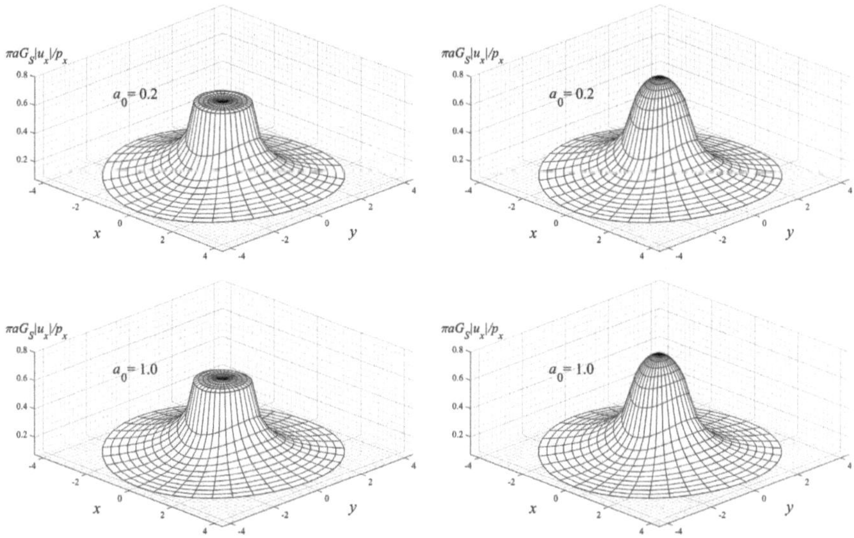

Fig. 6. The radial displacement field distribution for the Boussinesq-type (left) and uniform (right) harmonic excitations

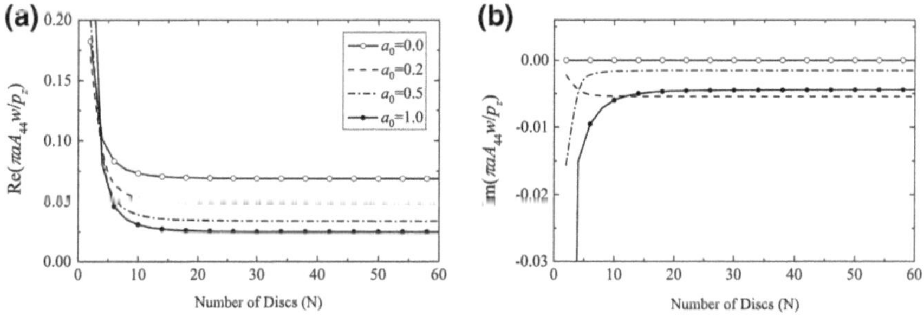

Fig. 7. The convergence of the real and imaginary parts of the vertical displacement field with increasing the number of rigid disks

Table 1. Properties of the Transversely Isotropic Materials

Material	$E^S/\bar{G}^S$	$\bar{E}^S/\bar{G}^S$	υ^s	$\bar{\upsilon}^S$	$\zeta_{c_4}/\omega\sqrt{\rho^S/A_{44}}$
I	2.5	2.5	0.25	0.25	1.08766
II	5	2.5	0.25	0.25	1.03780
III	1.25	2.5	0.25	0.25	1.20145
IV	2.5	5	0.25	0.25	1.05006
V	2.5	1.25	0.25	0.25	1.14367

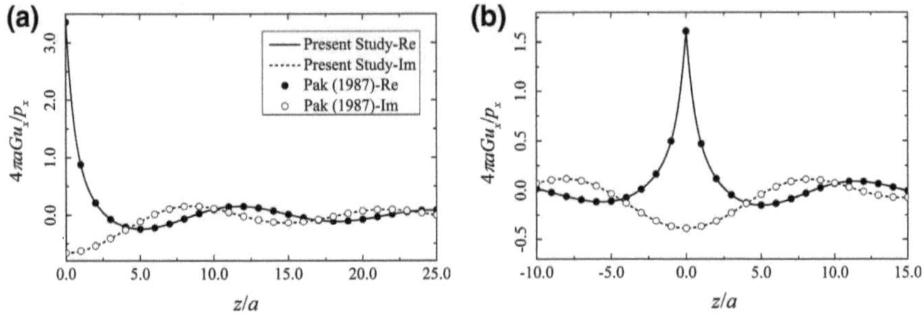

Fig. 8. Comparison of the radial displacement response in the present model with the results in Apsel and Luco (1983)

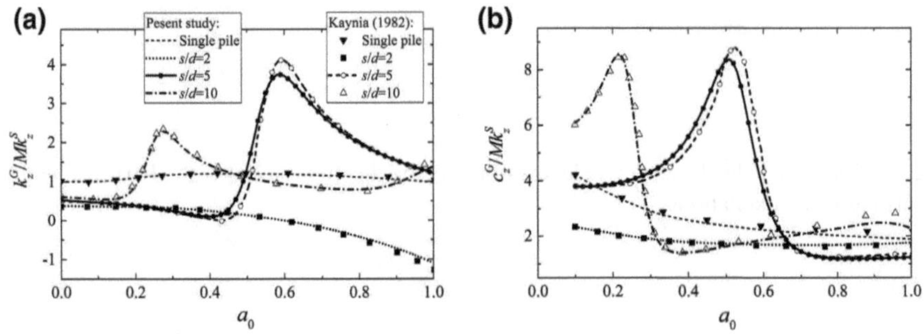

Fig. 9. Comparison of the present model with the results in Apsel and Luco (1983) for the pile groups

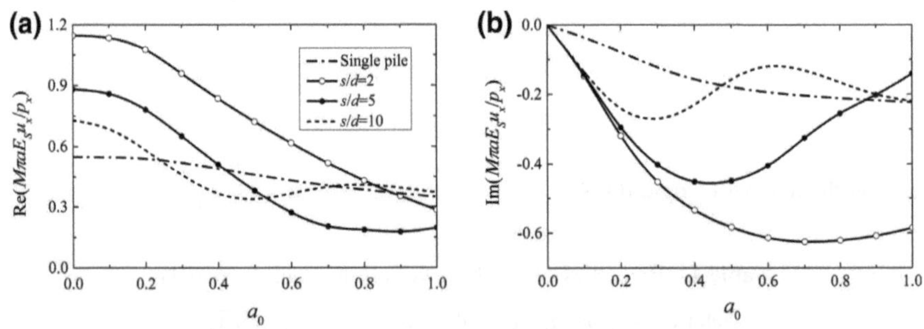

Fig. 10. The effect of the pile spacing on the dynamic response of the pile groups vs the normalized frequency in the transversely isotropic medium

4 Conclusions

A novel semi analytical-numerical computational model is developed in this paper to solve pile group dynamic interactions. The medium is assumed to a linear transversely isotropic half-space. For inhomogeneous and layered soils, the effective transversely isotropic soil parameters can be estimated for an equivalent transversely isotropic and elastic behavior therefore the present model is still valid. In layered or inhomogeneous soils, the wave speed varies with depth, the stress waves refract in the layer interfaces and thus analytical solutions are not available for general cases. Therefore, one is constrained to rely on finite element or boundary element simulations for modeling the pile dynamics. The present model is very efficient in modeling pile dynamics compared to the above numerical methods since it only discretizes the piles along the depth and alleviates the necessities of discretizing the piles' cross sections and the soil. Comparing the results of the present model with the reference solutions proves the accuracy of this model and helps engineers to solve relevant soil dynamics problem more efficiently and accurately.

References

Apsel, R.J., Luco, J.E.: On the Green's functions for a layered half space. part II. Bull. Seism. Soc. Am. **73**(4), 931–951 (1983)

Buchwald, V.: Rayleigh waves in transversely isotropic media. Quart J. Mech. Appl. Math. **14** (3), 293–318 (1961)

Chen, L., Polous, J.: Analysis of pile-soil interaction under lateral loading using infinite and finite elements. Comput. Geotech. **15**(4), 189–220 (1993)

Kashtalyan, M., Rushchitsky, J.J.: Revisiting displacement functions in three-dimensional elasticity of inhomogeneous media. Int. J. Solids Struct. **46**, 3463–3470 (2009)

Khodar, R., Abdel-Mohti, A.: Numerical analysis of pile-soil interaction under axial and lateral loads. Int. J. Concr. Struct. Mater. **8**(3), 239–249 (2014)

Kukukarsalan, S., Banerjie, P.: Inelastic analysis of pile-soil interaction. J. Geotech. Geoenviron. Eng. **130**(11), 1152–1157 (2004)

Liu, W., Novak, M.: Dynamic response of single piles embedded in transversely isotropic layered media. Earthq. Eng. Struct. Dynam. **73**(4), 931–951 (1994)

Padron, L.A., Aznarez, J., Maeso, M.: BEM–FEM coupling model for the dynamic analysis of piles and pile groups. Eng. Anal. Boundary Elem. **31**(16), 473–484 (2007)

Synge, J.: Elastic waves in anisotropic media. J. Math. Phys. **35**(35), 323–334 (1957)

Waas, G., Riggs, H., Werkle, H.: Displacement solutions for dynamic loads in transversely isotropic stratified media. Earthq. Eng. Struct. Dynam. **13**(2), 173–193 (1985)

Wang, C.D., Pan, E., Tzeng, C.S., Han, F., Liao, J.J.: Displacements and stresses due to a uniform vertical circular load in an inhomogeneous cross-anisotropic half-space. Int. J. Geomech. **6**(1), 1–10 (2006)

Yaghmaie, R., Noorzad, A.: 3D dynamic soil-foundations interactions in a transversely isotropic medium-impedance characteristics. In: Engineering Mechanics Institute Conference, University of Southern California, Los Angeles, CA., USA, 8–11 Aug 2010. https://doi.org/10.13140/rg.2.1.1988.0803

Yaghmaie, R., Guo, S., Ghosh, S.: Simulating coupled transient electromagnetic-large deformation dynamical mechanical systems using a novel multi-time scaling method. In: Engineering Mechanics Institute Conference, Northwestern University, Evanston, IL, USA, 4–7 Aug 2013. https://doi.org/10.13140/rg.2.1.4744.3041

Yaghmaie, R., Guo, S., Ghosh, S.: Wavelet transformation based multi-time scaling method for FE simulations of coupled transient electromagnetic nonlinear dynamical systems. In: 17th

US National Congress on Theoretical and Applied Mechanics, Michigan State University, East Lansing, MI, USA. 15–20 June 2014. https://doi.org/10.13140/rg.2.1.4744.3041

Yaghmaie, R., Ghosh, S.: Wavelet transformation induced multi-time scaling (WATMUS) model for coupled transient electro-magnetic and structural dynamics finite element analysis. In: 13th U.S. National Congress on Computational Mechanics, San Diego, CA, USA, 27–30 July 2015

Yaghmaie, R., Asgari, H.: Dynamic Reissner-Sagoci problem for a transversely-isotropic bi-material half-space. In: Fourth Geo-China International Conference, Shandong, China, 25–27 July 2016, pp. 17–24 (2016a). https://doi.org/10.1061/9780784480069

Yaghmaie, R., Asgari, H.: The elastodynamic interaction of a rigid circular foundation embedded in a functionally graded transversely isotropic half-space. In: Fourth Geo-China International Conference, Shandong, China, 25–27 July 2016, pp. 57–64 (2016b). https://doi.org/10.1061/9780784480021

Yaghmaie, R., Asgari, H.: Axisymmetric vibration of an elastic circular plate in an inhomogeneous half-space. In: Fourth Geo-China International Conference, Shandong, China, 25–27 July 2016, pp. 65–72 (2016c). https://doi.org/10.1061/9780784480021

Yaghmaie, R., Guo, S., Ghosh, S.: Multi-time scale coupled transient electro-magnetic and structural dynamics finite element analysis for antenna simulations. In: Engineering Mechanics Institute and Probabilistic Mechanics and Reliability Conference, Vanderbilt University, Nashville, TN, USA. 22–25 May 2016

Yaghmaie, R., Guo, S., Ghosh, S.: Wavelet transformation induced multi-time scaling (WATMUS) model for coupled transient electro-magnetic and structural dynamics finite element analysis. Comput. Methods Appl. Mech. Eng. **303**, 341–373 (2016). https://doi.org/10.1016/j.cma.2016.01.016

Yaghmaie, R., Ghosh, S.: Multi-time scaling based modeling of transient electro-magnetic fields in vibrating media with antenna applications. Comput. Mech. **60**(1), 117–141 (2017). https://doi.org/10.1007/s00466-017-1396-1

Yaghmaie, R., Ghosh, S.: Computational modeling of finite deformation piezoelectric material behavior coupling transient electrical and mechanical fields. J. Comput. Phys. (2018a, in press)

Yaghmaie, R., Ghosh, S.: Two scale wavelet transformation based multi-time scale (WATMUS) method for simulating coupled transient electro-mechanical fields with evolving damage. In: 18th US National Congress on Theoretical and Applied Mechanics, Northwestern University, Evanston, IL, USA. 9–4 June 2018b

Yaghmaie, R., Ghosh, S.: 2-scale multi-time scale computational framework for simulating finite strain transient piezoelectricity with evolving damage (2018c, in press)

Yaghmaie, R., Ghosh, S.: Multi-time scale simulations of coupled transient electro-mechanical fields for modeling finite strain piezoelectricity. In: Engineering Mechanics Institute Conference, Massachusetts Institute of Technology, Cambridge, MA, USA. May 29–June 1, 2018d

The Spectral Finite Element Method for Simulating Wave Propagation in Viscoelastic Soils

Reza Yaghmaie(✉)

Department of Civil Engineering, Johns Hopkins University, 3400 N. Charles Street, Baltimore, MD 21218, USA
ryaghma2@jhu.edu

Abstract. The soil material behavior has been widely modeled with viscoelastic material models; however, choosing the actual material properties within a viscoelastic model is not trivial. This paper presents a rigorous methodology for obtaining the damping and stiffness terms in viscoelastic material model for modeling wave propagation in soils. The viscoelastic material model is calibrated for various deformation modes. The calibrated model is then utilized for simulating wave propagation in soils. The boundary conditions in the far field are treated with non-reflecting boundaries. The present model with calibrated data has a great potential in modeling wave propagations in soils.

1 Introduction

Many geotechnical applications such as structures exposed to dynamic loading such as impact loading, earthquake loads, bridges, machinery foundations, rock mechanics and oil extractions are governed by wave propagations in visoelastic media. In these applications, dynamic properties of soils, e.g. damping, dynamic stiffness and inertial forces govern the actual soil behavior. Hence, accurate viscoelastic material models that can mimic the actual soils behavior and can experimentally be validated are required. Furthermore, modeling the wave propagation in soils is a challenging phenomenon due to the anisotropy of the soil, the interaction of the structure with the soil and vice versa, challenges in defining the size of the no-reflecting boundary and also the computational expenses in discretizing the soil in 3D utilizing finite element method (FEM).

Viscoelastic models have been widely used for wave propagation simulations in soils (see for example Vyalov (1986), Carcione et al. (1988), Lakes (2009), Yaghmaie and Noorzad (2010) and Yaghmaie et al. (2013, 2014, 2015, 2016a)). The most common type of viscoelastic model for wave propagation utilized is the Maxwell model (see for example Tschoegl (1989), Harris and Piersol (2002)). Analytical solutions to elastic waves in anisotropic media is developed by Synge (1957). Rayleigh waves in transversely isotropic media is provided by Buchwald (1961). A viscoelastic model for the dynamic behavior of saturated poroelastic soils is investigated by Bardet (1992). Makris and Zhang (2000) developed a time-domain viscoelastic analysis of

H. Khabbaz et al. (eds.), *New Prospects in Geotechnical Engineering Aspects of Civil Infrastructures*, Sustainable Civil Infrastructures,
https://doi.org/10.1007/978-3-319-95771-5_18

earth structures subjected to earthquake loads. Finite element method (FEM) has been widely used in wave propagations in soils. Stucky and Lord (2001) developed a finite element modeling of transient ultrasonic waves in linear viscoelastic media. Deterministic and stochastic analyses of a nonlinear system with a Biot visco-elastic element is investigated by Spanos and Tsavachidis (2001). Muscolino et al. (2005) investigated the time-domain response of linear hysteretic systems to deterministic and random excitations. Dwivedi et al. (2011) simulated dynamic response of partially sealed circular tunnel in viscoelastic soil condition. Lee et al. (2012) modeled shear waves through a viscoelastic medium induced by acoustic radiation force. The wave propagation in transversely isotropic elastic media is investigated by Yaghmaie and Asgari (2016a, b, c).

One major problem with general viscoelastic models is the computational challenge to calibrate all the material parameters. In many cases in the literature, curve fitting is used however the validity of such approximations for general deformation modes of vibration are unclear. In this paper, a Maxwell based viscoelastic model is introduced to model the behavior of the soil for wave propagation problems. The model is calibrated for longitudinal, flexural and torsional deformation modes. The resulting wave propagation problem is modeled using the finite element model in Yaghmaie et al. (2016b) and Yaghmaie and Ghosh (2017, 2018a, b) where the boundary conditions are treated as non-reflecting boundaries. A few numerical examples are provided to show the applicability of the present constitutive model behavior.

2 Constitutive Material Behavior

The material behavior is assumed to be viscoelastic. In viscoelastic materials, the resulting stresses due to an applied strain field are shifted in the frequency domain. The phase lag for a harmonic stress-strain response at time t is expressed as (see e.g. Lakes (2009))

$$\gamma(t) = \gamma_0 e^{i\omega t}, \quad \tau(t) = \tau_0 e^{i(\omega t + \phi)} \tag{1}$$

where γ_0 and τ_0 are the harmonic amplitudes of the shear strain and stress corresponding to the angular frequency ω and ϕ is the phase lag. The shear modulus of the viscoelastic material consists of an elastic and viscous terms defined as

$$G = G_{elastic} + iG_{viscous}, \quad where\ G_{elastic} = \frac{\tau_0}{\gamma_0}\cos(\phi),\ G_{viscous} = \frac{\tau_0}{\gamma_0}\sin(\phi) \tag{2}$$

where $G_{elastic}$ and $G_{viscous}$ are respectively the elastic and viscous moduli. The loss factor of the material corresponds to the ratio of the vicous to elastic modulus defined as $\eta = \frac{G_{viscous}}{G_{elastic}} = \tan(\phi)$. The Kelvin–Voigt, Maxwell and Standard Linear Solid are commonly used to demonstrate the dilation and time dependent deformations in the constitutive material models and behaviors such as creep and relaxation. The Kelvin–Voigt induces the following deformation (see e.g. Lakes (2009))

$$\gamma(t) = \frac{\tau(t)}{G}\left(1 - e^{-t/t_c}\right) \tag{3}$$

where t_c is the retardation time. In relaxation deformations, the shear stress progressively decreases due to a uniform strain field written as

$$\tau(t) = G\gamma_0 e^{-t/t_r} \tag{4}$$

Where t_r is the relaxation time. The Kelvin–Voigt is very well used for modeling creep deformations however the Maxwell model is more broadly used for modeling the damping in soil. Consider an infinite number of Maxwell models are attached in parallel. The effective shear modulus of the soil can be represented in a series expansion, also known as the prony series as (see e.g. Lakes (2009))

$$G(t) = G_\infty(t) + \sum_{i=1}^{N} G_i(t)e^{-t/t_r^i} \tag{5}$$

where $G_\infty(t)$, $G_i(t)$ and t_r^i are the shear modulus at $t = \infty$, shear modulus and relaxation time of the i-th Maxwell model. The second term in the prony series reduces with progressing time and therefore the $G(t = \infty) = G_\infty(t = \infty)$. Numerous number of terms in the prony series allow the correct representation of the soil however calibration of multiple terms with correct experimental data is required which is not available easily. A shorter version of Eq. (5) using only two shear moduli is written as

$$G(t) = G_\infty(t) + (G_0 - G_\infty)e^{-t/t_r} \tag{6}$$

The calibration of the model is performed using the dynamic finite element module developed by Yaghmaie et al. (2016b) and Yaghmaie and Ghosh (2017, 2018a). The calibrated parameters are shown in Yaghmaie and Asgari (2016a, b, c). The rigorous finite element model developed in Yaghmaie et al. (2016b) and Yaghmaie and Ghosh (2017, 2018a) is designed for implicitly solving dynamic and transient multi-physics problems in the time domain. The code is fully parallelized for solving large number of degrees of freedom (dof). Using this model, the wave propagation in the soil is simulated. The shear strain rate in the Maxwell model is then written as (see the derivations in Yaghmaie and Asgari (2016b))

$$\frac{d\gamma}{dt} = \frac{1}{G}\frac{d\tau}{dt} + \frac{\tau}{Gt_r} \tag{7}$$

Substituting Eqs. (1) into (7) leads to

$$i\gamma_0\omega e^{i\omega t} = \left(\frac{i\tau_0\omega}{G} + \frac{\tau_0}{Gt_r}\right)e^{i(\omega t + \phi)} \tag{8}$$

Dividing the both sides with $\gamma_0 e^{i\omega t}$ leads to

$$i\omega = \left(\frac{i\omega}{G} + \frac{1}{Gt_r}\right)G^* \tag{9}$$

The modulus G^* is obtained by rearranging the terms as

$$G^* = Gt_r^2 + i\omega Gt_r \tag{10}$$

The above dynamic shear complex modulus is utilized in the finite element model in Yaghmaie et al. (2016b) and Yaghmaie and Ghosh (2017, 2018a) for dynamic simulations.

3 Results and Discussions

We modeled a few column tests for specimen with diameter of 20 mm and two lengths as 30 and 50 mm. Figure 1 illustrates the finite element mesh. The mesh consists of 128,000 quadrilateral elements. The bottom of the geometry is fixed from mechanical deformation. A few of the material properties are tabulated in Table 1. The general material properties an constitutive laws can be found in Yaghmaie et al. (2016b) and Yaghmaie and Ghosh (2017, 2018a, b). The spectral decomposition in the finite element model delivers the eigen frequencies for resonant cases. These frequencies are used to calibrate the modulus. The frequency response constitutes of the longitudinal and the flexural modes. The flexural frequencies are generally of the form (see Yaghmaie and Ghosh (2017))

$$\omega = \sqrt{\frac{3EI}{L^3\left(\frac{3}{14}m + \sum_{i=1}^{N} m_i h_i\right)}} \tag{11}$$

where I is the moment of inertia, h_i is the height of each layer and m is the mass of the specimen. The longitudinal frequencies are of the form (see Yaghmaie and Ghosh (2017))

$$\omega = \sqrt{\frac{3AE}{Lm_i}} \tag{12}$$

where A is the cross sectional area. A damping parameter is defined as (see Yaghmaie and Ghosh (2017))

$$\xi = \sqrt{\frac{\delta}{4\pi^2 + \delta^2}}, \quad \delta = \frac{1}{N}\ln\left(\frac{A_i}{A_{i+N}}\right) \tag{13}$$

The values i and N are integer numbers and A is the amplitude. Figure 2 shows the finite element model for verifying the viscoelastic model. In this simulation, the natural

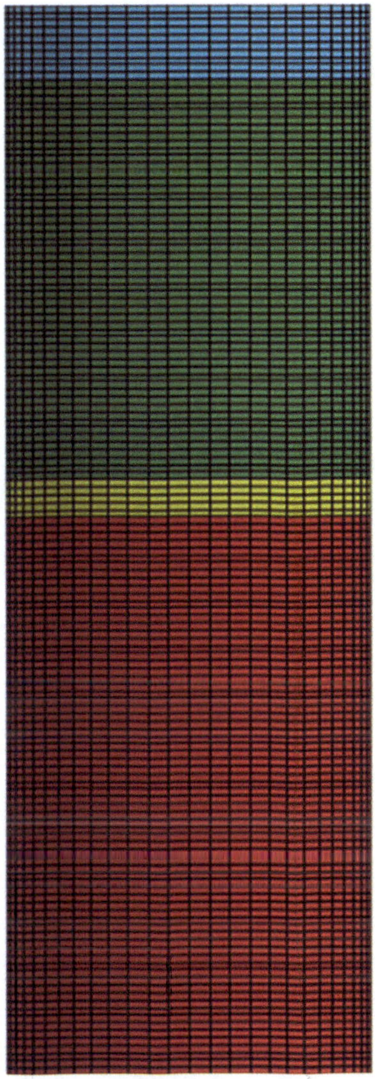

Fig. 1. The mesh for the finite element verification example. Different colors represent (from the top to the bottom) respectively: forcing part, cap, plastic ring and the specimen

Table 1. Properties of the soil in finite element modeling

Density	Bulk modulus	Initial shear modulus	Infinity shear modulus
$\rho\ (\mathrm{kg/m^3})$	K (MPa)	G_0 (MPa)	G_∞ (MPa)
1924	68.54	32.46	0.0

frequencies of the various modes are increased based on the parameters chosen in Table 1. The trapezoidal rule time integration scheme is used to obtain the torsional, flexural, longitudinal and shear frequency modes. Accordingly, harmonic excitations with time periods corresponding to these frequencies for 50 cycles is used to solve the dynamic problem. The response amplitude decays after the steady state solution is achieved. The displacements along the y-directions, z-direction and in the xy-plane are analyzed for all the deformation modes. The damping and shear moduli G_0 and G_∞ in the Maxwell models are varied consistently to investigate their effects on the amplitude decay for various dimensions of the specimen. Numerical investigations proved that the natural frequencies are a function of only two parameters, i.e. the damping and initial shear modulus G_0.

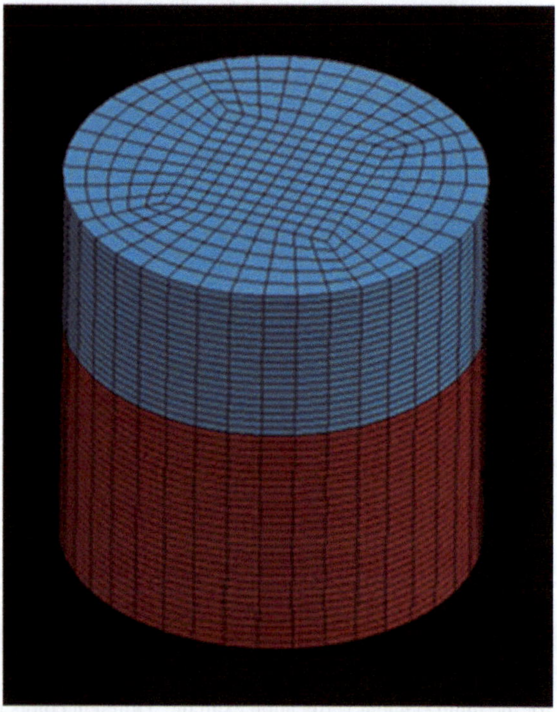

Fig. 2. Finite element mesh for the example considering the damping of the amplitude

Figure 3 shows the nodal displacement. The comparison of the damping ratios for three deformation modes is shown in Fig. 4. The end damping ratios for the flexural and longitudinal modes are smaller than the input damping ratios. These results correspond to the shear modulus representation in Eq. (6). The damping ratios are increased by 20%. The comparisons of the end and input damping ratios for various modes is plotted in Fig. 5. The results in the previous figures are plotted to Fig. 6 to

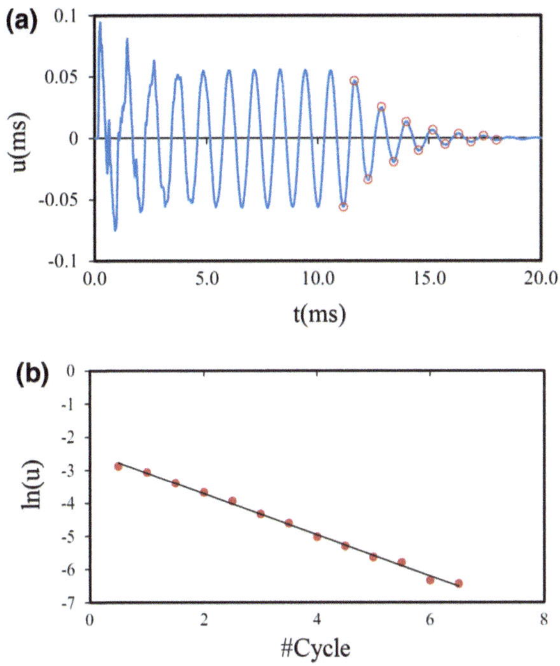

Fig. 3. **a** Time history response of the displacement field, and **b** the displacement amplitude decay with cycles in logarithmic scale. Fitted curve is ln(u) = −0.52N −2.65

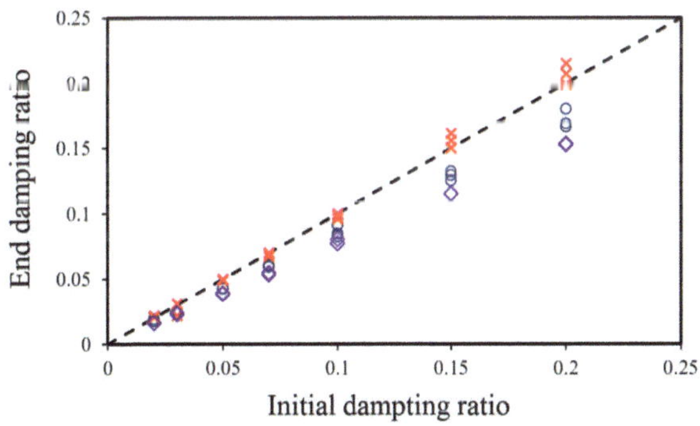

Fig. 4. End versus initial damping ratios for deformation modes including: flexural (circle), torsional (cross) and longitudinal (diamond)

provide a range of variation in the correlation of the output-input damping ratios. Fitting the data to the average curve, leads to the following linear form

$$\xi = 0.47\frac{\beta}{\omega} \tag{14}$$

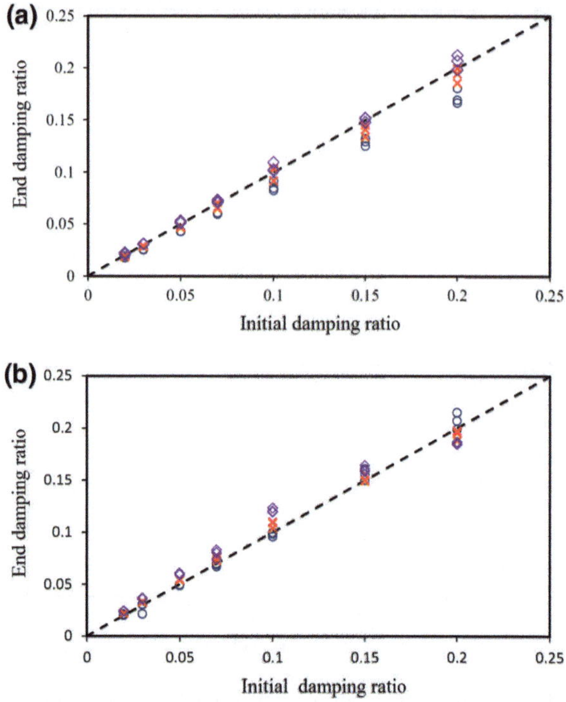

Fig. 5. End versus initial damping ratio for **a** flexural, and **b** longitudinal mode. Damping increased by 0% (circle), 5% (cross) and 20% (diamond)

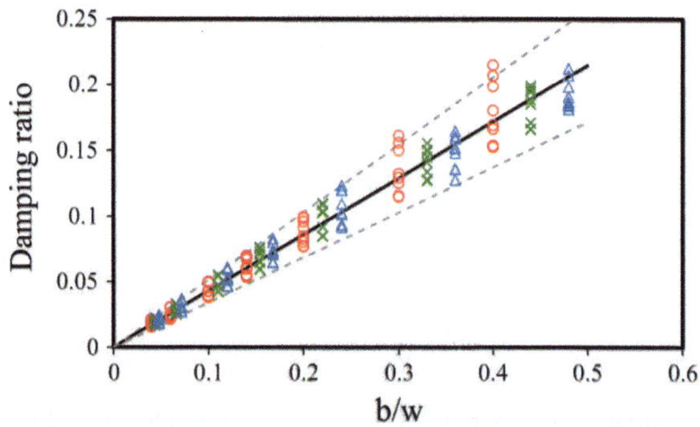

Fig. 6. Variation of damping ratio with β/ω. β increased by 0% (circle), 5% (cross) and 20% (triangle)

The above damping ratio fitted curve corresponds tp the Maxwell material model. Damping ratio is plotted versus $\frac{\beta}{\omega}$ for standard Linear Solid material model in Fig. 7. This viscoelastic response in this model is independent of G_0.

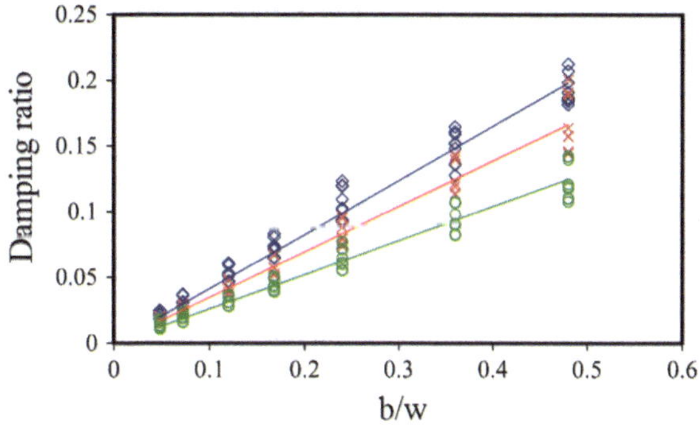

Fig. 7. Variation of damping ratio with β/ω. β increased by 0% (circle), 5% (cross) and 20% (triangle)

The previous mentioned model is used to solve the wave propagation in a viscoelastic medium using the finite element model in Yaghmaie et al. (2016b) and Yaghmaie and Ghosh (2018b, c, d). Figure 8 shows the finite element mesh consisting of 3600 quadrilateral elements. Only the r-z plane is modeled due to the symmetry in the θ-direction. The bottom and the right side of the domain are modeled with non-reflecting boundaries while the left boundary is constrained with rollers.

Fig. 8. The finite element mesh consisting of 3600 quadrilateral elements

A sinusoidal point excitation with frequency of 2π is applied on the middle of top surface. Numerical results for the damping ratio with z-direction is shown in Fig. 9. The damping ratio decreases in the medium with increasing depth.

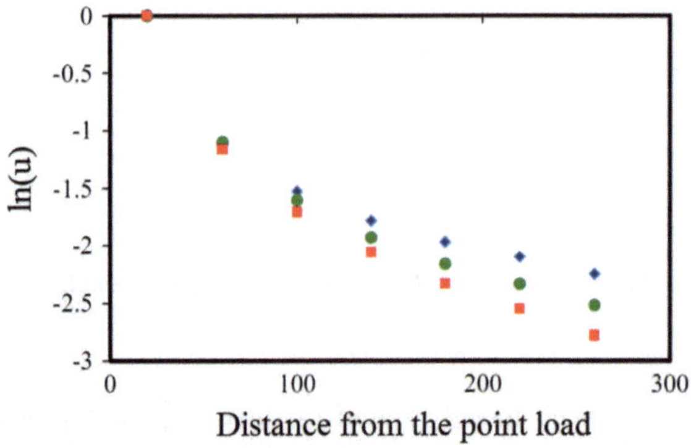

Fig. 9. The displacement amplitude decay with distance from the point load. Zero damping (blue), 5% damping (green) and 20% damping (red)

4 Conclusions

In this paper, a variety of viscoelastic models are reviewed to represent the actual viscoelastic nature of the soil under the dynamic loads. The most appropriate viscoelastic model is calibrated with the available experimental data. For inhomogeneous and layered soils, the effective viscoelastic soil parameters can be estimated equivalently through the proposed parametrized damping relations therefore the present model is still valid. In order to model the wave propagation in the soil medium, the finite element method is utilized. The boundary conditions in the far field are treated with non-reflecting boundaries. Numerical simulations proved that the viscoelastic response is independent of G_0. This model has great potential for solving wave propagation problems in actual soils.

References

Bardet, J.: A viscoelastic model for the dynamic behavior of saturated poroelastic soils. J. Appl. Mech. **59**(1), 128–135 (1992)

Buchwald, V.: Rayleigh waves in transversely isotropic media. Quart J. Mech. Appl. Math. **14** (3), 293–318 (1961)

Carcione, J.M., Kosloff, D., Kosloff, R.: Wave propagation simulation in a linear viscoacoustic medium. Geophys. J. **93**(2), 393–401 (1988)

Dwivedi, J., Singh, V., Lal, R.: Dynamic response of partially sealed circular tunnel in viscoelastic soil condition. Bull. Appl. Mech. **7**(26), 37–45 (2011)

Harris, C.M., Piersol, A.G.: Shock and Vibration Handbook Harris. Fifth ed. Mcgraw-Hill Professional, New York, (2002)

Lakes, R.: Viscoelastic Materials. Cambridge University Press, New York (2009)

Lee, K., Szajewski, A., Hah, Z., Parker, K., Maniatty, A.: Modelling shear waves through a viscoelastic medium induced by acoustic radiation force. Int. J. Numer. Methods Biomed. Eng. **28**(6–7), 678–696 (2012)

Makris, N., Zhang, J.: Time-domain viscoelastic analysis of earth structures. Earthq. Eng. Struct. Dyn. **29**(6), 745–768 (2000)

Muscolino, G., Palmeri, A., Ricciardelli, F.: Time-domain response of linear hysteretic systems to deterministic and random excitations. Earthq. Eng. Struct. Dyn. **34**(9), 1129–1147 (2005)

Spanos, D., Tsavachidis, S.: Deterministic and stochastic analyses of a nonlinear system with a Biot visco-elastic element. Earthq. Eng. Struct. Dyn. **30**(4), 595–612 (2001)

Stucky, P., Lord, W.R.: Finite element modeling of transient ultrasonic waves in linear viscoelastic media. IEEE Trans. Ultrason. Ferroelectr. Freq. Control **48**(1), 6–16 (2001)

Synge, J.: Elastic waves in anisotropic media. J. Math. Phys. **35**(35), 323–334 (1957)

Tschoegl, N.: The Phenomenological Theory of Linear Viscoelastic Behavior. Springer, Berlin Heidelberg (1989)

Vyalov, S.: Developments in geotechnical engineering. Volume 36: Rheological Fundamentals of Soil Mechanics. Elsevier, New York (1986)

Yaghmaie, R., Noorzad, A.: 3D dynamic soil-foundations interactions in a transversely isotropic medium-impedance characteristics. In: Engineering Mechanics Institute Conference, University of Southern California, Los Angeles, CA., USA, 8–11 Aug 2010. https://doi.org/10.13140/rg.2.1.1988.0803

Yaghmaie, R., Guo, S., Ghosh, S.: Simulating coupled transient electromagnetic-large deformation dynamical mechanical systems using a novel multi-time scaling method. In: Engineering Mechanics Institute Conference, Northwestern University, Evanston, IL, USA. 4–7 Aug 2013. https://doi.org/10.13140/rg.2.1.4744.3041

Yaghmaie, R., Guo, S., Ghosh, S. Wavelet transformation based multi-time scaling method for FE simulations of coupled transient electromagnetic nonlinear dynamical systems. In: 17th US National Congress on Theoretical and Applied Mechanics, Michigan State University, East Lansing, MI, USA. 15–20 June 2014. https://doi.org/10.13140/rg.2.1.4744.3041

Yaghmaie, R., Ghosh, S.: Wavelet transformation induced multi-time scaling (WATMUS) model for coupled transient electro-magnetic and structural dynamics finite element analysis. In: 13th U.S. National Congress on Computational Mechanics, San Diego, CA, USA, 27–30 July 2015

Yaghmaie, R., Asgari, H.: Dynamic Reissner-Sagoci problem for a transversely-isotropic bi-material half-space. In: Fourth Geo-China International Conference, Shandong, China, 25–27 July 2016, pp. 17–24 (2016a). https://doi.org/10.1061/9780784480069

Yaghmaie, R., Asgari, H.: The elastodynamic interaction of a rigid circular foundation embedded in a functionally graded transversely isotropic half-space. In: Fourth Geo-China International Conference, Shandong, China, 25–27 July, 2016, pp. 57–64 (2016b). https://doi.org/10.1061/9780784480021

Yaghmaie, R., Asgari, H.: Axisymmetric vibration of an elastic circular plate in an inhomogeneous half-space. In: Fourth Geo-China International Conference, Shandong, China, 25–27 July 2016, pp. 65–72 (2016c). https://doi.org/10.1061/9780784480021

Yaghmaie, R. and Guo, S., Ghosh, S.: Multi-time scale coupled transient electro-magnetic and structural dynamics finite element analysis for antenna simulations. In: Engineering Mechanics Institute and Probabilistic Mechanics and Reliability Conference, Vanderbilt University, Nashville, TN, USA, 22–25 May 2016 (2016a)

Yaghmaie, R., Guo, S., Ghosh, S.: Wavelet transformation induced multi-time scaling (WATMUS) model for coupled transient electro-magnetic and structural dynamics finite element analysis. Comput. Methods Appl. Mech. Eng. **303**, 341–373 (2016b). https://doi.org/10.1016/j.cma.2016.01.016

Yaghmaie, R., Ghosh, S.: Multi-time scaling based modeling of transient electro-magnetic fields in vibrating media with antenna applications. Comput. Mech. **60**(1), 117–141 (2017). https://doi.org/10.1007/s00466-017-1396-1

Yaghmaie, R., Ghosh, S.: A computational model coupling transient electro-mechanical fields for analyzing finite deformation piezoelectric material behavior. J. Comput. Phys. (2018a, in press)

Yaghmaie, R., Ghosh, S.: Two scale wavelet transformation based multi-time scale (WATMUS) method for simulating coupled transient electro-mechanical fields with evolving damage. In: 18th US National Congress on Theoretical and Applied Mechanics, Northwestern University, Evanston, IL, USA, 9–4 June 2018 (2018b)

Yaghmaie, R., Ghosh, S.: 2-scale multi-time scale computational framework for simulating finite strain transient piezoelectricity with evolving damage (2018c, in press)

Yaghmaie, R., Ghosh, S.: Multi-time scale simulations of coupled transient electro-mechanical fields for modeling finite strain piezoelectricity. In: 12-th World Congress in Computational Mechanics (WCCM), New York, NY, USA, 22–27 July 2018 (2018d)

Train Internal Noise Due to Wheel-Rail Interaction

Wenjun Luo[1(✉)], Junnan Jiang[1], Wennian Yu[2], and Chris K. Mechefske[2]

[1] Engineering Research Center of Railway Environment Vibration and Noise, Ministry of Education, East China Jiaotong University, Nanchang 330013, China
lwj06051979@163.com

[2] Department of Mechanical and Materials Engineering, Queen's University, Kingston, ON K7L 3N6, Canada

Abstract. In order to investigate different potentially effective methods to decrease the noise inside passenger trains, this paper employs the acoustic-solid coupling theory and the Finite Element(FE)-Statistical Energy Analysis (SEA) hybrid method to study the vibration response of the train body under wheel-rail excitation and the internal noise response caused by the vibration of the train body. The contribution of plates to the noise inside the train is also analyzed. The results show that the vibration of the floor has the greatest influence on the noise inside the train. Furthermore, compared with the FE method alone, the FE-SEA hybrid method shortens the computation time and improves the efficiency of the calculation.

Keywords: Vibration · Noise inside trains · Acoustic-solid coupling
Wheel-rail excitation · Hybrid methods

1 Introduction

With the continued development and widespread use of high-speed railways, scholars and engineers have intensively investigated the vibration and noise response of these systems. Results show that low and intermediate frequency internal noise caused by structure-acoustic coupling has the greatest influence on the human body and comfort. In addition, this problem is difficult to solve based on the noise transmission route. In order to forecast the internal noise of high-speed trains efficiently, many scholars (Craggs 1972; Shen et al. 2011; Ivanov et al. 2017) have explored new research methods.

The methods of numerical analysis for structure and noise include the Finite Element Method (FEM), the Boundary Element Method (BEM), Statistical Energy Analysis (SEA) and FE-SEA hybrid methods. Zheng (2015) analyzed the low frequency internal noise of the train by FEM software. The Acoustic Transmission Vector (ATV) was introduced to analyze the contribution of the plate to the sound pressure point. Wu and Ge (2014) used the FEM to simulate the low-frequency sound field in

H. Khabbaz et al. (eds.), *New Prospects in Geotechnical Engineering Aspects of Civil Infrastructures*, Sustainable Civil Infrastructures,
https://doi.org/10.1007/978-3-319-95771-5_19

high speed train compartments. The FE models of the train compartments with and without racks were established respectively, and the sound pressure level of the standard point and sound field distribution in these two cases were compared. FEM and BEM are usually employed to analyze the global vibrations of trains at low frequencies and are not appropriate to analyze high frequency local vibration because of inefficiency. Normally these two methods are employed for the case when the vibration frequency is less than 200 Hz. SEA is used to analyze the vibration and noise of structures under high frequency excitation. Liu et al. (2011) calculated the high frequency internal aerodynamic noise under the fluctuating pressure excitation based on the SEA theory. The result showed that the noise in the cab and passenger compartment of the train is mainly from the windows and the floor. Muto et al. (2017) established a SEA model to predict the internal noise in a passenger cabin of a high-speed train and used the approximated power injection method (APIM) to estimate SEA parameters. The calculation results were compared with the experimentally measured internal noise in a moving train.

This paper analyzes the train body-acoustic cavity coupling system, which can be divided into several subsystems. The modal density of each subsystem is different, and it is therefore difficult to ensure that the modal density of each subsystem is suitable for FEM or SEA. Because of this situation, the FE-SEA hybrid model, which combines the respective advantages of each method, can be adopted to achieve better results. The FE-SEA hybrid method is the combination of FEM, employed to the low frequency calculations, and SEA, used for the high frequency calculations. This method can be employed in the analysis of a complex structure over the entire frequency range of interest (Luo et al. 2013). In fact, Cotoni et al. (2005) have already used the FE-SEA hybrid method to simulate the structure noise in the intermediate frequency range, and proved the effectiveness and accuracy of this method.

In this paper, the FE-SEA hybrid method is employed to predict internal noise of carriages (also known as China Railway High-speed train) under the secondary suspension force. It is designed to provide a fast and accurate method for the prediction of structure-borne noise in a train. Firstly, the modal density of each part of the train body is solved, and the FE and SEA subsystems are classified according to the modal density. Then, the coupled model of train body-acoustic cavity is established with FE-SEA hybrid method. Within the frequency range of 0–200 Hz, the train body is established using the FE model. Within the frequency range of 200–500 Hz, the roof, sidewall and end wall are established using a SEA model. The glass, sidewall between glass, floor and chassis are established using FE model. The internal acoustic cavity is established using the SEA model. Based on the theory of acoustic and solid coupling, the junctions between structure and structure, structure and acoustic cavity, acoustic cavity and acoustic cavity are created. The secondary suspension force is obtained by the analysis of the train-track model, which is used as the excitation input of the train body and internal acoustic cavity coupling model, and the vibration of the train body and the internal noise are solved in the frequency domain.

2 Internal Structure Noise Prediction Model Based on the FE-SEA Hybrid Method

2.1 Principle of the FE-SEA Hybrid Method

The FE-SEA hybrid method was first introduced (Langley and Cordioli 2009). This method is based on the wave coupling theory. The complex system is divided into two kinds of subsystem to reduce the calculation time. The total degrees of freedom contained in the FE subsystem are expressed as the column vector $\boldsymbol{q}$, and $\boldsymbol{f}$ is a column vector composed of external harmonic excitation forces acting on the respective degrees of freedom of the FE subsystem. Therefore, the equation of motion of the FE subsystem can be expressed as (Langley et al. 2005)

$$\boldsymbol{D}_{tot}\boldsymbol{q} = \boldsymbol{f} + \sum_{k=1}^{N} \boldsymbol{f}_{rev}^{(k)} \tag{1}$$

The aggregate stiffness matrix of the FE subsystem is expressed as

$$\boldsymbol{D}_{tot} = \boldsymbol{D}_{d} + \sum_{k=1}^{N} \boldsymbol{D}_{dir}^{(k)} \tag{2}$$

where N is the total number of SEA subsystems; $\mathbf{f}_{rev}^{(k)}$ is the reverberation field load imposed by the kth SEA subsystem at the connection boundary; $\boldsymbol{D}_d$ is the dynamic stiffness matrix of the FE subsystem at the connection boundary; and $\boldsymbol{D}_{dir}^{(k)}$ is the direct field dynamic stiffness matrix of the kth SEA subsystem. The reverberation load on the coupling boundary is

$$\boldsymbol{S}_{ff}^{(k)} = \left(\frac{4E_k}{\pi\omega n_k}\right) \mathrm{Im}\left\{\boldsymbol{D}_{dir}^{(k)}\right\} \tag{3}$$

The average power flow of the direct field input to the jth SEA subsystem can be expressed as

$$P_{in,j} = P_{in,j}^{ext} + \sum_{k=1}^{N} P_{in,k\to j} \tag{4}$$

$$P_{in,j}^{ext} = \frac{\omega}{2} \sum_{rs} \left[\mathrm{Im}\left\{\boldsymbol{D}_{dir.rs}^{(j)}\right\} \left(\boldsymbol{D}_{tot}^{-1} \boldsymbol{S}_{ff} \boldsymbol{D}_{tot}^{-H}\right)_{rs} \right] \tag{5}$$

$$P_{in,k\to j} = \omega \eta_{jk} n_j \frac{E_k}{n_k} \tag{6}$$

The output power flow can be expressed as

$$P_{out,j} = P_{tran,j \to d} + \sum_{k=1}^{N} P_{tran,j \to k} \tag{7}$$

$$P_{\mathrm{tran},j \to k} = \omega \eta_{kj} n_k \frac{E_j}{n_j} \tag{8}$$

$$P_{\mathrm{tran},j \to d} = \omega \eta_{d,j} E_j \tag{9}$$

The average power flow consumed by the subsystem itself is

$$P_{diss,j} = \omega \eta_j E_j \tag{10}$$

Thus, the energy balance equation of the subsystem can be expressed as

$$P_{in,j}^{ext} + \sum_{k=1}^{N} \left[\omega \eta_{jk} n_j \frac{E_k}{n_k} \right] = \omega \eta_j E_j + \sum_{k=1}^{N} \left[\omega \eta_{kj} n_k \frac{E_j}{n_j} \right] + \omega \eta_{d,j} E_j \tag{11}$$

Due to symmetry: $n_j \eta_{jk} = n_k \eta_{kj}$, Eq. (11) can be expressed as

$$\frac{\omega}{2} \sum_{rs} \left[\mathrm{Im}\left\{ \boldsymbol{D}_{dir.rs}^{(j)} \right\} \left(\boldsymbol{D}_{tot}^{-1} \boldsymbol{S}_{ff} \boldsymbol{D}_{tot}^{-H} \right)_{rs} \right] = \omega \left(\eta_j + \eta_{d,j} \right) E_j + \sum_{k=1}^{N} \omega \eta_{jk} n_j \left(\frac{E_j}{n_j} - \frac{E_k}{n_k} \right) \tag{12}$$

where $P_{in,j}^{ext}$ is the total power outside input to the jth subsystem; r, s represent the degree of freedom at the deterministic boundary of the jth SEA subsystem; η_j is the damping loss factor for the jth SEA subsystem; $\eta_{d,j}$ is the coupling loss factor between the FE subsystem and the jth SEA subsystem; E_j, E_k represent the statistical energy response of the jth and kth SEA subsystem respectively; n_j, n_k represent the modal density of the jth and kth SEA subsystem respectively.

The acoustic power radiation of the ith subsystem is

$$W_i = \rho_{air} c_{air} \sigma_i S_i \left\langle v_i^{-2} \right\rangle \tag{13}$$

where ρ_{air} is the air density, c_{air} is the speed of sound in air, σ_i is the radiation efficiency of the ith subsystem, S_i is the surface area of the ith subsystem, $\left\langle v_i^{-2} \right\rangle$ is the mean square velocity.

2.2 Analysis Process

Figure 1 shows the analysis process of train internal structure noise. First, the wheel-rail force in time domain can be obtained by solving the FE model of train-track coupling vibration. The wheel-rail force in the time domain is transformed into the frequency domain by Fast Fourier Transform (FFT). Second, the FE-SEA model of the

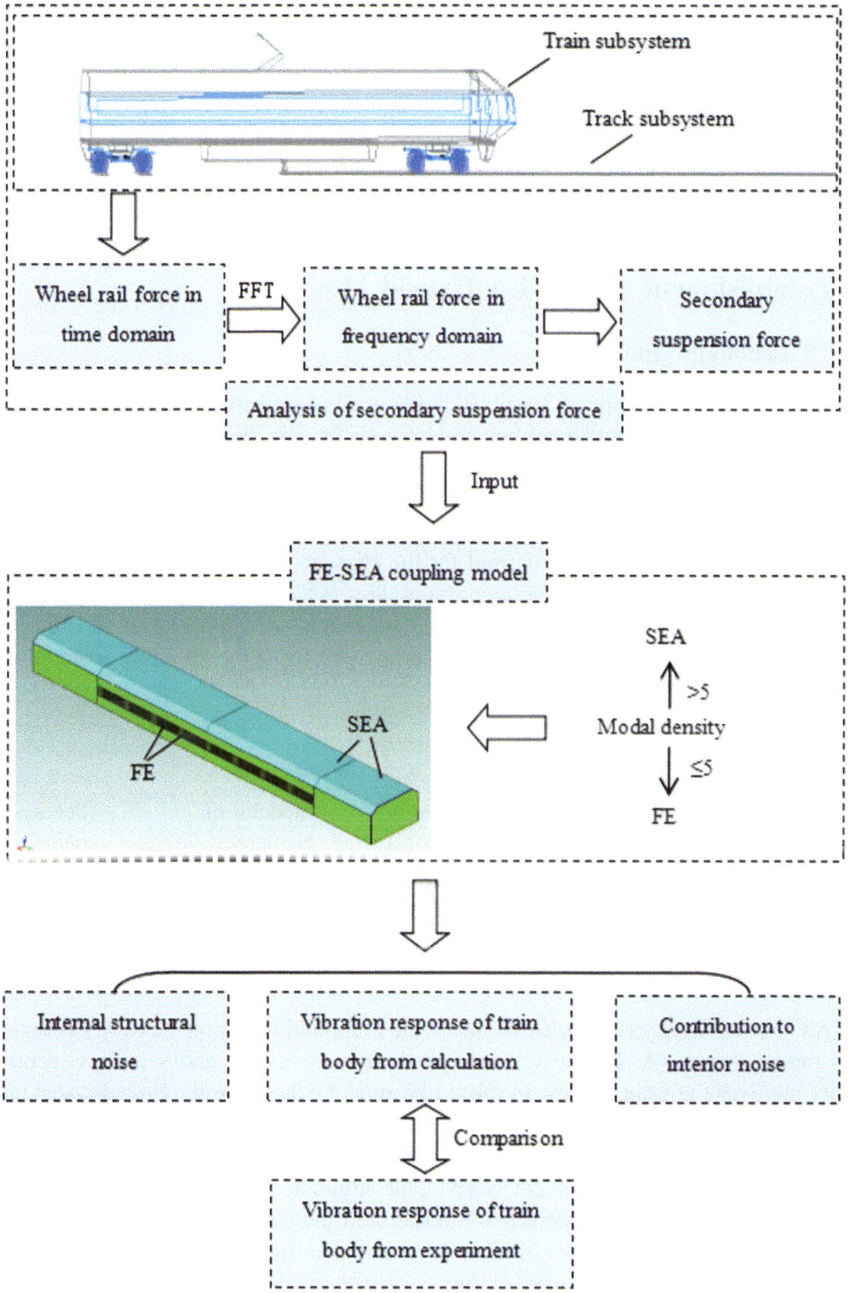

Fig. 1. Analysis process of train internal structure noise

noise in the train according to modal density is established, and the secondary suspension force transformed by wheel-rail force is applied to the train-acoustic cavity coupling model to calculate the vibration response of the train body structure. Third, the calculation results are compared with the experimental results to verify the accuracy of the model. Then the internal structure noise contribution of the noise in the train is analyzed based on the FE-SEA model. Finally, the measures to reduce the structure noise in the train are put forward.

3 Establishment of FE-SEA Hybrid Model

3.1 Calculation Model

A China Railway High-speed 2 train (CRH2) is employed to establish the train body and internal acoustic cavity coupling model with the FE-SEA hybrid method in this paper. A previous SEA model for a train body was established in the *VA One* software. A single curved shell element is simulated for the roof. Plate elements are simulated for the sidewall, end wall, floor and glass windows. A beam element is simulated for the chassis. Aluminum alloy 7N01 is used for the chassis. Aluminum alloy 6N01 is used for the sidewall and roof, and the aluminum alloy 5083 is used for the end wall. The material properties of these three aluminum materials (density, elastic modulus and Poisson's ratio) are almost the same. The properties for some structure materials are shown in Table 1.

Table 1. Properties for some structure materials

Structure	Type	Material	Density (kg/m^3)	Modulus of elasticity (GPa)	Poisson's ratio
Sidewall, roof	Sandwich	Aluminum	2700	71	0.3
Window	Laminated	Glass	2500	48.5	0.2398

An acoustic cavity is established inside the train, and this acoustic cavity is divided into three sections (A, B, and C). Section A acoustic cavity and section C acoustic cavity are over the train bogies. In these two sections there are the boiled water room and the washroom, respectively. Therefore, the acoustic cavity here should not be divided vertically. In section B, the acoustic cavity is divided into three layers vertically, i.e. the standing acoustic cavity (B1), the sitting acoustic cavity (B2) and the leg acoustic cavity (B3). They represent the position of the ears when passengers stand, sit and squat. The SEA models for the train body and acoustic cavity are shown in Figs. 2 and 3, respectively.

The relationship between the modal density and frequency is shown in Fig. 4, which indicates that the modal density increases with increasing frequency. Similar to previous research (Liu 2015; Yu 2012), when the modal density of the subsystems is over 5, these subsystems are estimated using the SEA model, and the remaining

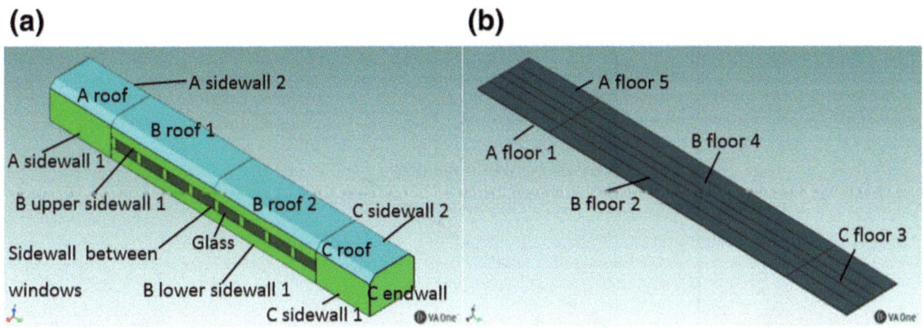

Fig. 2. SEA model for train body: **a** sidewall and roof, **b** floor

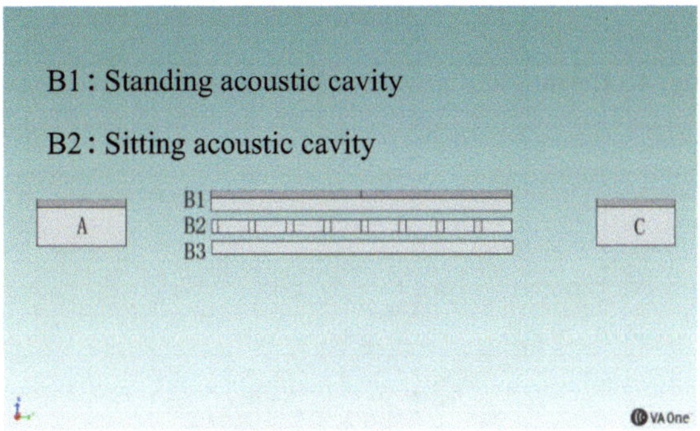

Fig. 3. SEA model for acoustic cavity

subsystems are estimated using the FE model. The modal density of the floor, roof and sidewall are over 5 at 200 Hz, and the modal density of the glass and the sidewall between the windows are less than 5. So calculation should be divided into two parts. The cutoff point is 200 Hz.

Within the frequency range of 0–200 Hz, the modal density of all the plates is less than 5, and they are established using the FE model and the desired number of elements per wavelength is 6. The total number of elements is 31,040. The FE model is shown in Fig. 5. Then the modes of the train body are analyzed. The top 6 order modes are rigid modes, which are not considered. The 7th–16th order modal frequencies are shown in Table 2. The 7th order frequency is 10.6 Hz, and the mode is torsion of the middle of the sidewall and roof. The 8th order frequency is 11.6 Hz, and the mode is bending of the whole structure in the vertical direction. The 9th order frequency is 12.4 Hz, and the mode is vibration of the middle of the sidewall in the horizontal direction, and vibration of the roof and floor in the vertical direction. The 10th order frequency is 14.6 Hz, and the mode is the second order vibration of the middle of the sidewall in the horizontal direction, and the second order vibration of the roof and the floor in the vertical direction.

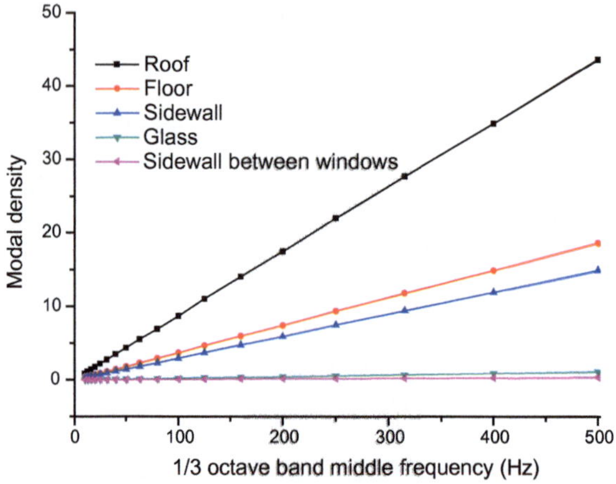

Fig. 4. Relationship between modal density and frequency

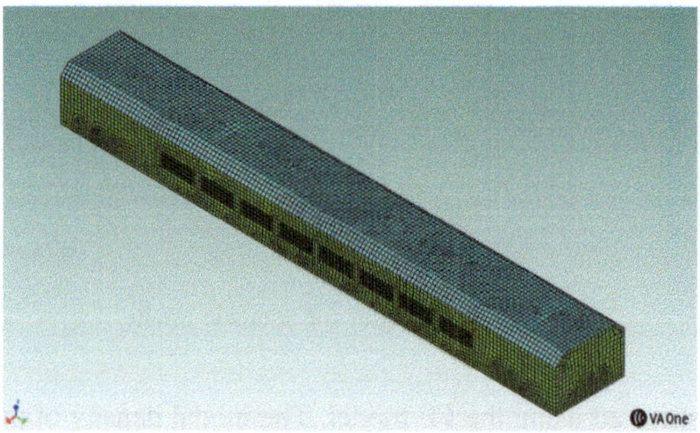

Fig. 5. FE model

Table 2. 7th–16th order model frequency

Order	Frequency	Order	Frequency
7th	10.6	12th	15.9
8th	11.6	13th	16.4
9th	12.4	14th	16.8
10th	14.6	15th	17.3
11th	14.8	16th	17.5

Within the frequency range of 200–500 Hz, the roof, sidewall and end wall are established using the SEA models as their modal densities are more than 5. The glass, sidewall between glass panels, floor and chassis are established using FE models as their modal densities are less than 5. The FE-SEA hybrid model is shown in Fig. 6. The junctions between structure and structure, between structure and acoustic cavity, and

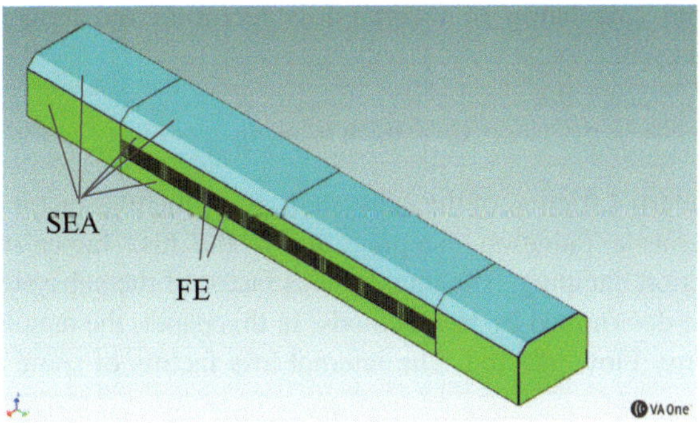

Fig. 6. FE-SEA hybrid model

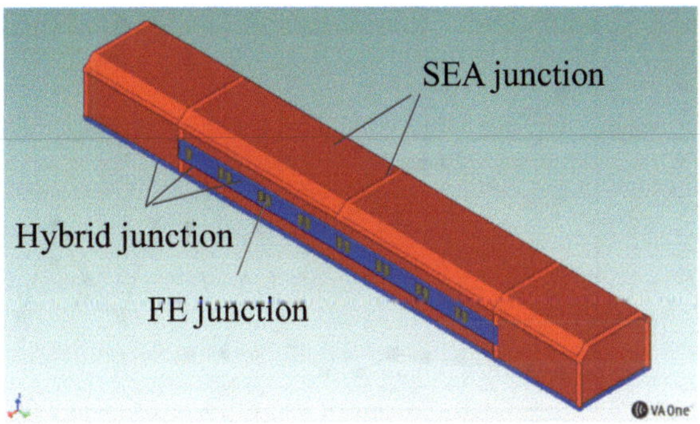

Fig. 7. Hybrid junctions

between acoustic cavity and acoustic cavity are created, with the purpose of transmitting different kinds of energies among subsystems. The hybrid junctions are shown in Fig. 7. The red portions represent the SEA junctions. The blue portions represent the hybrid junctions and the gray portions represent the FE junctions.

3.2 Parameter Calculation

In the *VA One* software, model parameters include modal density, internal loss factor and coupling loss factor. Internal loss factor is the ratio of the energy loss of the subsystem to the average energy stored in a unit time in a unit frequency. Since the energy loss is determined by the subsystem damping characteristics; the internal loss factor is also called the damping loss factor. The coupling loss factor is used to represent the amount of power flow or damping effect when one subsystem is connected to another; it is an important parameter of energy exchange between coupling subsystems. *VA One* software can automatically calculate the modal density and the coupling loss factor (Blanchet and Matla 2010). Therefore, only the internal loss factor is calculated in this research, including the calculation of the internal loss factor in the train body and internal acoustic cavity.

The theoretical calculation of internal loss factor of the train body adopts the following formula:

$$\eta_i = \eta_{is} + \eta_{ir} + \eta_{ib} \tag{14}$$

where η_{is} is the internal friction of the subsystem; η_{ir} is the loss factor of the subsystem vibration and acoustic radiation damping; η_{ib} is the loss factor of the subsystem boundary connection damping. The internal loss factor of the subsystem with complex structure could be determined by test methods. In this paper, the data is obtained by the Steady-state Energy Flow Method. The internal loss factors of some plates are shown in Fig. 8.

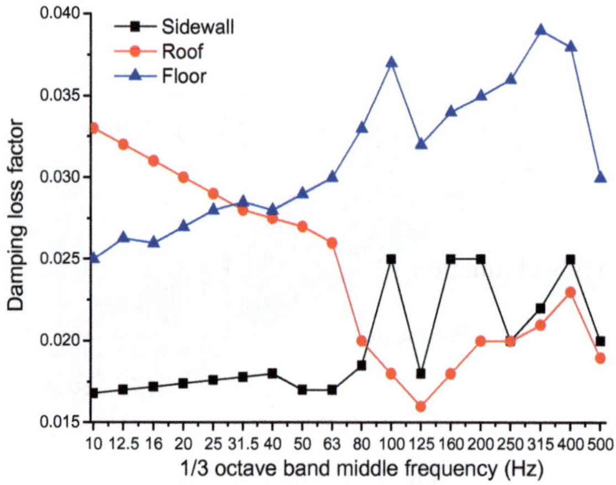

Fig. 8. Damping loss factor of some plates

The internal acoustic cavity loss factor is calculated by theoretical calculation (Ouis 2002; Mao 2015):

$$\eta = \frac{13.82}{T_{60}\omega} \tag{15}$$

where T_{60} is the reverberation time, referring to the time required for the sound pressure level to attenuate by about 60 dB from the steady state level, when the sound field reaches a steady state and the sound source is stopped. Some acoustic cavity damping loss factors are shown in Fig. 9.

3.3 Excitation Calculation

In this paper, the train track coupling model with the CRH2 model as reference is established by using SIMPACK software. During the modeling process, the train body is assumed as a rigid body, whose parameters include train body mass, rolling inertia, nodding inertia, shaking head inertia of train body, vertical stiffness of primary spring,

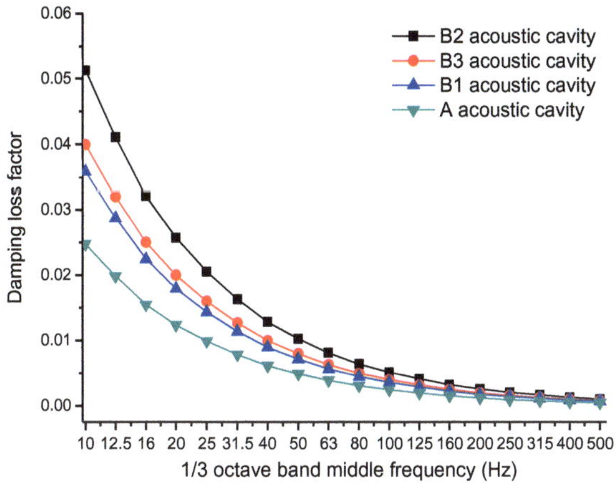

Fig. 9. Damping loss factor of some acoustic cavities

vertical damping of primary shock absorber and so on. The values of these parameters are selected based on an actual train. The train parameters are listed in Table 3. The train-track coupling model is shown Fig. 10. When the train runs in a uniform speed, the wheel rail excitation is mainly in the vertical direction. Therefore, only the vertical cxcitation is considcrcd in this papcr.

Table 3. Parameters for train

Parameters	Values	Parameters	Values
Train body mass	33.93 t	Shaking head inertia of bogie	1979 t*m^2
Rolling inertia of train body	109.9 t*m^2	Height of barycentre of bogie	–0.575 m
Nodding inertia of train body	1636.2 t*m^2	Wheel set mass	1627 kg
Shaking head inertia of train body	1569.1 t*m^2	Vertical stiffness of primary spring	919.8 kN/m
Height of barycentre of train body	1.66 m	Vertical damping of primary shock absorber	10 kN*s/m
Mass of bogie	1790 kg	Vertical stiffness of secondary spring	195 kN/m
Rolling inertia of bogie	1261 t*m^2	Vertical damping of secondary shock absorber	10 kN*s/m
Nodding inertia of bogie	844 t*m^2		

60 kg/m track is adopted for the rails of high-speed railway in China, and the speed of the train is usually more than 300 km/h. China Academy of Railway Sciences tested a lot on such rails and get the power spectrum density formula of the track irregularity

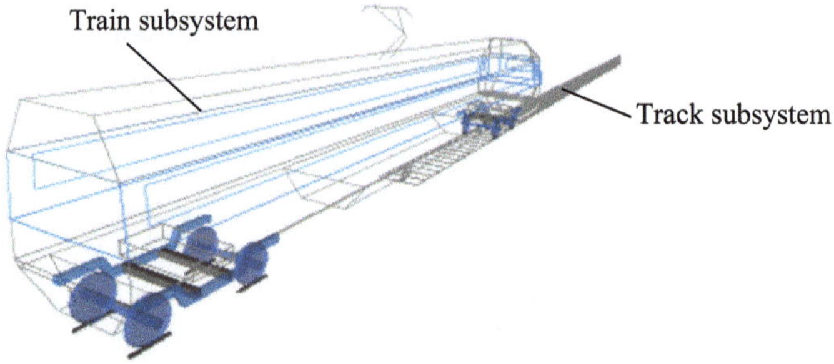

Fig. 10. Train-track coupling model

reflecting the characteristics of such rails. Power spectrum density for Chinese rails of 60 kg/m track can be expressed as:

$$S(f) = \frac{0.127(f^2 - 2.1531f + 1.5503)}{f^4 + 4.9835f^3 + 1.3891f^2 - 0.0327f + 0.0018} \tag{16}$$

where f is the spatial frequency (cycle/m).

Under the excitation of wheel-rail irregularity (Steinwolf 2014a; Steinwolf 2014b), the wheel-rail force in the vertical direction is calculated at the speed of 350 km/h. The wheel-rail force, decayed by the bogie suspension system, forms the secondary suspension force on the corbel at the end of air spring. The train body vibration responses are caused by the secondary suspension force transmitted to the chassis through the corbel. Figure 11 shows the secondary suspension vertical force spectrum at the speed of 350 km/h.

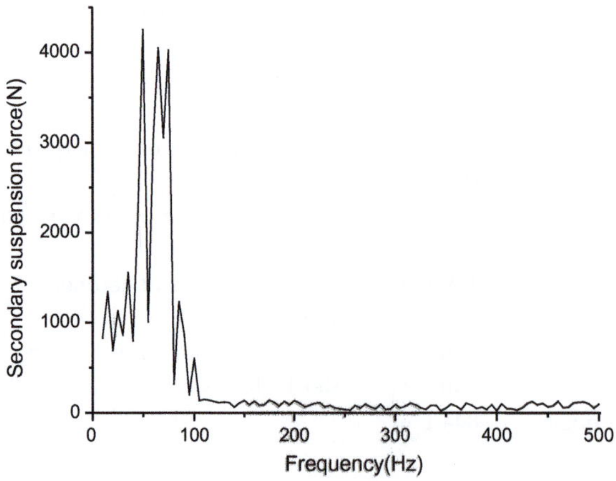

Fig. 11. Secondary suspension force spectrum

3.4 Model Validation

The vertical excitation is applied in the model to calculate the vibration of the train plates. Then the vibration acceleration level of some plates at the frequency of 50–500 Hz can be obtained, as shown in Fig. 12, which shows a good agreement with the experimental results provided in reference (Yu 2012), except that the vibration acceleration level of the sidewall obtained from model is larger than that from the experiment. The reason is that the aerodynamic noise outside causes the vibration of the side wall plate during the experiment, while in this paper only secondary suspension force is considered. The comparison in Fig. 12 can verify the accuracy of the model.

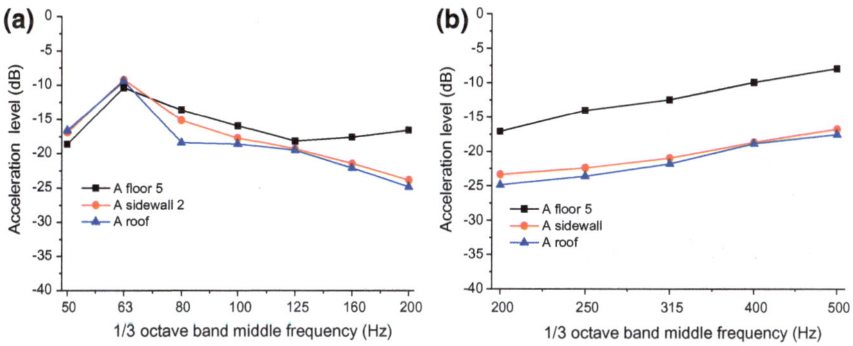

Fig. 12. Acceleration spectrum for comparison: **a** 0–200 Hz, **b** 200–500 Hz

In this paper, the highest analysis frequency is 500 Hz. When this model is calculated based on FEM, the number of elements is over 150,000 after being meshed. In order to investigate the computational efficiency, the computation time of this method is compared with that of the traditional method. It was found that the modal analysis within the frequency range of 0–200 Hz described above needs only 23 min of calculation time. Compared with the computation time of 2 h in reference (Zheng 2015), this method greatly shortens the calculation time, and improves the efficiency.

4 Vibration Analysis of Train Body

The vibration velocity level of some plates is shown in Fig. 13. It can be seen that the velocity level decreases sharply within the frequency range of 63–125 Hz. The reason is that this is the frequency of the secondary suspension force. Meanwhile, for most frequencies, the maximum vibration is at the floor, followed by the sidewall, and the velocity level of the roof is lowest.

The comparison of the velocity level calculated by the two models (FE model and SEA model) at 200 Hz is shown in Table 4. It can be seen that the difference is lower than 4 dB, meaning that the result is in good agreement at the cutoff point. Therefore, when the plates of train body structure are classified as FE or SEA, the proposed classification strategy is reasonable.

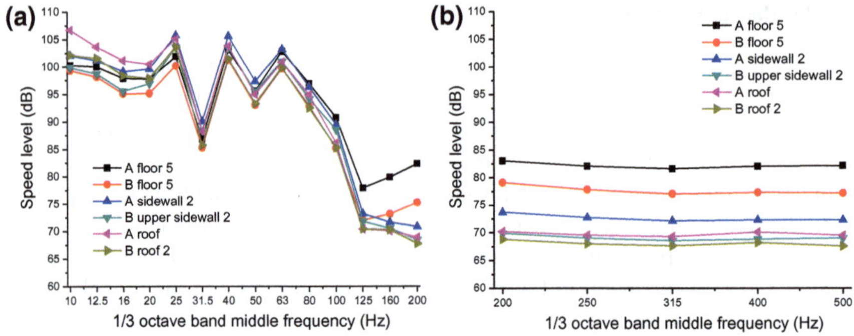

Fig. 13. Speed level spectrum: **a** 0–200 Hz, **b** 200–500 Hz

Table 4. Comparison of the speed level of the two models at 200 Hz

Speed level (dB)	A floor 5	A sidewall 2	A roof	B floor 5	B upper sidewall 2	B roof 2
0–200 Hz model	82.4	70.8	68.9	75.3	68.5	67.8
200–500 Hz model	83.0	73.7	70.2	79.1	70.0	68.9
Difference	−0.7	−2.9	−1.3	−3.9	−1.5	−1.1

The contour plots of the overall speed level of plates at different frequency ranges are shown in Fig. 14. It can be seen that the vibration speed level of the end plate is larger than that of the middle plate. This is because that the secondary suspension force is at the end of the train body and the vibration energy is transmitted from the end plate to the middle plate by coupling junction. Moreover, the vibration speed level of each plate within 0–200 Hz is higher than that within 200–500 Hz. It indicates that the second suspension force has greater influence on low frequency vibration.

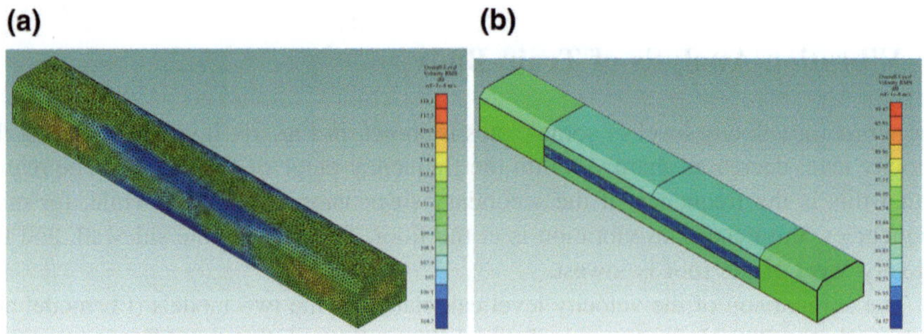

Fig. 14. Contour for Overall Speed Level: **a** 0–200 Hz, **b** 200–500 Hz

5 Analysis of Internal Structure Noise

5.1 Analysis of Sound Pressure Level

When the train runs at the speed of 350 km/h, the sound pressure level of the internal noise spectrum obtained by simulation is shown in Fig. 15. It can be seen that the difference of the sound pressure level calculated by the two models is less than 2 dB at 200 Hz. Besides, the variation trend of internal noise is almost the same as the variation trend of the secondary suspension force, which indicates that the vibration of the train body caused by the track irregularities is the source of the internal noise. The greater the excitation load, the more significant of the vibration, and the louder of the noise generated. Therefore, decreasing the wheel-rail force is an effective way to decrease the internal noise.

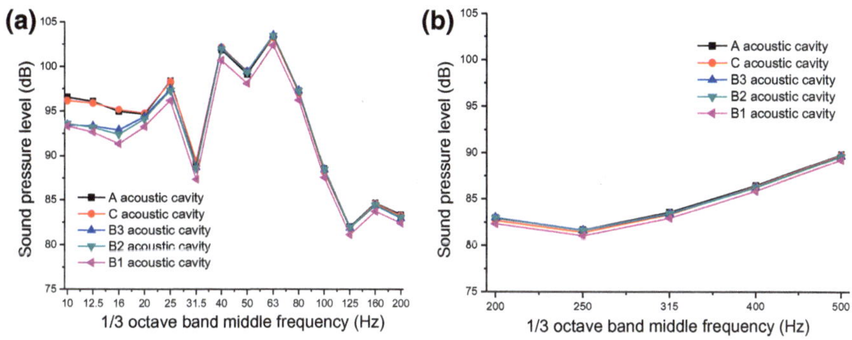

Fig. 15. Internal noise spectrum: **a** 0–200 Hz, **b** 200–500 Hz

The contour plots for overall sound pressure level at different frequency ranges are shown in Fig. 16. It can be seen that the difference of noise at different positions is negligible. The reason is that when under the secondary suspension force, the difference of vibration of each plate is low, thus the difference of structure noise at different positions caused by vibration is negligible. Meanwhile, acoustic cavities are coupled with each other and the acoustic energy is flowing among them. Moreover, sound pressure level of each acoustic cavity within 0–200 Hz is higher than that within 200–500 Hz. This is because that the vibration caused by the second suspension force at low frequency is more significant.

5.2 Analysis on the Contribution of Internal Structure Noise

The middle of the train is the passenger compartment. In this section the acoustic cavity is divided into three parts, namely, the standing acoustic cavity (B1), the sitting acoustic cavity (B2) and the leg acoustic cavity (B3). Under normal circumstances, the ears of passengers are in the sitting acoustic cavity position. Therefore, the contribution to the sitting acoustic cavity (B2) is of utmost concern. Figures 17, 18, 19, 20 show the main power level contribution to these acoustic cavities.

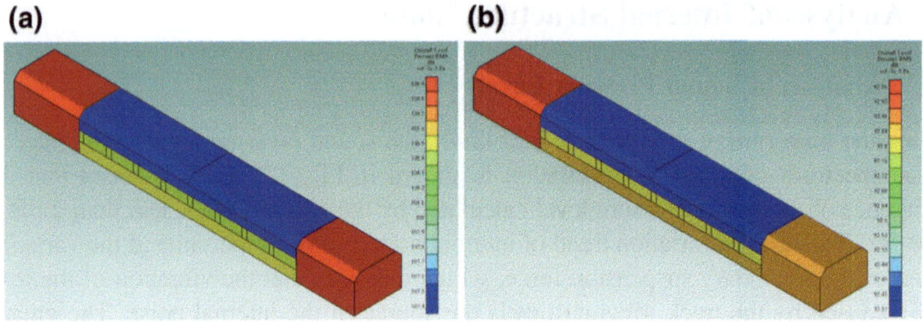

Fig. 16. Contour for overall sound pressure level: **a** 0–200 Hz, **b** 200–500 Hz

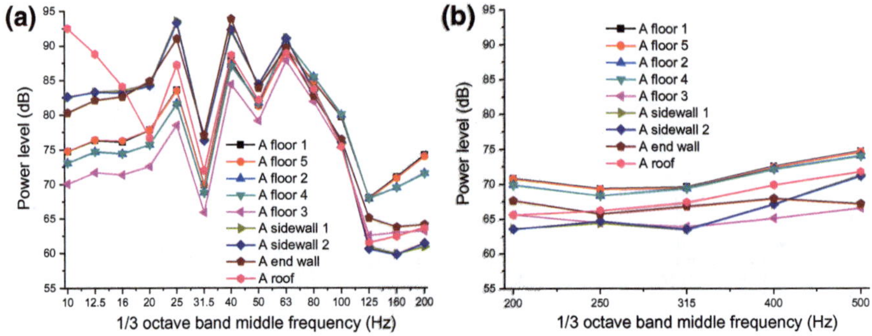

Fig. 17. Contribution to the A Acoustic Cavity: **a** 0–200 Hz, **b** 200–500 Hz

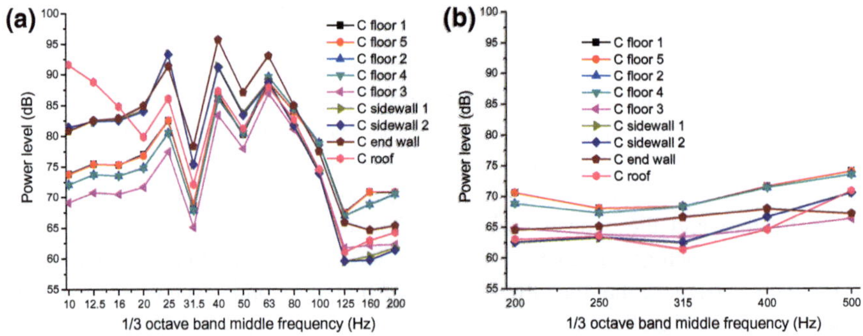

Fig. 18. Contribution to the C Acoustic Cavity: **a** 0–200 Hz, **b** 200–500 Hz

Power level contributions to the end acoustic cavities (A, C) are shown in Figs. 17 and 18. It can be seen that the noise in A acoustic cavity and C acoustic cavity mainly come from the vibration of the floors, followed by the vibration of the end walls. The power level contribution to the leg acoustic cavity (B3) is shown in Fig. 19. It can be seen that the noise in the leg acoustic cavity (B3) is mainly from the vibration of floors and the inflow of acoustic energy from the end acoustic cavities (A, C).

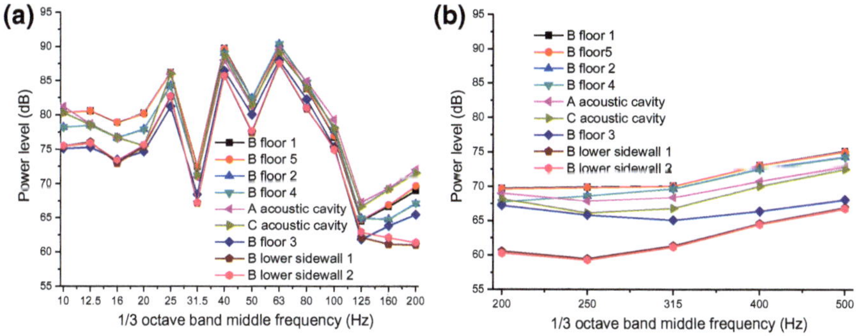

Fig. 19. Contribution to the B3 Acoustic Cavity: **a** 0–200 Hz, **b** 200–500 Hz

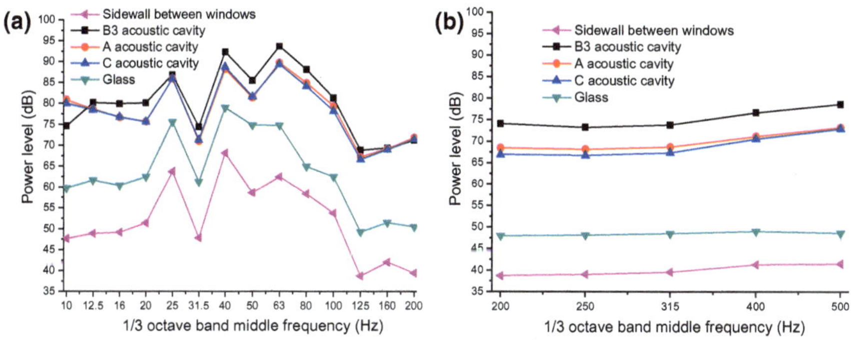

Fig. 20. Contribution to the B2 Acoustic Cavity: **a** 0–200 Hz, **b** 200–500 Hz

Power level contributions to the sitting acoustic cavity (B2) are shown in Fig. 20. It can be seen that the leg acoustic cavity (B3) has the highest contribution to the sitting acoustic cavity (B2), followed by the inflow of acoustic energy from the end acoustic cavities (A, C), the contribution to the sitting acoustic cavity (B2) from the vibration of the glass and the sidewalls between the windows are low. Therefore, the contribution to the sitting acoustic cavity (B2) is mainly considered as the inflow of acoustic energy from the leg acoustic cavity (B3) and the end acoustic cavities (A, C). The contribution to the leg acoustic cavity (B3) and the end acoustic cavities (A, C) are mainly the vibration of the floors.

In summary, the main contribution to the sitting cavity (B2) is the vibration of floors and the energy is flowing through the leg acoustic cavity (B3) and the end acoustic cavities (A, C) into the sitting cavity (B2). Thus, reducing the vibration of the floors to reduce the noise of the sitting acoustic cavity (B2) is the most effective measure. Besides, it can improve the track state and wheel-rail contact state, reduce the wheel-rail interaction and optimize the train body structure. Meanwhile, reducing the acoustic energy flowing into the sitting acoustic cavity (B2) from the lower part and the end of the train is advisable. That is, measures should be taken to reduce noise transmission, such as adding noise absorption material under the passenger seats or sound proofing treatments at the end of the passenger compartment.

6 Conclusions

An FE-SEA hybrid method was employed to analyze the relationship between the noise inside a train and the frequency response. According to the modal density of subsystems and the condition of sufficient accuracy, the structure is established using a FE-SEA hybrid model. Compared with the method that the whole structure is established using a FE model, the proposed method greatly shortens the computation time, and improves the efficiency. Therefore, the FE-SEA hybrid method is an effective method to forecast the noise inside a train.

Under the vertical secondary suspension force, the difference between noise levels at different positions is low, and the variation trend of internal noise is almost the same as the variation trend of the secondary suspension force. The vibration of the train body caused by track irregularity is shown to have significant influence on the internal noise and decreasing the wheel-rail force is an effective way to decrease the internal noise.

Vibration of the train body was analyzed at 200 Hz. The velocity level calculated by the two models is almost the same, with a difference of less than 4 dB. Compared with other plates, the vibration velocity level of the floors is the most significant. In addition, the second suspension force has greater influence on low frequency vibration.

Through the analysis of the contribution of the acoustic cavities inside the train, it was found that floors have the greatest influence on the internal noise, and therefore relevant measures of floor vibration reduction strategies can effectively reduce the noise inside the train. The control of the noise around the passenger's ears can also be done by taking measures such as applying noise absorption material under the passenger's seats or adding soundproofing treatment at the end of the passenger compartment.

Acknowledgements. This research was partly supported by National Natural Science Foundation of China (51468021), National Natural Science Foundation of China (51768022), Jiangxi Province Outstanding Young Funding Scheme (20162BCB23048), Jiangxi Province Natural Science Foundation (20161BAB206160) and Key Young Foundation of Jiangxi Province (20171ACB21037).

Appendix

See Table 5.

Table 5. A list of notation

$\boldsymbol{q}$	A column vector of total degrees of freedom contained in the FE subsystem
f	A column vector composed of external harmonic excitation forces acting on the respective degrees of freedom of the FE subsystem
$f_{rev}^{(k)}$	Reverberation field load imposed by the *k*th SEA subsystem at the connection boundary
$\boldsymbol{D}_{tot}$	Aggregate stiffness matrix of the FE subsystem
$\boldsymbol{D}_d$	Dynamic stiffness matrix of the FE subsystem at the connection boundary
$\boldsymbol{D}_{dir}^{(k)}$	Direct field dynamic stiffness matrix of the *k*th SEA subsystem
$\boldsymbol{S}_{ff}^{(k)}$	Reverberation load on the coupling boundary
$P_{in,j}$	Average power flow of the direct field input to the *j*th SEA subsystem
$P_{in,j}^{ext}$	Total power outside input to the *j*th subsystem
r, s	Degree of freedom at the deterministic boundary of the SEA subsystem
η_j	Damping loss factor for the *j*th SEA subsystem
$\eta_{d,j}$	Coupling loss factor between the FE subsystem and the *j*th SEA subsystem
E_j	Statistical energy response of the *j*th SEA subsystem respectively
n_j	Modal density of the *j*th SEA subsystem respectively
W_i	Acoustic power radiation of the *i*th subsystem
ρ_{air}	Air density
c_{air}	Speed of sound in air
σ_i	Radiation efficiency of the *i*th subsystem
S_i	Surface area of the *i*th subsystem
v_i	Velocity
η_{is}	Internal friction of the subsystem
η_{ir}	Loss factor of the subsystem vibration and acoustic radiation damping
η_{ib}	Loss factor of the subsystem boundary connection damping
η	Internal acoustic cavity loss factor
T_{60}	Reverberation time
f	Spatial frequency

References

Blanchet, D., Matla, S.: Building SEA predictive models to support vibro-noise ship design. Presented at the ESI global forum, Munich, Germany, 19–20 May 2010

Cotoni, V., Gardner, B., Shorter, P. et al.: Demonstration of hybrid FE-SEA analysis of structure-borne noise in the mid frequency range. SAE Technical Paper 2005-1-2331 (2005)

Craggs, A.: The use of simple three-dimensional acoustic finite elements for determining the natural modes and frequencies of complex shaped enclosures. J. Sound Vib. **23**(4), 331–339 (1972)

Ivanov, N.I., Boiko, I.S., Shashurin, A.E.: The problem of high-speed railway noise prediction and reduction. Procedia Eng. **189**, 539–546 (2017)

Langley, R.S., Cordioli, J.A.: Hybrid deterministic-statistical analysis of vibro-acoustic systems with domain couplings on statistical components. J. Sound Vib. **321**(3–5), 893–912 (2009)

Langley, R.S., Shorter, P.J., Cotoni, V.: A hybrid FE-SEA method for the analysis of complex vibro-acoustic systems. J. Acoust. Soc. America **117**(1), 85–95 (2005)
Liu, J.L., Zhang, J.Y., Zhang, W.H.: Calculation method of interior aerodynamic noises with middle and high frequencies for high-speed train. J. Traffic Transp. Eng. **11**(3), 55–60 (2011)
Liu, Q.M.: Prediction of structure-borne noise from railway composite bridge and suppression study of constrained layer damping. Dissertation, Southwest Jiaotong University, Chengdu, China (2015) (in Chinese)
Luo, W.J., Lei, X.Y., Lian, S.L.: Analysis on vibration of ballastless track-bridge system based on hybrid FE-SEA method. J. China Railway Soc. **34**(8), 94–101 (2013). (in Chinese)
Mao, J.: High-speed train internal full-spectrum noise prediction and sound quality optimization under multi-physical-field coupling excitation. Dissertation, Zhejiang University, Hangzhou. (in Chinese) (2015)
Muto, D., Takano, Y., Takeichi, M.: Statistical energy analysis of interior noise in a high-speed train. Presented at the proceedings of international symposium on seed-up and service technology for Railway and maglev systems: STECH. The Japan Society of Mechanical Engineers, pp. 136–140 (2017)
Ouis, D.: An acoustical technique for determining the loss factor of solid materials. J. Test. Eval. **30**(6), 497–500 (2002)
Shen, H.M., Zhang, Y.M., Xiao, X.B.: Low-noise optimization design of external corrugated floor for high-speed train. J. Traffic Transp. Eng. **11**(2), 65–71 (2011)
Steinwolf, A.: Vibration testing by non-Gaussian random excitations with specified kurtosis. Part I: Discussion and methods. J. Test. Eval. **42**(3), p JTE20120277, p13 (2014a)
Steinwolf, A.: Vibration testing by non-Gaussian random excitations with specified kurtosis. Part II: Numerical and experimental results. J. Test. Eval. **42**(3) p JTE20120278, p15. (2014b)
Wu, D., Ge, J.M.: Analysis of the influence of racks on high speed train internal noise using finite element method. Appl. Mech. Mater. **675–677**, 257–260 (2014)
Yu, Y.: Prediction and control of internal noise of super-high-speed train based on hybrid FE-SEA method. Dissertation, Southwest Jiaotong University, Chengdu, China. (in Chinese) (2012)
Zheng, J.H.: Research on the analysis and prediction of the internal noise of high-speed train in low frequency domain. Dissertation, Shandong University, Jinan, China. (in Chinese) (2015)

Sustainable Use of Reclaimed Ballast Rejects for Construction of Rail Corridor Access Road-an Australian Experience

M. Mirzababaei[1(✉)], T. Decourcy[2], and B. Fatahi[3]

[1] Central Queensland University, Melbourne, Australia
m.mirzababaei@cqu.edu.au
[2] Aurizon, Rockhampton, Australia
troy.decourcy@aurizon.com.au
[3] University of Technology Sydney (UTS), Sydney, Australia
behzad.fatahi@uts.edu.au

Abstract. Ballast cleaning practice results in depositing large quantities of fouled ballast in the rail corridors on and around existing access roads, drains and embankments. Therefore, such stockpiles of waste ballast may result in safety issues for the staff driving on rail access roads and moreover, may damage the drainage paths around the railway embankments leading to reduce the service life of the railway formation. Once ballast is cleaned, remaining materials with a particle size of smaller than 30 mm (i.e., called fine ballast) are discarded and therefore, the accumulation of such waste is challenging. In this study, the potential re-use application of fine ballast to construct access roads is investigated. This helps to remove the waste stockpiles and therefore, reduce the access road construction costs. A series of Modified Proctor compaction tests and CBR tests carried out on the fine ballast samples collected from stockpiles along Central Queensland coal network (CQCN) in Rockhampton area. The results indicated that the sample with a particle size of smaller than 26.5 mm performs satisfactorily in soaked CBR tests. An access road was constructed at the Waitara rail yard on the Goonyella line using the fine ballast (<26.5 mm) and the post-construction static plate load tests showed less deflection of the compacted base compared to that of the roads constructed with standard materials (<19.0 mm). Therefore, the waste fine ballast provided the confidence that it can perform well as an access road material.

1 Introduction

The design origins of gravel roads as they are today is traced back to the Scottish Engineer John McAdam who developed unsealed roads in early 1800s. In his design, a layer of single-sized crushed rock laid on top of the existing ground compared to the earlier design by Thomas Telford who used a stone foundation of 450 mm thickness (Pavement Interactive 2008). The proposed pavement system could resist the applied vehicles' load while minimising the road surface wear and protect the soil underneath from being failed. Present day's gravel roads are similar to the Macadam's design,

H. Khabbaz et al. (eds.), *New Prospects in Geotechnical Engineering Aspects of Civil Infrastructures*, Sustainable Civil Infrastructures,
https://doi.org/10.1007/978-3-319-95771-5_20

although they incorporate a mixture of larger aggregates as well as finer materials to improve binding and prevention of water penetration.

Aurizon as the owner and operator of one of the world's largest coal rail networks, linking approximately 50 mines with three major ports in Queensland, is Australia's largest rail freight operator which transports more than 250 million tonnes of Australian commodities annually in Australia. Aurizon network has approximately 2670 km of heavy haul railway track throughout the Central Queensland area. This track is generally constructed of ballasted track whereby the rail and sleeper assembly are laid on a bed made of 19–63 mm granular material that is in turn laid above the constructed formation. The ballast bed's thickness on most coal lines ranges from 250 to 300 mm when constructed new (Fig. 1). The ballast is employed to provide a number of functions including the following (Mirzababaei et al. 2014):

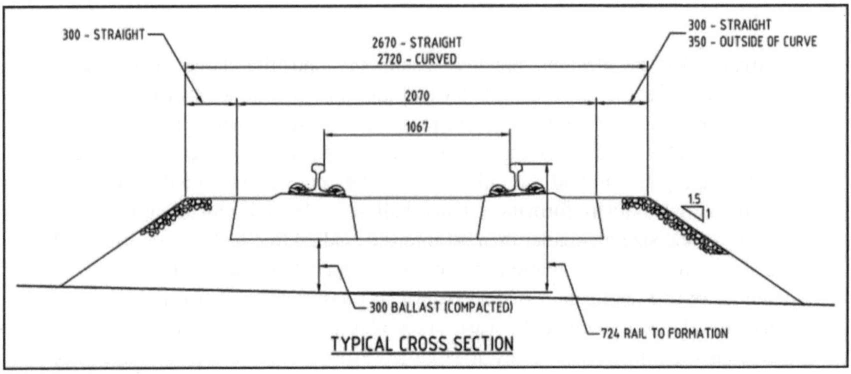

Fig. 1. Typical Aurizon heavy haul track cross-section (from Aurizon standard drawing 10673)

i. Providing lateral restraint of the rail and sleepers to prevent track buckles;
ii. Evenly distributing train axle loading to prevent formation damage due to the track "punching" into it; and
iii. Providing drainage during wet weather events.

Ballast layer deteriorates progressively due to the heavy train cyclic loads and this may end up in differential settlement of the rail track and therefore, a poor lateral drainage (Fatahi and Khabbaz 2011). Throughout the ballast bed's lifecycle, it gradually collects parasitic coal dust and other fine materials. The other fines are generally derived from:

iv. Dropped sand (from locomotive) onto the railhead;
v. Smaller particles and fines from the ballast itself as it degrades, chips and grinds;
vi. Powdered material from the concrete sleepers;
vii. Formation material that finds its way up into the ballast bed;
viii. Dust from the surrounding landscape; and
ix. Organic material from grasses and weeds on the railway corridor.

Accumulation of trapped fines results in losing the ability of the ballast to drain effectively and therefore, the track structure begins to ‘pump’ as trains pass over and subsequently, coal fines drop from the coal wagons which then trap more water, and this cycle continues. Once this process reaches a certain stage, it will become unsafe for normal line speed and the ballast will have to be renewed.

Aurizon annually renews/cleans the equivalent of approximately 152 km of standard railway track ballast materials. On average 30% of the cleaned ballast is rejected as being unsuitable for re-use back into track and is, therefore, replaced with new and clean materials. Current heavy haul rail standards for Aurizon require 2500 m^3 of ballast for every lineal kilometre of track. If the track has experienced alignment issues, however, maintenance activities can raise this figure to 3900 m^3 of ballast per kilometre. This equates to 114,500–229,000 m^3 of rejected ballast being discarded in railway corridors annually.

Fouled ballast that is stockpiled close to the existing waterways is considered to contain quantities of coal fines and a certain quantity of fine soil that has the potential to contaminate the waterways via runoff during the rain events. The coal fines and soil particles could produce a high suspended solid level in water and increase the turbidity in the stream and can slow down the process of photosynthesis by limiting the light transmission through water. This can have a negative impact on the waterway’s aquatic ecosystem. Moreover, if fines do settle in the water, they also tend to create an additional problem by smothering the aquatic plant life and organisms that live on the stream bed (Aurizon 2016). Based on the Aurizon’s policies, it is illegal to actively deposit the contaminants in form of fouled ballast, sand, silt and mud closer than 50 m to the bank of nearby waterways.

Aurizon has been undertaking a program over the past 25 years to clean ballasted tracks, using undercutting process in order to ensure the track structure performs at its optimum level. The amount of fouled ballast already stored in the rail corridors has become substantial. For this reason in addition to the desire for reducing the operating costs, Aurizon has created the need to clean and re-use as much of this material as possible back on track. Therefore, the rejected fouled ballast is processed through a mobile grading plant and then the suitably sized materials (i.e., 30–63 mm) are thoroughly washed and stockpiled ready for re-use back into the track as fresh ballast materials. Any material less than 30 mm is discarded and set aside in a rejected material stockpile as it is not suitable for re-use in operational track conditions. Such waste, coal-fouled ballast being discarded in the rail corridor has been a growing concern to the point where it is now becoming a major issue to both safety of staff and the performance of the Aurizon infrastructure.

A substantial quantity of access roads is required for periodic quality assurance and maintaining the minor and major failures throughout the railway network. The current approved method for constructing access roads on Aurizon land is based on a simple design of a 200 mm thick layer of imported material that matches the specification of material required for constructing a railway formation with a minimum CBR ratio of 45%. Figure 2 depicts the grading curve required for constructing the Aurizon’s formation roads. The ‘B’ grading material is of higher quality and therefore, more expensive to produce and as such should only be utilised in track formation situations. The ‘C’ graded material has better lateral drainage properties that are needed when no final seal is provided and is, therefore, the preferred type.

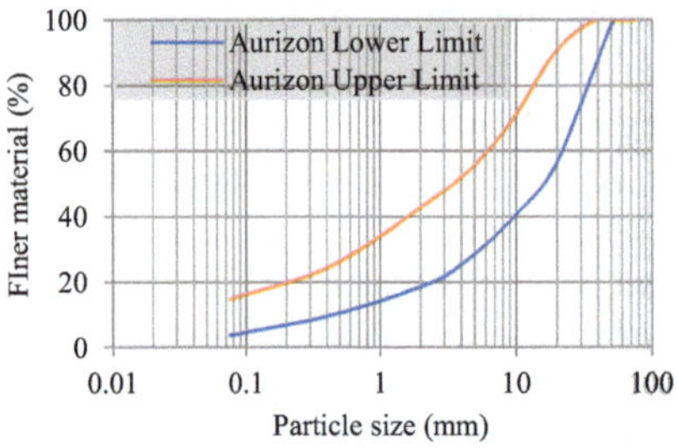

(a) Grade 'B' Aurizon standard formation material

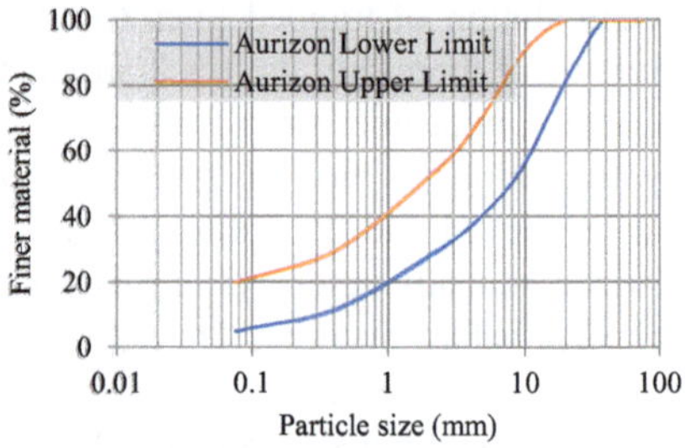

(b) Grade 'C' Aurizon standard formation material

Fig. 2. Grading 'B' and 'C' particle size criteria for Aurizon's formation

Recycled materials in road construction appear to be quite common with both Department of Transport & Main Roads (DTMR) Queensland and Austroads providing documentation to cover their use. Recycled materials such as crushed concrete, blast furnace slag and crushed stone are often used however, they are generally constructed as part of the sub-base and mostly are covered by a surface seal/wearing surface. For example, DTMR requires a minimum of 100 mm thick surface to be constructed above any recycled materials and requires that they are not used on any road constructed as an unbound pavement (DTMR MRTS35 2014).

The New South Wales transport department also has a standard that outlines the construction requirements for their access roads in rail corridors. This standard advises that an access road can be constructed using spent ballast however, there is no commentary on whether the spent ballast material is regraded to remove larger than 26.5 mm material or whether fouling by coal fines should be considered (ASA Access Roads Standard 2014). The use of fines recovered from spent railway ballast or even the entire product does not appear to have had any notable studies undertaken to assess their performance as an unsealed road construction medium. There are studies that involve using crushed, waste concrete and waste asphalt (Hoy et al. 2017; Yaowarat et al. 2018), however, there is no documentation on utilising fines from recycled and washed ballast.

Therefore, in this study, the authors have investigated the suitability of rejected material smaller than 30 mm for use as access road construction materials. The roads are presently constructed using material specified for railway formation performance under heavy and frequent train loading of 26.5 tonnes per axle. As access roads experience significantly lower traffic tasks (i.e., less than 100 per day), a material of only slightly lesser quality would be expected to perform adequately, especially if the total life cost is reduced through the use of essentially 'free' material.

1.1 Methodology

A quantity of rejected materials was taken from the existing stockpiles at a trial waste ballast cleaning site along CQRN. The samples were gathered directly from two stockpiles and from multiple locations on each stockpile in accordance with AS1141.3.1 (2012) to ensure that an average representative sample of the entire material was analysed. The collected sample was riffled in accordance with AS1141.11.1 (2009) for the sieve analysis. The dry sieve analysis result of the soil sample based on AS1141.11.1 (2009) is shown in Fig. 3. The soil was categorised as a well-graded gravel (GW) based on the USCS classification.

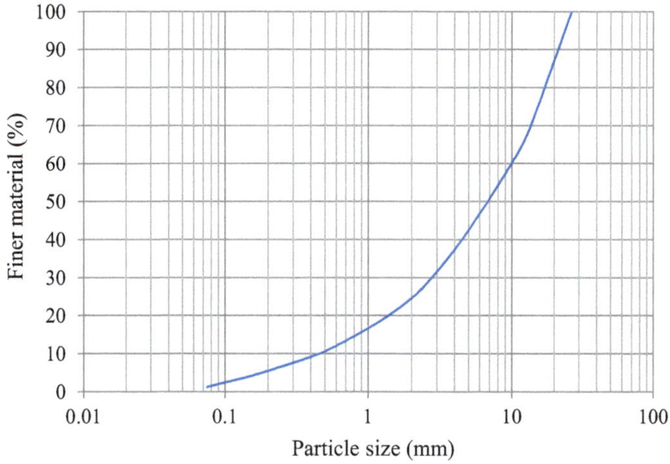

Fig. 3. Particle size analysis of the representative soil sample from stockpiles

In order to investigate the proficiency of the different particle size bands of the rejected waste material, the collected materials were further divided into 4 categories of (a) less than 26.5 mm (grade I), (b) less than 19.0 mm (grade II), (c) less than 13.2 mm (grade III), and (d) less than 9.5 mm (grade IV). The maximum dry density (MDD) and optimum moisture content (OMC) of each grade were measured in accordance to AS 1289.5.2.1 (2003). Figures 4 and 5 show the compaction curves of the soil samples. The maximum dry density of grades I to IV were determined as 3.03, 2.25, 2.18 and 2.16 t/m^3, respectively with optimum moisture contents of 4.30, 5.05, 5.70 and 6.44%, respectively.

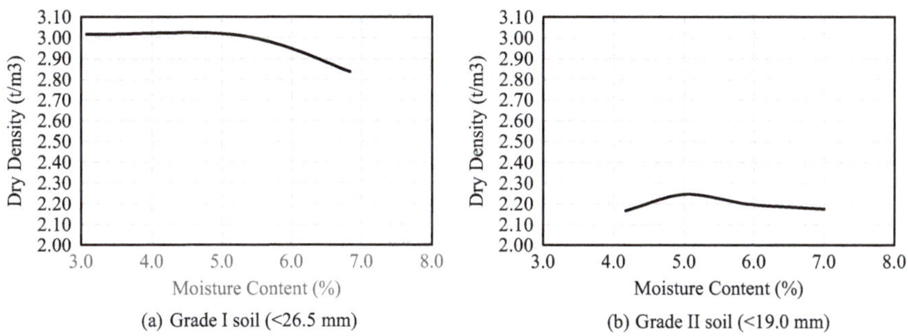

Fig. 4. Modified Proctor compaction curves of soil grades I and II

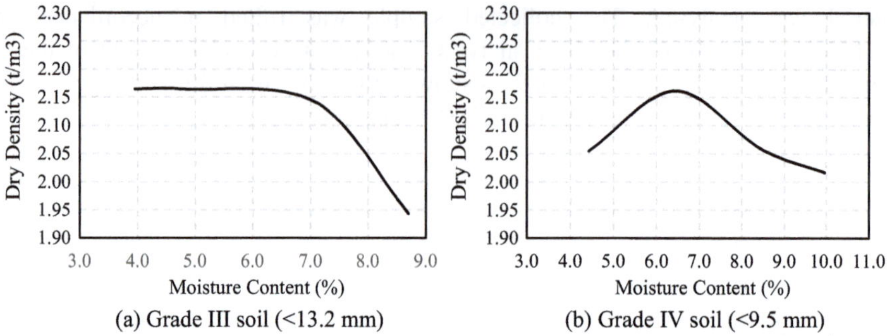

Fig. 5. Modified Proctor compaction curves of soil grades III and IV

A series of soaked and unsoaked California bearing ratio (CBR) tests were carried out on the soil samples based on AS 1289.6.1.1 (2014) to determine the strength of rejected ballast material compacted to the maximum dry density and optimum moisture content.

1.2 Laboratory Test Results

Figure 6 compares the particle size distribution of the collected rejected ballast with the upper and lower limits of the Aurizon's standard formation grade 'B' material. The collected sample seems to have a suitable grading for the larger particle range however, it appears to lack suitable quantities of finer sized material.

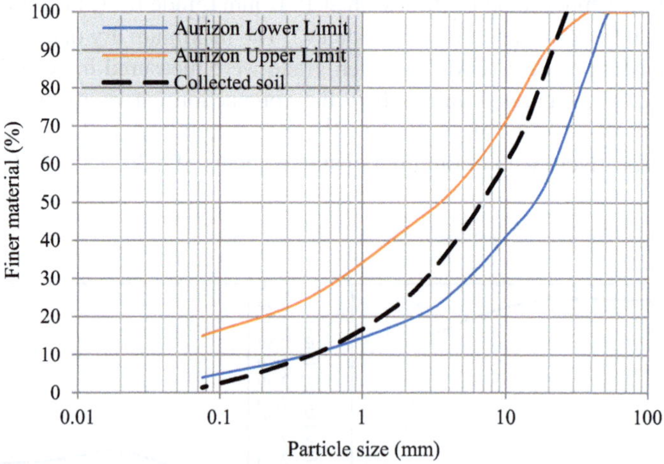

Fig. 6. Particle size distribution of the collected sample compared to the boundaries of grade 'B' Aurizon's materials

Figure 7 shows the stress-strain behaviour of unsoaked CBR tests for the four soil grades (I to IV) and Table 1 reports the compaction data and CBR ratio of unsoaked samples. All un-soaked CBR tests demonstrated acceptable results with the exception of the grade III (<13.2 mm) sample that gained small CBR ratios of 9–26.8% at penetration values of 2.5–5.0 mm, respectively. The low CBR ratio of the grade III soil could be due to the moisture ratio of the prepared sample that was 124.9% of its optimum moisture content and resulted in a weak resistance of the sample during CBR piston penetration. This was most likely due to the unsoaked sample had an existing moisture content slightly above zero even though it was air-dried prior to adding the required water quantities for compaction. Although the density and moisture ratios of grade I soil was also not close enough to its optimum condition, the resulted CBR ratio could have significantly improved if it has been compacted properly. Whilst the CBR tests were not performed at OMC for all samples, the Austroads Guide to Pavement Technology, Part 2 (2012) provides guidance that suggests for zones with a median rainfall greater than 800 mm per annum, the designer can consider using a moisture content of up to 1.15 × OMC of the soil. Majority of the networks in Queensland receive a minimum rainfall of 600 mm per annum or greater. At this higher moisture content, the CBR value is expected to be slightly lower than at that at OMC and this provides a more realistic service condition of roads. Access roads are maintained less frequently and drainage is not always ideal and therefore, their actual daily moisture content would be expected to be higher than public roads.

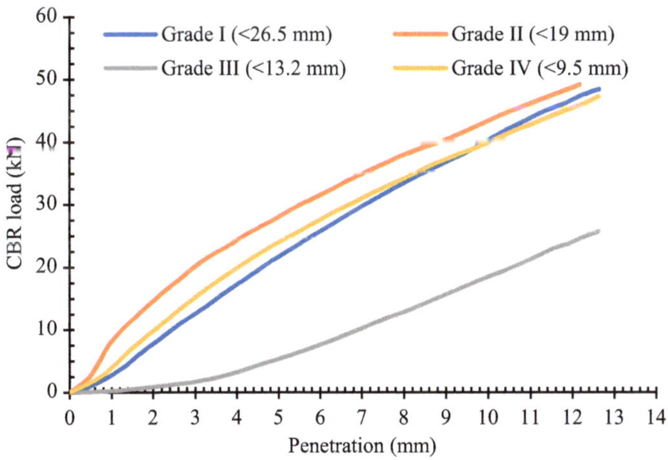

Fig. 7. Unsoaked CBR of soil grades I to IV

In order to investigate the CBR ratio of soil samples under severe rainfall and near saturation state, a new set of samples were cured in a water bath for 4 days followed by conducting soaked CBR. Figure 8 and Table 2 summaries the results. All soaked CBR tests achieved promising results apart from the grade IV sample which its CBR ratio at 2.5 mm penetration was relatively low (i.e., 12.3%). This, however, could be discarded

Table 1. Unsoaked CBR of soil grades I–IV

Soil grade		Grade I (<26.5 mm)	Grade II (<19.0 mm)	Grade III (<13.2 mm)	Grade IV (<9.5 mm)
Maximum dry density, $\rho_{dry-max}(\text{t/m}^3)$		3.03	2.25	2.18	2.16
Optimum moisture content (%)		4.30	5.05	5.70	6.44
Lab. density ratio, $\rho_{dry}/\rho_{dry-max}$ (%)		70.5	94.8	96.9	96.7
Lab. moisture ratio ω/ω_{opt} (%)		124.0	83.0	124.9	94.3
2.5 mm penetration	CBR load (kN)	10.3	17.5	1.2	12.7
	CBR ratio (%)	78.3	132.3	9.0	96.1
5.0 mm penetration	CBR load (kN)	21.8	28.6	5.3	24.0
	CBR ratio (%)	109.9	144.4	26.8	121.3

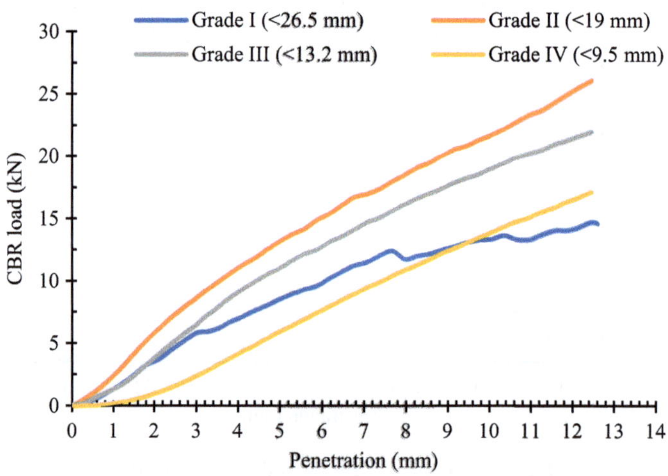

Fig. 8. Soaked CBR of soil grades I to IV

as at 5 mm penetration it provided a more acceptable CBR ratio of 30.2%. Apart from the grade IV, the lowest CBR ratio value obtained for grade I soil with a CBR ratio of 36.2% at 2.5 mm penetration. AS1289.6.1.1 allows the larger of the two readings (i.e., CBR ratio at 2.5 mm or 5.0 mm penetration) to be reported for the CBR of the sample

Table 2. Soaked CBR of soil grades I to IV

Soil grade		Grade I (<26.5 mm)	Grade II (<19.0 mm)	Grade III (<13.2 mm)	Grade IV (<9.5 mm)
Maximum dry density, $\rho_{dry-max}(\mathrm{t/m^3})$		3.03	2.25	2.18	2.16
Optimum moisture content (%)		4.30	5.05	5.70	6.44
Lab. density ratio, $\rho_{dry}/\rho_{dry-max}$ (%)		66.3	92.0	93.7	93.8
Lab. moisture ratio ω/ω_{opt} (%)		124.0	113.9	128.9	117.6
2.5 mm penetration	CBR load (kN)	4.8	7.3	5.2	1.6
	CBR ratio (%)	36.2	55.0	39.2	12.3
5.0 mm penetration	CBR load (kN)	8.6	13.2	11.1	6.0
	CBR ratio (%)	43.2	66.6	55.9	30.2

and therefore, grade I to IV soaked samples gained CBR ratios of 43.2, 66.6, 55.9 and 30.2%, respectively. Therefore, grade I material seems suitable as a replacement for Aurizon present formation material that requires a CBR ratio of at least 45%.

1.3 Field Test Results

A short section of a trial access road construction was undertaken at the Waitara rail yard on the Goonyella line close to the location of the waste material pile (see Fig. 9). The site was selected in part to reduce the material transportation costs for trial but also to comply with the current environmental approvals/restrictions with regard to moving the fouled ballast outside the rail corridor. The selected access road was positioned on the same parcel of land as the waste material stockpiles, therefore, no additional approvals were required. The section of the trial access road was approximately 50 m long and 4 m wide and was in a section of the access road that was generally heavily trafficked and had previous issues with pooling water leading to degradation of the surface. The bearing capacity of the constructed road was examined using a plate load test and the result was compared with the plate load test results on an access road that was constructed in the past using Aurizon's standard materials.

As the grade I soil performed satisfactorily in the soaked CBR test, it was decided that the access road would be constructed from an unmodified portion of the rejected waste ballast material (i.e. no specific refined mix or additives). This would ensure that the maximum quantity of waste material can be utilised and therefore, no waste ballast

Fig. 9. Location of stockpiles and the proposed access road (Map data @2015 Google)

requires disposal at an additional cost. Static plate load test (i.e., using a 350 mm diameter plate) was undertaken on the constructed road and also on a previously constructed road of Aurizon's existing design for a comparison purpose. Figure 10 shows the setup of the plate load tests along the access roads. The settlement of the plate was plotted against the applied load and the values were measured at staged intervals.

(a) Static Plate test on previously constructed road

(b) Static Plate test on recently constructed road using waste ballast

Photos were taken by Troy Decourcy

Fig. 10. Plate load test on the access road

Figure 11 shows the results of plate load tests. This figure shows that at an ultimate bearing pressure 300 kPa (i.e., equivalent to an allowable bearing pressure of 100 for a safety factor of 3), the settlements of the plate for the Aurizon's standard access road and ballast reject access road were 2.20–1.42 mm, respectively. This clearly shows that the rejected waste ballast could be stronger than the current Aurizon's standard material and can significantly save the incurred cost for constructing and maintaining access roads along the network.

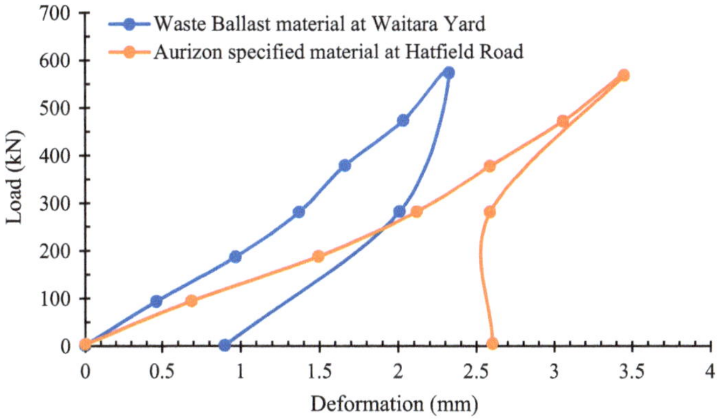

Fig. 11. Plate load test results on a previously constructed access road and ballast reject access road

Given using the same quantity of material for a 200 mm thick unsealed pavement, the costs savings that may be achieved from repairing or constructing railway access roads using the rejected fouled waste ballast are very likely to be only due to the material costs. Based on the undertaken cost analysis, for a 100 m road with a width of 4 m, a total of saving of around $3100 is expected.

After a short period of four weeks of use, a visual inspection of the trial access road section was undertaken. The area where the road was constructed had received some rainfall during the 4-week period and the road was always available for traffic use. Unfortunately, no exact rainfall readings were able to be obtained for the area however, coastal towns of Mackay and Sarina (close to the location of the access road) experienced 8–32 mm of rainfall, respectively. The road did not appear to have any significant damage or deterioration that could be contributed to normal vehicle use (Fig. 12). There was no evidence of any rutting having occurred or likely to occur in the near future and no signs of any sections likely to fail drastically or rapidly. This road has been in operation for approximately 15 months and during a significant wet season (cyclone Debbie), road resisted the severe environmental effects well.

(a) After four weeks
Photo was taken by Troy Decourcy

(b) After 15 months (after cyclone Debbie)
Photo was taken by Mitchell Lees

Fig. 12. The condition of trial access road after construction

2 Conclusions

In order to reduce the cost of access road construction and prevent the discarded rejected waste ballast material from becoming an additional problem, in this study, it was attempted to construct access roads from the waste <26.5 mm material. The project was undertaken purely to determine if discarded, smaller sized waste ballast material would be suitable for the construction of railway corridor access roads in the Central Queensland Coal Network. The desired grading of <26.5 mm material provided a soaked CBR ratio of 43% which was almost compliant with the Aurizon present formation material that requires a CBR ratio of at least 45%. Given that access roads experience significantly lower axle loading than rail formations, the obtained slightly lower CBR value should provide a suitable performance.

A static plate load test was undertaken on the completed waste ballast access road and this resulted in a deflection of 1.42 mm under an applied pressure of 300 kPa. The observed performance was better than the test results obtained for a similar road that was constructed using Aurizon formation specified material (i.e., 2.20 mm settlement at 300 kPa pressure). Overall, the material showed a genuine potential as a low-cost alternative for construction of Aurizon access roads. Whilst quantities of material available would be quite small, it would be possible to utilise these to repair and upgrade the local access roads in locations where ballast recycling is being performed. This would help to reduce capital costs and assist the ballast recycling program to improve conditions with the rail corridors.

Acknowledgements. The authors acknowledge the support of Aurizon Network assets management which provided a financial support for constructing the access road in Aurizon's land. The authors also would like to appreciate the technical support of the Aurizon's engineering staff who continually supported this research and provided guidance on Aurizon policies, procedures and standards to ensure the research and outcomes were valuable and had a suitable application to the railway industry.

References

Aurizon: Aurizon Network, Fouled Ballast Management Guideline N03-GUI-002-ENV (2016)

Fatahi, B., Khabbaz, H.: Enhancement of ballasted rail track performance using geosynthetics. In: Advances in pile foundations, geosynthetics, geoinvestigations, and foundation failure analysis and repairs, pp. 222–230 (2011)

Hoy, M., Rachan, R., Horpibulsuk, S., Arulrajah, A., Mirzababaei, M.: Effect of wetting—drying cycles on compressive strength and microstructure of recycled asphalt pavement–fly ash geopolymer. Constr. Build. Mater. **144**, 624–634 (2017)

Jameson, G., Harrison, C.: Austroads guide to pavement technology. Part 2, pavement structural design. Austroads, Sydney (Updated Feb 2012)

Mirzababaei, M., Egwurube, J., Gyasi-Agyei, Y., Foun, D., Hammond, A., Keleher, P., Nissen, D.: Classification of common geotechnical failure types occurring in Queensland's heavy haul rail network. In: Proceedings of the Second International Conference on Railway Technology: Research, Development and Maintenance, Ajaccio, Corsica, France (2014)

NSW Government, Transport NSW: Access roads standard THR Ci 12200 ST Version 1.0 (2014)

Pavement interactive: Pavement history. (2016). [ONLINE] Available at: http://www.pavementinteractive.org/pavement-history/. Accessed 21 Jan 2018

Standards Australia: Methods for testing soils for engineering purposes—Method 5.2.1: Soil compaction and density tests—determination of the dry density/moisture content relation of a soil using modified compactive effort (AS1289.5.2.1) (2003)

Standards Australia: Methods for sampling and testing aggregates method 11.1: particle size distribution—sieving method (AS1141.11.1) (2009)

Standards Australia: Methods for sampling and testing aggregates method 3.1: sampling—aggregates (AS1141.3.1) (2012)

Standards Australia: Methods for testing soils for engineering purposes—Method 6.1.1: soil strength and consolidation tests—determination of California bearing capacity of a soil—standard laboratory method for a remoulded specimen (AS1289.6.1.1) (2014)

State of Queensland Department of Transport & Main Roads (DTMR): Technical specification, MRTS35 recycled materials for pavements (2014)

Yaowarat, T., Horpibulsuk, S., Arulrajah, A., Mirzababaei, M., Rashid, A.: Compressive and flexural strength of polyvinyl alcohol modified pavement concrete using recycled concrete aggregates, ASCE Mater. Civil Eng. (2018, in Press)

Potassium Aluminate Geopolymerisation of Acidic Gold Mine Tailings

Felix Ndubisi Okonta[1(✉)], Thabo Falayi[1], and Freeman Ntuli[2]

[1] Department of Civil Engineering Science, University of Johannesburg, Johannesburg 2006, South Africa
fnokonta@uj.ac.za
[2] Department of Chemical Engineering, University of Johannesburg, Johannesburg 2006, South Africa

Abstract. Acidic gold mine tailings were alkaline activated using KOH. The effect of potassium aluminate (KA) on the strength and durability of the geopolymers was investigated. A 2.8 KA:KOH geopolymer had a UCS of 18.10 MPa after curing for 5 days at 100 °C. There was an increase in UCS with an increase in loading of KA up to a ratio of 2.8. Beyond the KA:KOH ratio of 2.8, there was a 48% drop in UCS due to excess K^+ ions in the system which resulted in the loss of charge balance of the system leading to reduction of UCS. It is worth mentioning that the KA:KOH ratio of 2.8 represented a Si/Al ratio of 1.02. This study showed that KA activation of acidic gold mine tailings is an attractive route to stabilise/solidify hazardous tailing material. Though there is use of elevated temperature to achieve high strength for the KA based geopolymer, this pales in comparison to energy requirements of cement manufacturing and clay brick firing.

Keywords: Gold mine tailings · Geopolymers · Alkaline activation
Toxicity leaching characteristic procedure · Durability

1 Introduction

Mine tailings are waste material left after the extraction of valuable minerals from the ore body. Storage of mine tailings is an environmental concern in that it can lead to loss of agrarian land, pollution of environment and underground water (Ahmari and Zhang 2013). Gold mine tailings (GMT) are of particular concern in that they contain the mineral pyrite and thus a potential source for the formation of acid mine drainage (Nehdi and Tariq 2007). About 600,000 tonnes GMT are found around the Gauteng province of South Africa exposing people to pollution effects of these tailings dams (Winde 2010). Cement stabilisation of GMT has been used to produce paste backfill (Ercikdi et al. 2009) but the continued use of cement is discouraged due to its CO_2 production and energy intensive production process. Geopolymerisation is an alternative route for the stabilisation of GMT.

Geopolymers are inorganic three dimensional amorphous/semi crystalline structures formed when an aluminosilicate material reacts with an alkaline solution. Any material with silica and alumina can be a precursor for geopolymerisation although

H. Khabbaz et al. (eds.), *New Prospects in Geotechnical Engineering Aspects of Civil Infrastructures*, Sustainable Civil Infrastructures,
https://doi.org/10.1007/978-3-319-95771-5_21

silicate and aluminate solutions can be added to enhance the process (Ahmari et al. 2012). Geopolymers have been shown to have chemical, fire and thermal stabilities, high compressive strength, low density, shrinkage and permeability (Kiventera et al. 2016).

South African GMT has successfully been stabilised through geopolymerisation (Falayi et al. 2018), although a low unconfined compressive strength of 3.49 MPa was a major drawback.

This research article therefore investigated the effects of potassium aluminate on the geopolymerisation of acidic GMT in order to improve the strength of the geopolymer. The research article further looks at the potential uses of the geopolymer by subjecting it to two types of acidic environments. The influence of the geopolymer gel towards strength gain was also investigated. This research allows the opportunity of turning of a hazardous material into a useful green civil engineering material.

2 Materials

Potassium aluminate (KA) powders and potassium hydroxide (KOH) were supplied by Rochelle chemicals, South Africa. GMT was taken from a tailings dam in the West Rand of Gauteng province of South Africa.

3 Methods

The 10 M KOH solution was prepared by dissolving the appropriate amount of KOH in reverse osmosis water. The effect of KA on the geopolymerisation was investigated by adding appropriate amounts of KA respectively to the 10 M KOH solution. KA was added in the mass ratios of 1.3 to 3.1 to KOH. The alkaline solution was allowed to cool down before casting. Casting was done by mixing the appropriate alkali solution with GMT. A liquid/solid (L/S) ratio of 0.26 was employed. The mixture was then hand mixed for 15 min before being cast into a 50 * 50 * 50 mm mould. The paste was allowed to harden before demoulding and subsequent curing at different temperatures for 5 days. Unconfined compressive strength (UCS) was tested on all samples using a Cyber plus evolution compression machine at 0.25 MPa/s. The reported UCS was an average of 3 samples. The effect of KA on the microstructure of the geopolymers was investigated using the Scanning electron microscope (SEM).

4 Results and Discussion

4.1 Effect of Potassium Aluminate (KA)

Figure 1 shows the variation in UCS with the ratio of KA:KOH.

There was an increase in UCS with an increase in loading of KA up to a ratio of 2.8. Beyond the KA:KOH ratio of 2.8, there was a 48% drop in UCS due to excess K^+ ions in the system which resulted in the loss of charge balance of the system leading to

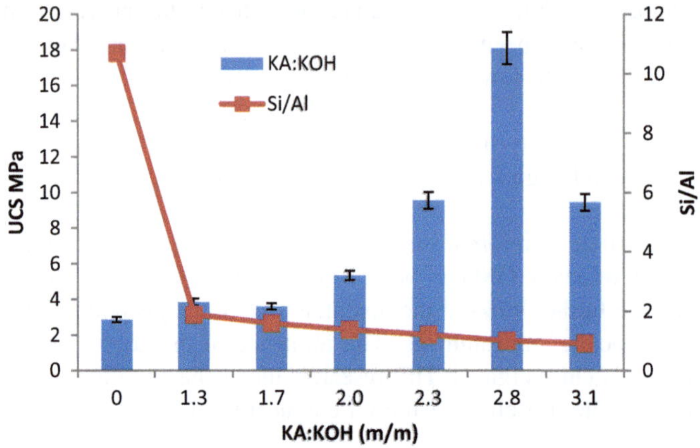

Fig. 1. Variation of UCS with ratio of KA:KOH and Si/Al ratio (all geopolymers synthesised with 10 M KOH and cured at 100 °C for 5 days)

reduction of UCS. It is worth mentioning that the KA:KOH ratio of 2.8 represented a Si/Al ratio of 1.02. This has been reported to be the optimum Si/Al ratio for geopolymerisation (Ahmari et al. 2012).

The geopolymerisation kinetics of the KA:KOH (2.8) geopolymer was investigated. The synthesised pastes were cured for a variation of 1–5 days. After the curing period was completed, the cured samples were crushed using a mortar bowl and pestle. 2 g of the ground geopolymer was mixed with 20 ml concentrated HCl. The slurry was left in a Thermoshaker for 24 h. After 24 h the slurry was filtered and the solid residue was dried in an oven for 24 h at 100 °C. The dried solid residue was weighed with the weight difference being used to calculate the % geopolymer in the cures sample.

Figure 2 shows the variation in geopolymer gel with curing time.

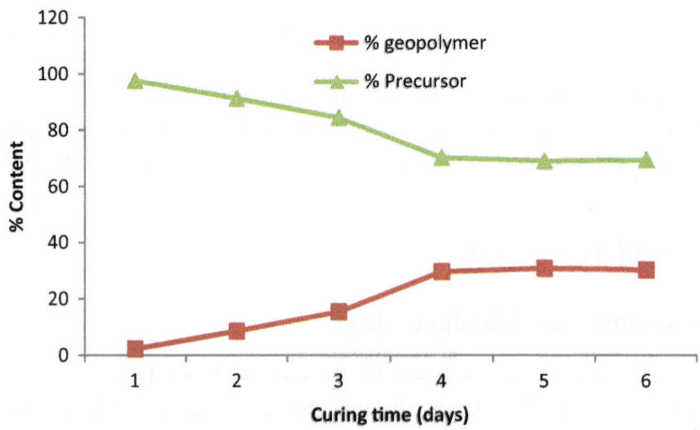

Fig. 2. Variation of geopolymer gel in the geopolymer with curing time, geopolymers synthesised with 10 M KOH and cured at 100 °C

The precursor material is GMT or ungeopolymerised material. Geopolymerisation was a slow process for the first 3 days as only the geopolymer gel consisted of only 15% of the geopolymer. The optimum curing time was 4 days as thereafter the increase in geopolymer gel was insignificant.

A correlation between geopolymer gel content and UCS was established (Table 1).

Table 1. Variation in UCS with geopolymer gel content

Curing days	Geopolymer gel (%)	UCS (MPa)
1	2.35	0.98
2	8.64	1.44
3	15.44	1.64
4	29.65	12.34
5	30.88	18.44
6	30.44	18.48

Though a higher geopolymer gel corresponded to an increase in UCS, the correlation coefficient was below 0.6 and therefore considered poor. This might be due to the delay in hardening of the gel to form an amorphous three dimensional structure.

The geopolymer cured for 5 days was exposed to Acetic acid solution at a pH of 2.88 ± 0.05) and sulphuric acid solution at a pH = 4.20 ± 0.05. These solution were used simulate landfill leachate and acid rain respectively (Phoungthong et al. 2016) respectively (Table 2).

Table 2. Variation in UCS with time in different acidic solution

Days	Acetic acid solution (MPa)	Sulphuric acid solution (MPa)
1	12.56	16.25
2	8.55	15.44
3	8.45	15.35
4	8.44	15.22
5	8.41	15.2

After 5 days there was a 54.4% reduction for in UCS for the acetic acid solution as compared to 17.6% for the sulphuric acid solution. This therefore means the geopolymer is not suited to be used for as a landfill liner but can be used in areas which receive acid rain. The reduction in UCS was due to the dissolution of the geopolymer gel by the acidic environment as shown in Fig. 3. There was a visible decrease in geopolymer particles after immersion.

Fig. 3. SEM of geopolymer **a** before immersion in acetic acid solution and **b** after

5 Conclusions

Gold mine tailings can successfully be stabilised via alkaline activation with the use of potassium aluminate as an activator. Geopolymerisation of gold mine tailings is accompanied by the increase in geopolymer gel in the material with only 30% of the original material converted. Geopolymerisation was a slow process for the first 3 days as only the geopolymer gel consisted of only 15% of the geopolymer. The optimum curing time was 4 days as thereafter the increase in geopolymer gel was insignificant. After 5 days there was a 54.4% reduction for in UCS for the acetic acid solution as compared to 17.6% for the sulphuric acid solution. This therefore means the geopolymer is not suited to be used for as a landfill liner but can be used in areas which receive acid rain. This research therefore provides opportunities of turning waste materials into useful construction materials satisfying the requirements of industrial ecology, though there is a need for activation processes which reduce the curing temperature.

References

Ahmari, S., Zhang, L., Zhang, J.: Effects of activator type/concentration and curing temperature on alkali activated binder based on copper mine tailings. J. Mater. Sci. **47**, 5933–5945 (2012)

Ahmari, S., Zhang, L.: Durability and leaching behavior of mine tailings-based geopolymer bricks. Constr. Build. Mater. **44**(2013), 743–750 (2013)

Falayi, T., Ntuli, F., Okonta, F.N.: Synthesis of paste backfill using acidic gold mine tailings. J. Solid Waste Technol. (2018, in press)

Nehdi, M., Tariq, A.: Stabilization of sulphidic mine tailings for prevention of metal release and acid drainage using cementitious materials: a review. J. Environ. Eng. Sci. **6**, 423–436 (2007)

Ercikdi, B., Kesimal, A., Cihangir, F., Deveci, H., Alp, I.: Cemented paste backfill of sulphide-rich tailings: importance of binder type and dosage. Cem. Concr. Compos. **31**, 268–274 (2009)

Kiventera, J., Golek, L., Yliniemi, J., Ferreira, V., Deja, J., Illikainen, M.: Utilization of sulphidic tailings from gold mine as raw materials in geopolymerisation. Int. J. Miner. Process. **149**, 104–110 (2016)

Phoungthong, K., Xia, Y., Zhang, H., Shao, L., He, P.: Leaching toxicity characteristics of municipal solid waste incineration bottom ash. Front. Environ. Sci. Eng. **10**, 399–411 (2016)

Winde, F.: Uranium pollution of the Wonderfonteinspruit, 1997–2008 part 1: uranium toxicity, regional background and mining-related sources of uranium pollution. Water Sa **36**(3), 239–256 (2010)

Author Index

H. Khabbaz et al. (eds.), *New Prospects in Geotechnical Engineering Aspects of Civil Infrastructures*, Sustainable Civil Infrastructures,
https://doi.org/10.1007/978-3-319-95771-5

MIX
Papier aus verantwortungsvollen Quellen
Paper from responsible sources
FSC® C105338

If you have any concerns about our products,
you can contact us on
ProductSafety@springernature.com

In case Publisher is established outside the EU,
the EU authorized representative is:
Springer Nature Customer Service Center GmbH
Europaplatz 3, 69115 Heidelberg, Germany

Printed by Libri Plureos GmbH
in Hamburg, Germany